Liquid Transportation Fuels from Coal and Biomass

TECHNOLOGICAL STATUS, COSTS, AND ENVIRONMENTAL IMPACTS

America's Energy Future Panel on Alternative Liquid Transportation Fuels

NATIONAL ACADEMY OF SCIENCES
NATIONAL ACADEMY OF ENGINEERING
NATIONAL RESEARCH COUNCIL
OF THE NATIONAL ACADEMIES

THE NATIONAL ACADEMIES PRESS
Washington, D.C.
www.nap.edu

THE NATIONAL ACADEMIES PRESS 500 Fifth Street, N.W. Washington, DC 20001

NOTICE: The project that is the subject of this report was approved by the Governing Board of the National Research Council, whose members are drawn from the councils of the National Academy of Sciences, the National Academy of Engineering, and the Institute of Medicine. The members of the panel responsible for the report were chosen for their special competences and with regard for appropriate balance.

Support for this project was provided by the Department of Energy under Grant Number DE-FG02-07-ER-15923 and by BP America, Dow Chemical Company Foundation, Fred Kavli and the Kavli Foundation, GE Energy, General Motors Corporation, Intel Corporation, and the W.M. Keck Foundation. Support was also provided by the Presidents' Circle Communications Initiative of the National Academies and by the National Academy of Sciences through the following endowed funds created to perpetually support the work of the National Research Council: Thomas Lincoln Casey Fund, Arthur L. Day Fund, W.K. Kellogg Foundation Fund, George and Cynthia Mitchell Endowment for Sustainability Science, and Frank Press Fund for Dissemination and Outreach. Any opinions, findings, conclusions, or recommendations expressed in this publication are those of the author(s) and do not necessarily reflect the views of the organizations that provided support for the project.

International Standard Book Number-13: 978-0-309-13712-6
International Standard Book Number-10: 0-309-13712-8
Library of Congress Catalog Card Number 2009937432

Limited copies of this report are available from:

Board on Energy and Environmental Systems
National Research Council
500 Fifth Street, NW, Keck 934
Washington, DC 20001
(202)334-3344

Additional copies of this report are available from:

The National Academies Press
500 Fifth Street, NW, Lockbox 285
Washington, DC 20055
(800) 624-6242 or (202) 334-3313 (in the Washington metropolitan area)
Internet: http://www.nap.edu

Printed on recycled stock

Printed in the United States of America

THE NATIONAL ACADEMIES
Advisers to the Nation on Science, Engineering, and Medicine

The **National Academy of Sciences** is a private, nonprofit, self-perpetuating society of distinguished scholars engaged in scientific and engineering research, dedicated to the furtherance of science and technology and to their use for the general welfare. Upon the authority of the charter granted to it by the Congress in 1863, the Academy has a mandate that requires it to advise the federal government on scientific and technical matters. Dr. Ralph J. Cicerone is president of the National Academy of Sciences.

The **National Academy of Engineering** was established in 1964, under the charter of the National Academy of Sciences, as a parallel organization of outstanding engineers. It is autonomous in its administration and in the selection of its members, sharing with the National Academy of Sciences the responsibility for advising the federal government. The National Academy of Engineering also sponsors engineering programs aimed at meeting national needs, encourages education and research, and recognizes the superior achievements of engineers. Dr. Charles M. Vest is president of the National Academy of Engineering.

The **Institute of Medicine** was established in 1970 by the National Academy of Sciences to secure the services of eminent members of appropriate professions in the examination of policy matters pertaining to the health of the public. The Institute acts under the responsibility given to the National Academy of Sciences by its congressional charter to be an adviser to the federal government and, upon its own initiative, to identify issues of medical care, research, and education. Dr. Harvey V. Fineberg is president of the Institute of Medicine.

The **National Research Council** was organized by the National Academy of Sciences in 1916 to associate the broad community of science and technology with the Academy's purposes of furthering knowledge and advising the federal government. Functioning in accordance with general policies determined by the Academy, the Council has become the principal operating agency of both the National Academy of Sciences and the National Academy of Engineering in providing services to the government, the public, and the scientific and engineering communities. The Council is administered jointly by both Academies and the Institute of Medicine. Dr. Ralph J. Cicerone and Dr. Charles M. Vest are chair and vice chair, respectively, of the National Research Council.

www.national-academies.org

PANEL ON ALTERNATIVE LIQUID TRANSPORTATION FUELS

MICHAEL P. RAMAGE, ExxonMobil Research and Engineering Company (retired), *Chair*
G. DAVID TILMAN, University of Minnesota, St. Paul, *Vice Chair*
DAVID GRAY, Nobilis, Inc.
ROBERT D. HALL, Amoco Corporation (retired)
EDWARD A. HILER, Texas A&M University (retired)
W.S. WINSTON HO, Ohio State University
DOUGLAS L. KARLEN, U.S. Department of Agriculture, Agricultural Research Service
JAMES R. KATZER, ExxonMobil Research and Engineering Company (retired)
MICHAEL R. LADISCH, Purdue University and Mascoma Corporation
JOHN A. MIRANOWSKI, Iowa State University
MICHAEL OPPENHEIMER, Princeton University
RONALD F. PROBSTEIN, Massachusetts Institute of Technology
HAROLD H. SCHOBERT, Pennsylvania State University
CHRISTOPHER R. SOMERVILLE, Energy BioSciences Institute
GREGORY STEPHANOPOULOS, Massachusetts Institute of Technology
JAMES L. SWEENEY, Stanford University

Liaisons from the Committee on America's Energy Future
CHRISTINE A. EHLIG-ECONOMIDES, Texas A&M University
JOHN B. HEYWOOD, Massachusetts Institute of Technology
ARISTIDES A.N. PATRINOS, Synthetic Genomics, Inc.

Consultants
ADRIAN A. FAY, Massachusetts Institute of Technology
SAMUEL FLEMING, Claremont Canyon Consultants
JASON HILL, University of Minnesota, St. Paul
SHELDON KRAMER, Independent Consultant, Grayslake, Illinois
THOMAS KREUTZ, Princeton University
ERIC LARSON, Princeton University
ROBERT WILLIAMS, Princeton University

America's Energy Future Project Director
PETER D. BLAIR, Executive Director, Division on Engineering and Physical Sciences

America's Energy Future Project Manager
JAMES ZUCCHETTO, Director, Board on Energy and Environmental Systems

Foreword

Energy, which has always played a critical role in our country's national security, economic prosperity, and environmental quality, has over the last two years been pushed to the forefront of national attention as a result of several factors:

- World demand for energy has increased steadily, especially in developing nations. China, for example, saw an extended period (prior to the current worldwide economic recession) of double-digit annual increases in economic growth and energy consumption.
- About 56 percent of the U.S. demand for oil is now met by depending on imports supplied by foreign sources, up from 40 percent in 1990.
- The long-term reliability of traditional sources of energy, especially oil, remains uncertain in the face of political instability and limitations on resources.
- Concerns are mounting about global climate change—a result, in large measure, of the fossil-fuel combustion that currently provides most of the world's energy.
- The volatility of energy prices has been unprecedented, climbing in mid-2008 to record levels and then dropping precipitously—in only a matter of months—in late 2008.
- Today, investments in the energy infrastructure and its needed technologies are modest, many alternative energy sources are receiving insufficient attention, and the nation's energy supply and distribution systems are increasingly vulnerable to natural disasters and acts of terrorism.

All of these factors are affected to a great degree by the policies of government, both here and abroad, but even with the most enlightened policies the overall energy enterprise, like a massive ship, will be slow to change course. Its complex mix of scientific, technical, economic, social, and political elements means that the necessary transformational change in how we generate, supply, distribute, and use energy will be an immense undertaking, requiring decades to complete.

To stimulate and inform a constructive national dialogue about our energy future, the National Academy of Sciences and the National Academy of Engineering initiated a major study in 2007, "America's Energy Future: Technology Opportunities, Risks, and Tradeoffs." The America's Energy Future (AEF) project was initiated in anticipation of major legislative interest in energy policy in the U.S. Congress and, as the effort proceeded, it was endorsed by Senate Energy and Natural Resources Committee Chair Jeff Bingaman and former Ranking Member Pete Domenici.

The AEF project evaluates current contributions and the likely future impacts, including estimated costs, of existing and new energy technologies. It was planned to serve as a foundation for subsequent policy studies, at the Academies and elsewhere, that will focus on energy research and development priorities, strategic energy technology development, and policy analysis.

The AEF project has produced a series of five reports, including this report on alternative liquid fuels for transportation, designed to inform key decisions as the nation begins this year a comprehensive examination of energy policy issues. Numerous studies conducted by diverse organizations have benefited the project, but many of those studies disagree about the potential of specific technologies, particularly those involving alternative sources of energy such as biomass, renewable resources for generation of electric power, advanced processes for generation from coal, and nuclear power. A key objective of the AEF series of reports is thus to help resolve conflicting analyses and to facilitate the charting of a new direction in the nation's energy enterprise.

The AEF project, outlined in Appendix A, included a study committee and three panels that together have produced an extensive analysis of energy technology options for consideration in an ongoing national dialogue. A milestone in the project was the March 2008 "National Academies Summit on America's Energy Future" at which principals of related recent studies provided input to the AEF study committee and helped to inform the panels' deliberations. A report chronicling the event, *The National Academies Summit on America's Energy Future:*

Summary of a Meeting (Washington, D.C.: The National Academies Press), was published in October 2008.

The AEF project was generously supported by the W.M. Keck Foundation, Fred Kavli and the Kavli Foundation, Intel Corporation, Dow Chemical Company Foundation, General Motors Corporation, GE Energy, BP America, U.S. Department of Energy, and our own academies.

Ralph J. Cicerone, President
National Academy of Sciences
Chair, National Research Council

Charles M. Vest, President
National Academy of Engineering
Vice Chair, National Research Council

Preface

Transportation plays a key role in the economies of industrialized societies, especially in light of increasing globalization. As in most countries, transportation in the United States has relied heavily on petroleum-based fuels. The influence of volatile oil prices on the U.S. economy, increasing U.S. dependence on imported oil and its effect on U.S. energy security, and recognition of the large contribution of transportation to emissions of greenhouse gases call for development of alternative transportation fuels from domestic sources that have lower greenhouse gas emissions than do petroleum-based fuels. Biofuels and coal-to-liquid fuels are options that can improve the nation's energy security inasmuch as biomass is a renewable resource and the United States has the world's largest known coal reserves. However, those options raise important questions about economic viability, carbon impact, and technology status. To assess the technological status, costs, and environmental effects of alternative liquid transportation fuels produced from coal and biomass, the National Research Council convened the Panel on Alternative Liquid Transportation Fuels. The panel's work was part of a larger study initiated by the National Academy of Sciences and the National Academy of Engineering—the America's Energy Future project (Appendix A).

In approaching its task (Appendix B), the 16-member panel of experts (Appendix C) began by reviewing the literature and also gathered input from invited speakers (Appendix D) on the production of biofuels and coal-to-liquid fuels. Because of the uncertainties and widely different opinions expressed in the literature, the panel decided to conduct its own analyses of the costs, potential supply, and life-cycle greenhouse gas emissions of alternative fuels produced from biomass, coal, or both. An advantage of conducting its own analyses was that the panel could use a consistent basis and assumptions to compare the costs and environmental effects of different alternative fuel options. As the panel was writing its report (from November 2007 to November 2008), the commodity prices and capital costs of building energy plants fluctuated widely. The panel therefore included sensitivity analyses of feedstock costs, capital costs, and oil prices to see how they might affect choices of fuels.

The panel concluded that alternative liquid fuel technology can be deployable and supply a substantial volume of clean fuels for U.S. transportation at a reasonable cost. Transforming the U.S. transportation fuel system from domination by petroleum-based fuels to supply by various domestic sources will take several decades. Sustained and aggressive efforts are needed to accelerate the further development and penetration of alternative liquid fuel technologies.

I thank the panel members and the liaisons from the Committee on America's Energy Future for dedicating much time to the study. We were on a tight schedule to complete a complex task. Each member devoted time and effort to the study because we recognized not only the importance of achieving energy security for the nation but also, and more importantly, the immediate need for demonstration of the technical feasibility and economic viability of alternative liquid transportation fuels from domestic sources.

> Michael P. Ramage, *Chair*
> Panel on Alternative Liquid Transportation Fuels

Acknowledgments

This report is a product of the cooperation and contributions of many people. The members of the panel thank the consultants and the following persons who provided input to the panel:

John Baker, U.S. Department of Agriculture, Agricultural Research Service

Gary M. Banowetz, U.S. Department of Agriculture, Agricultural Research Service

Dana Dinnes, U.S. Department of Agriculture, Agricultural Research Service

Curt R. Fischer, Massachusetts Institute of Technology

Jane M-F. Johnson, U.S. Department of Agriculture, Agricultural Research Service

Youngmi Kim, Purdue University

Daniel Klein-Marcuschamer, Massachusetts Institute of Technology

Alicia Rosburg, Iowa State University

Wallace W. Wilhelm *(deceased)*, U.S. Department of Agriculture, Agricultural Research Service

The members of the panel also thank all the speakers who provided briefings to the panel. (Appendix D contains a list of presentations to the panel.)

This report has been reviewed in draft form by persons chosen for their diverse perspectives and technical expertise in accordance with procedures approved by the National Research Council's Report Review Committee. The purpose of this independent review is to provide candid and critical comments that will assist the institution in making its published report as sound as possible and to ensure that the report meets institutional standards for objectivity, evidence, and responsiveness to the study charge. The review comments and draft manuscript remain confidential to protect the integrity of the deliberative process. We thank the following individuals for their review of this report:

Noubar Afeyan, Flagship Ventures
Douglas Chapin, MPR Associates, Inc.
Joel Darmstadter, Resources for the Future, Inc.
Christopher B. Field, Carnegie Institution of Washington
Richard Flavell, Ceres, Inc.
Kevin B. Fogash, Air Products and Chemicals, Inc.
Bruce C. Gates, University of California, Davis
Lester Lave, Carnegie Mellon University
Bruce A. McCarl, Texas A&M University
Jeffrey Peterson, Energy Resources Group
Timothy Searchinger, Princeton University
Richard Sheppard, Independent Consultant
Jeff Siirola, Eastman Chemical Company
Kenneth Vogel, U.S. Department of Agriculture, Agricultural Research
   Service
Charles E. Wyman, University of California, Riverside

Although the reviewers listed above have provided many constructive comments and suggestions, they were not asked to endorse the conclusions or recommendations, nor did they see the final draft of the report before its release. The review of the report was overseen by Elisabeth M. Drake, Massachusetts Institute of Technology, and Robert A. Frosch, Harvard University. Appointed by the National Research Council, they were responsible for making certain that an independent examination of this report was carried out in accordance with institutional procedures and that all of the review comments were carefully considered. Responsibility for the final content of the report rests entirely with the authoring panel and the institution.

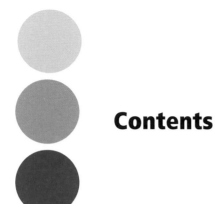

Contents

SYNOPSIS   **1**

SUMMARY   **7**

1 **LIQUID FUELS FOR TRANSPORTATION**   **51**

  Demand for Liquid Transportation Fuels, 52
  Alternative Transportation Fuels, 55
  Purpose of This Study, 56
  Context of Report, 58
  Structure of This Report, 60
  References, 61

2 **BIOMASS RESOURCES FOR LIQUID TRANSPORTATION FUELS**   **63**

  Current Biomass Production for Biofuels, 64
  Biomass Resources, 70
  Research and Development, 89
  Costs of Supplying Biofuel Feedstocks, 92
  Environmental Effects, 100
  Findings and Recommendations, 103
  References, 106

3 **BIOCHEMICAL CONVERSION OF BIOMASS**   **117**

  Technology Alternatives, 117
  Biochemical Conversion of Cellulosic Biomass, 121
  Cost and Performance, 132
  Technology Forecast, 145

Findings and Recommendations, 153
References, 157

4 THERMOCHEMICAL CONVERSION OF COAL AND BIOMASS   163

Status and Challenges of Technology Alternatives, 164
Indirect-Liquefaction Technologies, 167
Direct-Liquefaction Technologies, 209
Findings and Recommendations, 215
References, 221

5 DISTRIBUTION   225

Ethanol Transportation, 227
The Market for Biofuels, 235
Findings and Recommendations, 239
References, 240

6 COMPARISON OF OPTIONS AND MARKET PENETRATION   243

Comparison of Costs, Greenhouse Gas Emissions, and
Potential Fuel Supply, 244
Market Penetration, 259
Findings and Recommendations, 265
Reference, 267

7 OVERALL FINDINGS AND RECOMMENDATIONS   269

8 KEY CHALLENGES TO COMMERCIAL DEPLOYMENT   277

Challenge 1, 277
Challenge 2, 278
Challenge 3, 279
Challenge 4, 279
Challenge 5, 280

9 OTHER ALTERNATIVE FUEL OPTIONS   281

Compressed Natural Gas, 282
Alternative Diesel, 284
Methanol, 285
Dimethyl Ether, 285
Hydrogen, 286
References, 291

Appendixes

A   **America's Energy Future Project**   295

B   **Statement of Task**   301

C   **Panel Members' Biographical Sketches**   303

D   **Presentations to the Panel**   311

E   **Research Supporting a Landscape Vision of Production of**   313
     Biofuel Feedstock

F   **Estimating the Amount of Corn Stover That Can Be Harvested in a**   317
     Sustainable Manner

G   **Life-Cycle Inputs for Production of Biomass**   321

H   **Background Information on the Economic and Environmental**   325
     Assessment of Biomass Supply

I   **Modeling of Capital and Operating Costs and Carbon Emissions of**   339
     Ethanol Plants with Superpro Designer®

J   **Resource Requirements for Production of Microbial Biomass**   359

K   **Nonquantified Uncertainties That Could Influence the Costs of**   367
     Carbon Storage

Synopsis

The U.S. transportation sector consumes about 14 million barrels of oil per day, 9 million of which are used in light-duty vehicles. Total U.S. oil consumption is 21 million barrels per day, 12 million of which are imported. The nation can reduce its dependence on imported oil, increase its energy security, and potentially reduce greenhouse gas emissions by producing alternative liquid transportation fuels from domestically available resources to replace gasoline and diesel. Two abundant domestic resources with potential for producing liquid fuels are biomass and coal. Although abundant supplies of biomass and coal can be produced, each resource has its own set of limitations and challenges. Unlike liquid fuels from biomass, liquid fuels from coal cannot, even with the use of carbon capture and storage, offer any greenhouse gas benefit relative to gasoline. However, liquid fuels from coal are probably less expensive than those from biomass unless the costs of greenhouse gas emissions are included. A robust set of conversion technologies needs to be developed or demonstrated immediately and driven to commercial readiness to enable the use of the abundant biomass and coal in the development of suitable liquid transportation fuels. This report of the America's Energy Future Panel on Alternative Liquid Transportation Fuels addresses technological readiness for producing liquid fuels from coal and biomass, their life-cycle costs, and their environmental impacts.

COAL AND BIOMASS SUPPLY

The United States has at least 20 years' worth of coal reserves in active mines and probably sufficient resources to meet the nation's needs for well over 100 years at current rates of consumption. In contrast, biomass can be produced continuously over a long term if managed sustainably, but the amount that can be produced in a given period is limited by the natural resources required to support biomass production. Given that production of corn requires large amounts of fertilizer and that corn grain is also an important source of food and feed, the panel regards corn-grain ethanol as a transition to cellulosic biofuels or other biomass-based liquid hydrocarbon fuels, such as biobutanol and algal biodiesel. Cellulosic biomass—obtained from dedicated fuel crops, agricultural and forestry residues, and municipal solid wastes—could potentially be sustainably produced at about 400 million dry tons per year with today's technology and agricultural practices and with minimal adverse impacts on U.S. food and fiber production or on the environment. A key assumption underlying that estimate is that dedicated fuel crops will be grown on idle agricultural land in the U.S. Department of Agriculture's Conservation Reserve Program. By 2020, the amount of sustainably produced biomass could reach 550 million dry tons per year.

Ensuring a sustainable biomass supply requires that the resource base be assessed systematically to address multiple environmental, public, and private sector concerns simultaneously. Producers will probably need additional incentives to grow biofeedstocks so as to avoid direct and indirect competition with the food supply and also to avoid land-use practices that add substantially to net greenhouse gas emissions. Appropriate incentives can encourage sustainable approaches to the production of lignocellulosic biomass.

CONVERSION TECHNOLOGIES

Biochemical Conversion

Technology for biochemical conversion of the starch in grains to ethanol has been deployed commercially. Although grain-based ethanol has been important for initiating public awareness and the development of industrial infrastructure for fuel ethanol, cellulosic ethanol and other advanced cellulosic biofuels have much greater potential for reducing U.S. oil use and greenhouse gas emissions without affecting the food supply and costs. Processes for biochemical conversion of cellulosic biomass to ethanol are in the early stages of commercial development, and process improvements are expected over the next decade from evolutionary improvements through commercial experience and from economies of scale as production becomes more widespread and expands to an optimal size. The panel estimated that incremental improvements in biochemical conversion technologies can be expected to reduce nonfeedstock process costs by about 25 percent by 2020 and by 40 percent by 2035. An expanded distribution infrastructure will be required because ethanol cannot be transported in the pipelines used to transport petroleum. Studies should be conducted to identify the infrastructure needed to accommodate increasing volumes of ethanol and to identify and address the challenges of distributing and integrating larger volumes into the fuel system. Research on biochemical pathways for converting biomass to fuels that are more compatible with the current distribution infrastructure could lead to the development of relevant technologies over the next 10–15 years.

If all necessary conversion and distribution infrastructure is in place, 550 million dry tons of biomass can in theory be used to produce up to 2 million barrels per day (30 billion gallons) of gasoline-equivalent fuels. However, potential supply does not translate to actual fuel supply. When the production of corn-grain ethanol was commercialized, U.S. production capacity grew by 25 percent each year over a 6-year period. Assuming that the rate of building cellulosic-ethanol plants exceeds that of corn-grain-ethanol plants by 100 percent, alternative fuels could be added to the U.S. fuel portfolio at a rate of up to 0.5 million barrels of gasoline equivalent per day by 2020 (1 bbl of oil yields about 0.85 bbl of gasoline and diesel). By 2035, up to 1.7 million barrels per day could be produced in this manner, leading to about a 20 percent reduction in oil used for light-duty transportation at current consumption levels.

Thermochemical Conversion

Technologies for the indirect liquefaction of coal to transportation fuels (gasification with Fischer-Tropsch or a methanol-to-gasoline process) without geologic carbon dioxide (CO_2) storage are commercially deployable today, but CO_2 life-cycle emission will be more than twice that of petroleum-based fuels. Requiring geologic CO_2 storage with these processes would have a relatively small effect on engineering costs and efficiency. However, the viability of geologic CO_2 storage has yet to be adequately demonstrated on a large scale in the United States, and unanticipated costs could occur.

Liquid fuels produced from thermochemical plants that use only biomass feedstock are more expensive than fuels produced from coal but can have CO_2 life-cycle emission that is close to zero without geologic CO_2 storage or is highly negative with effective geologic CO_2 storage. To make such fuels competitive, the economic incentive for reducing CO_2 emission has to be sufficiently high. Gasification of biomass and coal together to produce liquid fuels allows operation on a larger scale than would be possible with biomass only and reduces capital costs per unit of capacity. When biomass and coal undergo thermochemical conversion together, overall CO_2 life-cycle emission is lower than emission from coal because the CO_2 emission from the coal is countered by the CO_2 uptake by biomass during its growth. If 550 million tons of biomass are combined with coal (60 percent coal and 40 percent biomass on an energy basis), production of 4 million barrels per day (60 billion gallons per year) of gasoline equivalent is technically feasible. That amount of fuel represents about 45 percent of the current volume of liquid fuels used annually for light-duty vehicles in the transportation sector. Conversion of combined coal and biomass to liquid fuels at that ratio without geologic CO_2 storage yields CO_2 life-cycle emission similar to that of gasoline; with geologic CO_2 storage, it yields close to zero CO_2 life-cycle emission. A program of aggressive support for establishment of first-mover commercial coal-to-liquid transportation-fuel plants and coal-and-biomass-to-liquid transportation-fuel plants with integrated geologic CO_2 storage will have to be undertaken immediately if commercial plants are to be deployed by 2020 to address U.S. energy security concerns and to provide fuels whose levels of greenhouse gas emissions are similar to or less than that of petroleum-based fuels.

For thermochemical conversion of coal or combined coal and biomass, the viability of geologic CO_2 storage is critical for commercial implementation. Large-scale demonstration and establishment of regulatory procedures have to be pursued aggressively in the next few years if thermochemical conversion of biomass and coal with geologic CO_2 storage is to be ready for commercial deployment in 2020 or sooner. The federal government should continue to partner with industry and independent researchers to determine the costs, safety, and effectiveness of geologic CO_2 storage on a commercial scale. If such demonstrations are initiated immediately and geologic CO_2 storage is proved viable and safe by 2015, the first commercial plants could be operational in 2020. Combined coal-and-biomass-to-liquid plants will have one-fifth the output of a nominal 50,000-bbl/d coal plant and will probably be sited in regions near coal and biomass supplies, and so build-out rates will be lower than those for the cellulosic-ethanol plants discussed above. The panel estimates that at a growth rate of 20 percent until 2035, 2.5 million barrels of gasoline equivalent would be produced per day in combined coal and biomass plants. Production at that level would consume about 300 million dry tons of biomass—less than the projected biomass availability—and about 250 million tons of coal per year.

Because of the vast coal resources in the United States, the actual supply of coal-to-liquid fuel will be limited not by feedstock availability but rather by market penetration. At a build rate of two to three plants per year, up to 3 million barrels of gasoline equivalent could be produced per day by 2035. Production at that level would consume about 580 million tons of coal per year. However, issues related to an increase in coal mining by 50 percent need to be considered. At a build rate of three plants starting up per year, five to six plants would be under construction at any time.

COSTS, BARRIERS, AND DEPLOYMENT

Production of alternative liquid transportation fuels from coal and biomass with technology commercially deployable by 2020 can play an important role in reducing U.S. oil consumption and greenhouse gas emissions. The various options have different greenhouse gas impacts, and the choices will most likely depend on U.S. carbon policy. The panel used a consistent set of assumptions to estimate the costs of cellulosic ethanol, coal-to-liquid fuels with and without geologic CO_2 storage, and coal-and-biomass-to-liquid fuels with and without geologic CO_2 storage (see Table Sy.1). Although the estimates do not represent predictions of prices, they allow comparisons of fuel costs relative to each other. Coal-to-liquid fuels with geologic CO_2 storage can be produced at a cost of $70/bbl of gasoline equivalent and thus are competitive with a $75/bbl gasoline equivalent. The costs of fuels produced from biomass without geologic CO_2 storage are competitive with a $115/bbl gasoline equivalent with biochemical conversion and a $140/bbl gasoline equivalent with thermochemical conversion. The costs of cellulosic ethanol and coal-and-biomass-to-liquid fuels with geologic CO_2 storage become more attractive if a CO_2-equivalent price of $50/tonne is included.

Attaining supplies of 1.7 million barrels of biofuels per day, 2.5 million barrels of coal-and-biomass-to-liquid fuels per day, or 3 million barrels of coal-

TABLE Sy.1 Estimated Costs of Fuel Products With and Without a CO_2 Equivalent Price of $50/tonne

Fuel Product	Cost Without CO_2 Equivalent Price ($/bbl of gasoline equivalent)	Cost With CO_2 Equivalent Price of $50/tonne ($/bbl of gasoline equivalent)
Gasoline at crude-oil price of $60/bbl	75	95
Gasoline at crude-oil price of $100/bbl	115	135
Cellulosic ethanol	115	110
Biomass-to-liquid fuels without carbon capture and storage	140	130
Biomass-to-liquid fuels with carbon capture and storage	150	115
Coal-to-liquid fuels without carbon capture and storage	65	120
Coal-to-liquid fuels with carbon capture and storage	70	90
Coal-and-biomass-to-liquid fuels without carbon capture and storage	95	120
Coal-and-biomass-to-liquid fuels with carbon capture and storage	110	100

Note: Numbers are rounded to the nearest $5.

to-liquid fuels per day will require the permitting and construction of tens to hundreds of conversion plants and the associated fuel transport and delivery infrastructure. It will take more than a decade for these alternative fuels to penetrate the U.S. market. In addition, investments in alternative fuels have to be protected against crude-oil price fluctuations.

Integrated geologic CO_2 storage is key to producing liquid fuels from coal with greenhouse gas life-cycle emissions comparable to those of gasoline. Commercial demonstrations of coal-to-liquid and coal-and-biomass-to-liquid fuel conversion technologies with integrated geologic CO_2 storage should proceed immediately if the goal is to deploy commercial plants by 2020. Detailed scenarios for market penetration of U.S. biofuels and coal-to-liquid fuels should be developed to clarify the hurdles and challenges facing full feedstock use and to establish the enduring policies required. Current government and industry programs should be evaluated to determine whether emerging biomass and coal conversion technologies can further reduce U.S. oil consumption and greenhouse gas emissions over the next decade.

Summary

Growing worldwide energy demand, high commodity prices, high economic growth in developing countries, and growing scientific evidence that atmospheric carbon dioxide (CO_2) is an important contributor to global climate change make it urgent to increase energy supply and reduce worldwide greenhouse gas emissions at the same time. Achieving the first goal will require increasingly efficient energy production and use and expanded development of alternative sources of energy supplies that have low greenhouse gas emissions. In the United States today, the transportation sector relies almost exclusively on oil. Although domestic energy sources can supply all U.S. electricity needs, the United States is unable by itself to satisfy transportation sector and petrochemical industry demand for oil and so currently imports about 56 percent of the petroleum used in the United States. Moreover, volatile crude-oil prices and recent tightening of global supplies relative to demand, combined with fears that oil production will peak in the next 10–20 years, have aggravated concerns over oil dependence. The second goal is reduction of greenhouse gas emissions from the transportation sector, which accounts for one-third of the total emissions in the United States. Those two objectives have motivated the search for new vehicle power trains and alternative domestic sources of liquid fuels that can substantially lower greenhouse gas emissions.

Coal and biomass are abundant in the United States and can be converted to liquid fuels that can be combusted in existing and future vehicles with internal-combustion and hybrid engines. Their abundance makes them attractive candidates to provide non-oil-based liquid fuels for the U.S. transportation system. However, there are important questions about their economic viability, carbon impact, and technology status. Coal liquefaction is a potentially important source of alternative liquid transportation fuels, but the technology is capital-intensive. More important, fuel from liquefaction produces about twice the amount of greenhouse gas emissions on a life-cycle basis[1] as does petroleum-based gasoline if the process CO_2 is vented to the atmosphere. Capture of the process CO_2 and its geologic storage in the subsurface, often referred to as carbon capture and storage (CCS), will be required for producing coal-based liquid fuels in a carbon-constrained world. Thus, the viability, costs, and safety of lifetime geologic CO_2 storage could be barriers to commercialization.

Biomass is a renewable resource and, if properly produced and converted, can yield biofuels that have lower greenhouse gas emissions than do petroleum-based gasoline and diesel. Biomass production on already-cleared fertile land might compete with food, feed, and fiber production. If ecosystems are cleared directly or indirectly to produce biomass for biofuels, the resulting release of greenhouse gases from the cleared lands could negate for decades to centuries any greenhouse gas benefits of using biofuels. Thus, there are questions about how much biomass could be used for fuel without competing with food, feed, and fiber production to an important degree and without having adverse environmental effects.

STUDY SCOPE AND APPROACH

As part of its America's Energy Future (AEF) study (see Appendix A), the National Research Council appointed the 16-member Panel on Alternative Liquid Transportation Fuels to assess the potential for using coal and biomass to produce liquid fuels in the United States; provide thorough and consistent analyses of technologies for the production of alternative liquid transportation fuels; and prepare a report addressing the potential for use of coal and biomass to substantially reduce

---

[1]Life-cycle analysis yields an estimate of the emissions that will occur over the life cycle of a fuel. For example, life-cycle estimates cover the period from the time when the resource for the fuel is obtained (from the oil well in the case of petroleum-based gasoline, from the coal mine in the case of coal-to-liquid fuel) to the time when the fuel is combusted. In the case of biomass, the life cycle starts with the growth of biomass in the field and ends when the fuel is combusted. Greenhouse gas emissions that result from indirect land-use change, however, are not included in the estimates of life-cycle greenhouse gas emissions presented in this report.

U.S. dependence on conventional crude oil and also reduce greenhouse gas emissions in the transportation sector. The full statement of task is given in Appendix B. Although the report is the product of this independent panel, the results it presents will contribute to the larger AEF study mentioned in Appendix A.

The panel focused on technologies for converting biomass and coal to alternative liquid fuels that will be commercially deployable by 2020. Technologies that will be deployable after 2020 were also evaluated, but in less depth because they are associated with greater uncertainty than are the more developed technologies. For the purpose of this study, commercially deployable technologies are ones that have been scaled up from research and development to pilot-plant scale and then to several commercial-size demonstrations. Thus, the capital and operating costs of plants using commercially deployable technologies have been optimized so that the technologies can compete with other options. Commercial deployment of a technology—the rate at which it penetrates the market—depends on market forces, capital and human resource availability, competitive technologies, public policy, and other factors.

Because the choices for alternative liquid fuels are so many and so complex, the panel was unable to assess every potential biomass or conversion technology in the time available for this study. Instead, it focused on biomass supply and technologies that could potentially be commercially deployable over the next 15 years, be cost-competitive with petroleum fuels, and result in substantial reductions in U.S. oil consumption and greenhouse gas emissions. Other potential alternative fuels are reviewed at the end of the report (Chapter 9).

This study was initiated at a time (November 2007) when the prices of fossil fuels and other raw materials and the capital costs for infrastructure were rising rapidly. As the study progressed, those prices reached a peak (for example, the crude-oil price reached $147/bbl on July 11, 2008) and then began to fall steeply. Currently, there is continuing uncertainty about some of the factors that will directly influence the rate of deployment of technologies and the costs of new transportation-fuel supplies. The panel also recognized early in its deliberations the extent of the considerable debate reported on coal and biomass conversion technologies and biomass feedstock potential.

To decrease the uncertainty in its analysis and to ensure consistency among models used for comparison, the panel—with input from the Princeton Environmental Institute, the Massachusetts Institute of Technology, Purdue University, the University of Minnesota, Iowa State University, and the Renewable Energy Assessment Project team of the U.S. Department of Agriculture's Agricultural Research

Service—developed methods for estimating the costs and greenhouse gas impacts of supplying biomass, biochemical conversion, thermochemical conversion, and the potential quantity of fuel supply. Because of pervasive levels of uncertainty, however, the energy supply and cost estimates provided in this report should be considered as important first-step assessments rather than forecasts. The panel's estimates of the total costs of fuel products—including the feedstock, technical, engineering, construction, and production costs—were derived on a consistent basis and on the basis of a single set of conditions.

U.S. public policies related to energy have been introduced over the years. The oil crises of the 1970s sparked a number of energy-policy changes at the federal, state, and local levels. Price controls and rationing were instituted nationally, along with a reduced speed limit, to save gasoline. The Energy Policy and Conservation Act of 1975 created the Strategic Petroleum Reserve and mandated the doubling of fuel efficiency for automobiles from 13 to 27.5 miles/gal according to the corporate average fuel economy (CAFE) standards. Alternative fuels have been promoted in several other government incentives and mandates, including the Synthetic Liquid Fuels Act of 1944, the Energy Security Act of 1980 (which contained the U.S. Synthetic Fuels Corporation Act), the Alternative Motor Fuels Act of 1988, the Energy Policy Act of 1992, the Energy Policy Act of 2005, and the recent Energy Independence and Security Act of 2007 (which aims to increase the use of renewable fuels to at least 36 billion gallons by 2022 and set a new CAFE standard of 35 miles/gal by 2020). In addition, the American Jobs Creation Act of 2004 provided a tax credit of $0.51/gal of ethanol blended to companies that blend gasoline and a tax credit of $0.50–$1.00/gal of biodiesel to biodiesel producers.

Even though many public policies have addressed transportation-energy supply and use over the last 60 years and large amounts of public money have been spent, the use of alternative transportation fuels in the U.S. market today is still proportionately small. Many factors are involved in this low market penetration, such as generally low oil prices, but the fact that many of the policies have not been durable and sustainable has played an important role.

In its report, the panel identifies what it judged to be "aggressive but achievable" deployment opportunities for alternative fuels. Over the course of its study, it became clear to the panel that given the costs of alternative fuels and the volatility of fuel prices, significant deployment of alternative fuels in the market will probably require some realignment of public policies and regulations and the implementation of other incentives, such as substantial investment by both the public and the private sectors.

This summary includes some of the panel's key findings and recommendations; details of the panel's assessment and additional findings and recommendations are presented in subsequent chapters of the report. Quantities are expressed in standard units that are commonly used in the United States, except that greenhouse gas emissions are expressed in tonnes of CO_2 equivalent (CO_2 eq), the common unit used by the Intergovernmental Panel on Climate Change.

TECHNICAL READINESS FOR 2020 DEPLOYMENT

Biomass Supply

Responsible development of feedstocks for biofuels and expansion of biofuel use in the transportation sector must be socially, economically, and environmentally sustainable. The social, economic, and environmental effects of producing and using domestic biofuels have been mixed. In 2007, the United States consumed about 6.8 billion gallons of ethanol, mostly made from corn grain, and 491 million gallons of biodiesel, mostly made from soybean. The combined total of those two biofuels is less than about 3 percent of the fuels consumed for U.S. transportation. Diverting corn, soybean, or other food crops to biofuel production induces competition among food, feed, and fuel. Producing corn-grain ethanol and soybean biodiesel involves substantial use of fossil-fuel and other resources, and the improvements in greenhouse gas emissions compared with emissions associated with petroleum-based gasoline are small at best. Thus, the panel judges that corn-grain ethanol and soybean biodiesel are intermediate fuels in the transition from oil to cellulosic biofuels or other biomass-based liquid hydrocarbon transportation fuels, such as biobutanol and algal biofuels. In contrast, liquid biofuels made from lignocellulosic biomass can offer major improvements in greenhouse gas emissions relative to those from petroleum-based fuels if the biomass feedstock is a residual product of some forestry and farming operations or if it is grown on marginal lands that are not used for food and feed production.

Lignocellulosic feedstocks can be derived from both forestry and farming operations, including some production on marginal lands where commodity production often results in increased environmental problems because of erosion, runoff, and nutrient leaching. Therefore, the panel focused on the lignocellulosic resources available for biofuel production and assessed the costs of different biomass feedstocks delivered to a biorefinery for conversion. It considered societal needs, using recent analyses that have examined tradeoffs between land use for biofuel production and land use for food, feed, fiber, and ecosystem services. Corn stover, wheat and seed-grass straws, hay crops, dedicated perennial grass crops, woody biomass, wastepaper and paperboard, and municipal solid waste are the biofuel feedstocks considered in this report.

The panel estimated the amount of cellulosic biomass that could be produced sustainably in the United States and result in fuels with substantially lower greenhouse gas emissions than petroleum-based fuels. For the purpose of this study, the panel considers biomass to be produced in a sustainable manner (1) if croplands would not be diverted for biofuels and land therefore would not be cleared elsewhere to grow crops displaced by fuel crops and (2) if growing and harvesting of cellulosic biomass would incur minimal or even reduce adverse environmental effects such as erosion, excessive water use, and nutrient runoff. The panel estimated that about 400 million dry tons per year of biomass can potentially be made available for production of liquid transportation fuels with the technologies and management practices of 2008 (Table S.1). The cellulosic-biomass supply could increase to about 550 million dry tons per year by 2020. Key assumptions in the analysis are that 18 million acres of land currently enrolled in the Conservation Reserve Program (CRP) would be used to grow perennial grasses or other perennial crops for biofuel production and that the acreage would increase to 24 million by 2020 as knowledge increases. Other key assumptions are that harvesting methods would be developed for efficient collection of forestry or agricultural residues; that improved management practices and harvesting technology would increase agricultural crop yield; that yield increases could continue at the historical rates seen for corn, wheat, and hay; and that all the cellulosic biomass estimated to be available for energy production would be used for liquid fuels (this leads to an estimate of the potential amount of fuels produced).

The panel presented a scenario in which 550 million dry tons of cellulosic feedstock could be harvested or produced sustainably in 2020. That estimate is not a prediction of what would be available for fuel production in 2020. The supply of biomass could exceed the panel's estimate if croplands are used more efficiently or if genetic improvement of dedicated fuel crops exceeds the panel's estimate. In contrast, the panel's estimate could be lower if producers decide not to harvest agricultural residues or not to grow dedicated fuel crops on their CRP land.

TABLE S.1 Estimated Amount of Lignocellulosic Feedstock That Could Be Produced for Biofuel in 2008 with Technologies Available in 2008 and 2020

	Millions of Tons	
Feedstock Type	With Technologies Available in 2008	With Technologies Available by 2020
Corn stover	76	112
Wheat and grass straw	15	18
Hay	15	18
Dedicated fuel crops	104[a]	164
Woody biomass	110	124
Animal manure	6	12
Municipal solid waste	90	100
Total	416	548

[a]CRP land has not been used for dedicated fuel-crop production as of 2008. The panel assumed that two-thirds of the potentially suitable CRP land would be used for dedicated fuel production as an illustration.

The panel also estimated the costs of biomass delivered to a conversion plant (Table S.2). In that analysis, the price that the farmer or supplier would be willing to accept was assumed to include (1) land rental cost and other forgone net returns from not selling or not using the cellulosic material for feed or bedding and (2) all other costs incurred in sustainably producing, harvesting, and storing the biomass and transporting it to the processing plant. The willingness-to-accept price or feedstock price is the long-run equilibrium price that would induce suppliers to deliver biomass to the processing plant. Because an established market for cellulosic biomass does not exist, the panel's analysis relied on published estimates. However, the panel's estimates are higher than those in published reports because transportation and land rental costs are included.

The geographic distribution of biomass supply is also an important factor in the potential for development of a biofuels industry in the United States. The panel estimated the quantities of biomass that could, for example, be available within a 40-mile radius (which is about a 50-mile driving distance) of a given fuel-conversion plant in the United States (Figure S.1). An estimated 290 sites could supply 1,500–10,000 tons of biomass per day (0.5 million–2.4 million dry tons per year) to conversion plants within a 40-mile radius. The wide variation in the geographic distribution of the biomass potentially available for processing at plants will affect processing-plant size and is a factor in the potential to optimize

TABLE S.2 Biomass Suppliers' Willingness-to-Accept Prices in 2007 Dollars for 1 Dry Ton of Delivered Cellulosic Material

Biomass	Willingness-to-Accept Price (dollars per ton)	
	Estimated in 2008	Projected in 2020
Corn stover	110	86
Switchgrass	151	118
Miscanthus	123	101
Prairie grasses	127	101
Woody biomass	85	72
Wheat straw	70	55

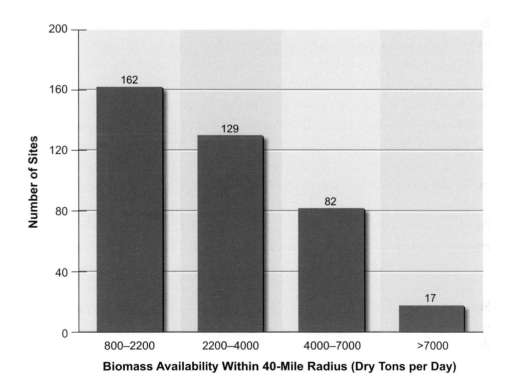

FIGURE S.1 *The number of sites in the United States with a potential to supply the indicated daily amounts of biomass within a 40-mile radius of each site.*

each conversion plant to decrease costs and maximize environmental benefits and supply in a given region. For example, increasing the distance of delivery could result in larger conversion plants with economies of scale that could lower fuel production costs.

To attain the panel's projected sustainable biomass supply, incentives would have to be provided to farmers and developers to use a systems approach for comprehensively addressing biofuel feedstock production; soil, water, and air quality; carbon sequestration; wildlife habitat; and rural development. The incentives would encourage farmers, foresters, biomass aggregators, and those operating biorefineries to work together to enhance technology development and ensure that the best management practices were used for different combinations of landscape and potential feedstock.

Finding S.1 (see Finding 2.1 in Chapter 2)

An estimated annual supply of 400 million dry tons of cellulosic biomass could be produced sustainably with technologies and management practices already available in 2008. The amount of biomass deliverable to conversion facilities could probably be increased to about 550 million dry tons by 2020. The panel judges that this quantity of biomass can be produced from dedicated energy crops, agricultural and forestry residues, and municipal solid wastes with minimal effects on U.S. food, feed, and fiber production and minimal adverse environmental effects.

Finding S.2 (see Finding 2.5 in Chapter 2)

Biomass availability could limit the size of a conversion facility and thereby influence the cost of fuel products from any facility that uses biomass irrespective of the conversion approach. Biomass is bulky and difficult to transport. The density of biomass growth will vary considerably from region to region in the United States, and the biomass supply available within 40 miles of a conversion plant will vary from less than 1,000 tons/day to 10,000 tons/day. Longer transportation distances could increase supply but would increase transportation costs and could magnify other logistical issues.

Recommendation S.1 (see Recommendation 2.3 in Chapter 2)

Technologies that increase the density of biomass in the field to decrease transportation cost and logistical issues should be developed. The densification of available biomass enabled by a technology such as field-scale pyrolysis could facilitate transportation of biomass to larger-scale regional conversion facilities.

Finding S.3 (see Finding 2.2 in Chapter 2)

Improvements in agricultural practices and in plant species and cultivars will be required to increase the sustainable production of cellulosic biomass and to achieve the full potential of biomass-based fuels. A sustained research and development (R&D) effort in increasing productivity, improving stress tolerance, managing diseases and weeds, and improving the efficiency of nutrient use will help to improve biomass yields.

Recommendation S.2 (see Recommendation 2.1)

The federal government should support focused research and development programs to provide the technical bases for improving agricultural practices and biomass growth to achieve the desired increase in sustainable production of cellulosic biomass. Focused attention should be directed toward plant breeding, agronomy, ecology, weed and pest science, disease management, hydrology, soil physics, agricultural engineering, economics, regional planning, field-to-wheel biofuel systems analysis, and related public policy.

Finding S.4 (see Finding 2.3 in Chapter 2)

Incentives and best agricultural practices will probably be needed to encourage sustainable production of biomass for production of biofuels. Producers need to grow biofuel feedstocks on degraded agricultural land to avoid direct and indirect competition with the food supply and also need to minimize land-use practices that result in substantial net greenhouse gas emissions. For example, continuation of CRP payments for CRP lands when they are used to produce perennial grass and wood crops for biomass feedstock in an environmentally sustainable manner might be an incentive.

Recommendation S.3 (see Recommendation 2.2 in Chapter 2)

A framework should be developed to assess the effects of cellulosic-feedstock production on various environmental characteristics and natural resources. Such an assessment framework should be developed with input from agronomists, ecologists, soil scientists, environmental scientists, and producers and should include, at a minimum, effects on greenhouse gas emissions and on water and soil resources. The framework would provide guidance to farmers on sustainable production of cellulosic feedstock and contribute to improvements in energy security and in the environmental sustainability of agriculture.

Coal Supply

Deployment of coal-to-liquids technologies would require the use of large quantities of coal and thus an expansion of the coal-mining industry. For example, a 50,000-barrels/day (50,000 bbl/d) plant will use about 7 million tons of coal per year, and 100 such plants producing liquid transportation fuels at 5 million bbl/d would use about 700 million tons of coal per year, which would mean a 70 percent increase in coal consumption. That would require major increases in coal-mining and transportation infrastructure for moving coal to the plants and moving fuel from the plants to the market. Those issues could represent major challenges, but they could be overcome. A key question is the availability of sufficient coal in the United States to support such increased use while supporting the coal-based power industry. A National Research Council evaluation (NRC, 2007) of domestic coal resources concluded as follows:

> Federal policy makers require accurate and complete estimates of national coal reserves to formulate coherent national energy policies. Despite significant uncertainties in existing reserve estimates, it is clear that there is sufficient coal at current rates of production to meet anticipated needs through 2030. Further into the future, there is probably sufficient coal to meet the nation's needs for more than 100 years at current rates of consumption. . . . A combination of increased rates of production with more detailed reserve analyses that take into account location, quality, recoverability, and transportation issues may substantially reduce the number of years of supply. Future policy will continue to be developed in the absence of accurate estimates until more detailed reserve analyses—which take into account the full suite of geographical, geological, economic, legal, and environmental characteristics—are completed. (p. 4)

Recently, the Energy Information Administration estimated the proven U.S. coal reserves to be about 260 billion tons. A key conclusion was that there are sufficient coal reserves in the United States to meet the nation's needs for over 100 years at current rates of consumption, and possibly even with increased rates of consumption. The primary issue probably is not the reserves but increased mining of coal and the opening of many new mines. Increased mining has numerous environmental effects that will need to be dealt with, and there will probably be public opposition to it. Increasing use of coal will undoubtedly increase the cost of coal, which is low relative to the cost of biomass.

Finding S.5 (see Finding 4.1 in Chapter 4)

Despite the vast coal resource in the United States, it is not a forgone conclusion that adequate coal will be mined and be available to meet the needs of a growing coal-to-fuels industry and the needs of the power industry.

Recommendation S.4 (see Recommendation 4.1 in Chapter 4)

The U.S. coal industry, the U.S. Environmental Protection Agency, the U.S. Department of Energy, and the U.S. Department of Transportation should assess the potential for a rapid expansion of the U.S. coal-supply industry and delineate the critical barriers to growth, environmental effects, and their effects on coal cost. The analysis should include several scenarios, one of which assumes that the United States will move rapidly toward increasing use of coal-based liquid fuels for transportation to improve energy security. An improved understanding of the immediate and long-term environmental effects of increased mining, transportation, and use of coal would be an important goal of the analysis.

Conversion Technologies

Two key technologies, biochemical and indirect thermochemical conversion, that are required for the conversion of biomass and coal to fuels are illustrated in Figure S.2. Biochemical conversion typically uses enzymes to transform starch (from grains) or lignocelluloses into sugars as intermediates (saccharification), and the sugars are converted to ethanol by microorganisms (fermentation). Indirect thermochemical conversion uses heat and steam to convert biomass or coal into primarily a mixture of carbon monoxide (CO) and hydrogen (H_2)—syngas—which can be cleaned and converted to have the right $CO:H_2$ ratio (now referred to as synthesis gas) and then be catalytically converted to liquid fuels, such as diesel

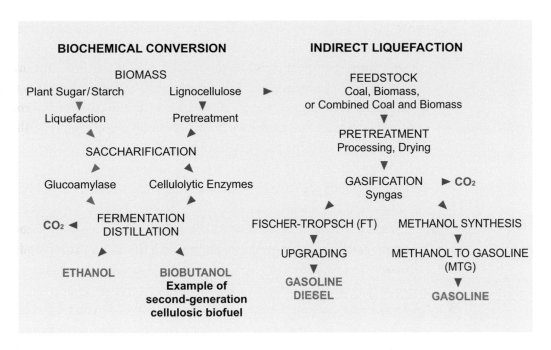

FIGURE S.2 *Steps involved in the biochemical conversion of biomass and the thermo-chemical conversion (indirect route only) of coal, biomass, and combined coal and biomass to liquid transportation fuels.*

and gasoline. The CO_2 from the fermentation process in biochemical conversion or from the off-gas streams of the thermochemical processes can be captured and stored geologically. Direct liquefaction of coal, which involves adding H_2 directly to slurried coal at high temperatures and pressures in the presence of suitable catalysts, represents another route from coal to liquid fuels but is less developed than indirect liquefaction. That route is not shown.

Biochemical Conversion

Biochemical conversion of starch from grains to ethanol (as shown on the left side of Figure S.2) has been deployed commercially. Grain-based ethanol, although important for initiating public awareness and industrial infrastructure for fuel ethanol, is considered by the panel to be a transition to cellulosic ethanol and other advanced cellulosic biofuels. The biomass supplies likely to be available by 2020 technically could be converted to ethanol by biochemical conversion, displace a substantial fraction of petroleum-based gasoline, and reduce greenhouse gas emissions, but the conversion technology has to be demonstrated first and be developed to a commercially deployable state.

Cellulosic ethanol could be the main biochemical route of converting biomass to fuels over the next decade or two. Further R&D could lead to commercial technologies that convert sugars to such other biofuels as butanol and alkanes, which have higher energy densities and could be distributed in the existing infrastructure. Although the panel focused on cellulosic ethanol as the most deployable technology for the next 10 years, it sees a long-term transition to cellulosic conversion to higher alcohols or hydrocarbons—so-called advanced biofuels—as having important long-term potential.

The challenge in biochemical conversion of biomass to fuels is first to break down the recalcitrant structure of the plant cell wall and then to break down the cellulose to five-carbon and six-carbon sugars that can be fermented by microorganisms. The effectiveness of this sugar generation is important for economical biofuel production. The process of production of cellulosic ethanol includes (see Figure S.2) preparation of feedstock to achieve size reduction by grinding or other means; pretreatment of feedstock with steam, hot water, or acid or base to release cellulose from the lignin shield; saccharification—cellulase to hydrolyze cellulose polymers to cellobiose (a disaccharide) and glucose (a monosaccharide) and hemicellulase to break down hemicellulose to monosaccharides; fermentation of the sugars to ethanol; and distillation to separate the ethanol. The CO_2 generated in the conversion process and in combustion of the fuel is mostly offset by the CO_2 taken up during the growth of the biomass. The unconverted materials are burned in a boiler to generate steam for the distillation; some excess electricity can thus be generated.

As of 2008, no commercial-scale cellulosic ethanol plants were operational. However, the Department of Energy announced in February 2007 that it would invest up to $385 million for six biorefinery projects (two based on gasification) over 4 years to move cellulosic ethanol to the market. When they are fully operational, the total production of the six plants would be 8,000 bbl/d. In addition, a number of private companies are actively pursuing commercialization of cellulosic-ethanol plants. Technologies for cellulosic ethanol will continue to evolve over the next 5–10 years as challenges are overcome and experience is gained in the first technology-demonstration and commercial-demonstration plants. The panel expects deployable, commercialized technology to be in place by 2020 if technology-demonstration plants continue to be built despite the current economic crisis and are rapidly followed by commercial-demonstration plants. Because of lack of commercial experience, the cost of initial commercial plants could well be higher than estimated by the modeling but decrease as commercial experience is gained.

An expanded transport and distribution infrastructure will be required to replace gasoline with a larger proportion of ethanol produced by biochemical conversion because ethanol cannot be transported in pipelines used for petroleum transport. Ethanol is currently transported by rail or barges and not by pipelines, because it is hygroscopic and can damage seals, gaskets, and other equipment and induce stress-corrosion cracking in high-stress areas. Gasoline vehicles can tolerate gasoline blends with up to 20 percent ethanol. If ethanol is to be used in a fuel at concentrations higher than 20 percent (for example, E85, which is a blend of 85 percent ethanol and 15 percent gasoline), the number of refueling stations will have to be increased to support alternative-fuel vehicles designed for alcohol fuels. The transport and distribution of synthetic diesel and gasoline produced with thermochemical conversion do not pose the same challenge because they are compatible with the existing infrastructure for petroleum-based fuels.

The key process-related challenges in R&D and demonstration that need to be addressed before widespread commercialization are as follows: to improve the effectiveness of pretreatment to remove and hydrolyze the hemicellulose, separate the cellulose from the lignin, and loosen the cellulose structure; to reduce the production cost of the enzymes for converting cellulose to sugars; to reduce operating costs by developing more effective enzymes and more efficient microorganisms for converting the sugar products of biomass deconstruction to biofuels; to demonstrate the biochemical conversion technology on a commercial scale; and to begin to optimize capital costs and operating costs. The size of a biorefinery will probably be limited by the supply of biomass available from the surrounding regions. That size limitation could result in loss of potential economies of scale that characterize large plants.

Finding S.6 (see Finding 3.2 in Chapter 3)

Process improvements in cellulosic-ethanol technology are expected to be able to reduce the plant-related costs associated with ethanol production by up to 40 percent over the next 25 years. Over the next decade, process improvements and cost reductions are expected to come from evolutionary developments in technology, from learning gained through commercial experience and increases in scale of operation, and from research and engineering in advanced chemical and biochemical catalysts that will enable their deployment on a large scale.

Recommendation S.5 (see Recommendation 3.2 in Chapter 3)

The federal government should continue to support research and development to advance cellulosic-ethanol technologies. R&D programs should be pursued to resolve the major technical challenges facing ethanol production from cellulosic biomass: pretreatment, enzymes, tolerance to toxic compounds and products, solids loading, engineering microorganisms, and novel separations for ethanol and other biofuels. A long-term perspective on the design of the programs and allocation of limited resources is needed; high priority should be placed on programs that address current problems at a fundamental level but with visible industrial goals.

Recommendation S.6 (see Recommendation 3.3 in Chapter 3)

The pilot and commercial-scale demonstrations of cellulosic-ethanol plants should be complemented by a closely coupled research and development program. R&D is necessary to resolve issues that are identified during demonstration and to reduce costs of sustainable feedstock acquisition. Industrial experience shows that such reductions typically occur as processes go through multiple phases of implementation and expansion.

Finding S.7 (see Finding 3.4 in Chapter 3)

Biochemical conversion processes, as configured in cellulosic-ethanol plants, produce a stream of relatively pure CO_2 from the fermenter that can be dried, compressed, and made ready for geologic storage or used in enhanced oil recovery with little additional cost. Geologic storage of the CO_2 from biochemical conversion of plant matter (such as cellulosic biomass) further reduces greenhouse gas life-cycle emissions from advanced biofuels, so their greenhouse gas life-cycle emissions would become highly negative.

Recommendation S.7 (see Recommendation 3.5 in Chapter 3)

Because geologic storage of CO_2 from biochemical conversion of biomass to fuels could be important in reducing greenhouse gas emissions in the transportation sector, it should be evaluated and demonstrated in parallel with the program of geologic storage of CO_2 from coal-based fuels.

Finding S.8 (see Finding 3.3 in Chapter 3)

Future improvements in cellulosic technology that entail invention of biocatalysts and biological processes could produce fuels that supplement ethanol production in the next 15 years. In addition to ethanol, advanced biofuels (such as lipids, higher alcohols, hydrocarbons, and other products that are easier to separate than ethanol) should be investigated because they could have higher energy content and would be less hygroscopic than ethanol and therefore could fit more smoothly into the current petroleum infrastructure than could ethanol.

Recommendation S.8 (see Recommendation 3.4 in Chapter 3)

The federal government should ensure that there is adequate research support to focus advances in bioengineering and the expanding biotechnologies on developing advanced biofuels. The research should focus on advanced biosciences—genomics, molecular biology, and genetics—and biotechnologies that could convert biomass directly to produce lipids, higher alcohols, and hydrocarbons fuels that can be directly integrated into the existing transportation infrastructure. The translation of those technologies into large-scale commercial practice poses many challenges that need to be resolved by R&D and demonstration if major effects on production of alternative liquid fuels from renewable resources are to be realized.

Finding S.9 (see Finding 5.1 in Chapter 5)

The need to expand the delivery infrastructure to meet a high volume of ethanol deployment could delay and limit the penetration of ethanol into the U.S. transportation-fuels market. Replacing a substantial proportion of transportation gasoline with ethanol will require a new infrastructure for its transport and distribution. Although the cost of delivery is a small fraction of the overall fuel-ethanol cost, the logistics and capital requirements for widespread expansion could present many hurdles if they are not planned for well.

Recommendation S.9 (see Recommendation 5.1 in Chapter 5)

The U.S. Department of Energy and the biofuels industry should conduct a comprehensive joint study to identify the infrastructure system requirements of, research and development needs in, and challenges facing the expanding biofuels

industry. Consideration should be given to the long-term potential of truck or barge delivery versus the potential of pipeline delivery that is needed to accommodate increasing volumes of ethanol. The timing and role of advanced biofuels that are compatible with the existing gasoline infrastructure should be factored into the analysis.

Thermochemical Conversion

Indirect liquefaction converts coal to liquid fuels (CTL), biomass to liquid fuels (BTL), or mixtures of coal and biomass to liquid fuels (CBTL) by gasifying the feedstocks to produce syngas, cleaning and adjusting the H_2:CO ratio, and then catalytically converting the synthesis gas with Fischer-Tropsch (FT) technology to high-cetane, clean diesel, and some naphtha (which can be upgraded to gasoline). The synthesis gas can also be converted to methanol with commercial technology, and methanol-to-gasoline (MTG) technology can then be used to produce high-octane gasoline from the methanol (see Figure S.2). Those technologies can be integrated with technologies that compress the CO_2 emitted during production and store it in Earth's subsurface (CCS), such as in deep saline aquifers. Unlike ethanol, the gasoline and diesel produced via FT and MTG are fully compatible with the existing infrastructure and vehicle fleet.

Gasification has been used commercially around the world for nearly a century by the chemical, refining, and fertilizer industries and for more than 10 years by the electric-power industry. More than 420 gasifiers are in use in some 140 facilities worldwide; 19 plants are operating in the United States. Coal gasification is commercially deployable today with any of several gasification systems that are being commercially used. Application in CTL fuels and other applications of gasification will lead to further improvements in the technology, and it will have become more robust and efficient by 2020. The improvements are part of the usual evolution of any new technology. Combined coal and biomass gasification is close to being commercially deployable, and further commercial application will make it more robust and efficient and enhance its ability to use higher fractions of biomass by 2020. Biomass gasification has been commercially demonstrated but requires more operational experience to make it robust and to allow well-informed designs.

FT technology was commercialized at the South African firm Sasol's complexes beginning in the middle 1950s. Sasol now produces transportation fuels from coal at more than 165,000 bbl/d, and large plants convert natural gas to synthesis gas, which is converted to diesel and gasoline with the FT process. FT synthesis technology can be considered commercially deployable today. Like several other ready-to-deploy technologies, it will undergo substantial process improvement by 2020, which will lead to more robust and efficient technology for producing liquid transportation fuels, and catalyst improvements for coal applications can be expected.

In technologies based on methanol synthesis, synthesis gas is converted to methanol with available commercial technology in plants as large as 6,000 tons/day. The methanol can be used directly or be upgraded into high-octane gasoline with a proprietary catalytic process developed by ExxonMobil and referred to as the MTG process, which was commercialized in New Zealand in the late 1980s. Standard MTG technology is considered by the panel to be commercially deployable today—and several projects are moving toward commercial deployment. Several variations of the technology are ready for commercial demonstration and could lead to improvements in the standard MTG technology.

Although the technologies involved in thermochemical conversion of coal have all been commercialized and have years of operating experience, geologic storage of CO_2 has not been adequately developed and demonstrated. In the case of power generation from coal, most of the costs for CCS are in the CO_2-capture part of the process, and this technology has been demonstrated on a large scale. However, geologic storage of CO_2 in the subsurface has not been developed and demonstrated to the point where there is sufficient confidence in its long-term efficiency and efficacy to embark on commercial application on the scale required. The CO_2 emissions from CTL technology are high because of the high heat needed for the process and the high carbon content of coal (which has about twice the carbon content of oil); even with geologic storage of CO_2, the well-to-wheel emissions from CTL are about the same as those of gasoline because, as with any hydrocarbon fuel, CO_2 is released when the fuel is combusted in vehicles.

Inclusion of biomass in the feedstock with coal decreases greenhouse gas life-cycle emissions because the biomass takes up atmospheric CO_2 during its growth. It is possible to optimize the biomass-plus-coal indirect liquefaction process to produce liquid fuels that have somewhat lower greenhouse gas life-cycle emissions than gasoline has and even to make carbon-neutral liquid fuels if geologic storage of CO_2 is used. Although the notion of gasifying mixtures of coal and biomass to produce liquid fuels is relatively new and commercial experience is small, several demonstration units are running in Europe. Gasifiers for biomass alone, designed around limited biomass availability, operate on a smaller scale than do those for coal and so will be more expensive because of the diseconomies of scale of small plants. However, the fuels produced from such plants can have greenhouse gas life-cycle emissions that are close to zero without geologic storage of CO_2. Thermochemical processes that use biomass only can therefore be carbon neutral, as is biochemical conversion, and can have highly negative carbon emissions if geologic storage of CO_2 is used. The panel judges that the technology for cofeeding biomass and coal is close to being ready for commercial deployment, but commercial experience with the technology is needed. Stand-alone biomass gasification technology is probably 5–8 years away from commercial scale-up.

The subject of greatest uncertainty in connection with conversion of coal and biomass to fuels is the geologic storage of CO_2. As of 2008, few commercial-scale geologic-storage demonstrations have been carried out or are ongoing. Well-monitored commercial-scale demonstrations are needed to gather data sufficient to assure industry and governments as to the long-term viability, costs, and safety of geologic CO_2 storage and to develop procedures for site choice, permitting, operation, regulation, and closure. The political and commercial acceptability of geologic storage of CO_2 is critical for the commercial viability of thermochemical technology. The estimates of potential CCS costs of \$10–\$15/tonne of CO_2 avoided are "bottom-up" largely on the basis of engineering estimates of expenses for transport, land purchase, permitting, drilling, required capital equipment, storing, capping wells, and monitoring for an additional 50 years. However, uncertainty about the regulatory environment arising from concerns of the general public and policy makers has the potential to raise storage costs and slow commercialization of thermochemical fuel technology (Appendix K). Ultimate requirements for design, monitoring, carbon-accounting procedures, liability for long-term monitoring of geologically stored CO_2, and associated regulatory frameworks depend on future commercial-scale demonstrations of geologic storage of CO_2. The commercial demonstrations will have to be pursued aggressively in the next few years if thermochemical conversion of biomass and coal with geologic storage of CO_2 is to be ready for commercial deployment by 2020.

As a first step toward accelerating the commercial demonstration of CTL and CBTL technology and addressing the CO_2-storage issue, commercial-scale demonstration plants could serve as sources of CO_2 for geologic-storage demonstration projects. So-called capture-ready plants that vent CO_2 would create liquid fuels with higher CO_2 emissions per unit of usable energy than petroleum-based fuels; their commercialization should not be encouraged unless the plants are integrated with geologic storage of CO_2 at their start-up.

Direct liquefaction of coal—which involves relatively high temperature, high hydrogen pressure, and liquid-phase conversion of coal directly to liquid products—has a long history, as does FT. Direct-liquefaction products generally are heavy liquids that require upgrading to liquid transportation fuels. The technology is not ready for commercial deployment. Furthermore, the panel's ability to estimate costs and performance was limited by the lack of recent detailed design studies in the available literature.

The three most important challenges in R&D and demonstration facing commercialization of thermochemical technologies are

- Immediate construction of a small number of commercial first-mover projects combined with geologic storage of CO_2 to move the technology toward reduced cost, improved performance, and robustness. The commercial first-mover projects would have a major R&D component to focus on solving issues and problems identified in the operation and to develop technology for specific improvements.
- R&D programs associated with commercial-scale geologic CO_2 storage demonstrations that involve detailed geologic analysis and a broad array of monitoring tools and techniques to provide the understanding and data on which future commercial projects will depend.
- R&D on gasification of biomass or combined biomass and coal, which has potential CO_2-reduction benefits, is critical to bring this technology to commercial deployment. In particular, penalties associated with the preprocessing of biomass, the choice of a best gasifier for a given biomass type, and the technical problems in feeding biomass to high-pressure gasification systems have to be resolved.

Finding S.10 (see Finding 4.2 in Chapter 4)

Technologies for the indirect liquefaction of coal to transportation fuels are commercially deployable today; but without geologic storage of the CO_2 produced in the conversion, greenhouse gas life-cycle emissions will be about twice those of petroleum-based fuels. With geologic storage of CO_2, CTL transportation fuels could have greenhouse gas life-cycle emissions equivalent to those of equivalent petroleum-derived fuels.

Finding S.11 (see Finding 4.7 in Chapter 4)

Technologies for the indirect liquefaction of coal to produce liquid transportation fuels with greenhouse gas life-cycle emissions equivalent to those of petroleum-based fuels can be commercially deployed before 2020 only if several first-mover plants are started up soon and if the safety and long-term viability of geologic storage of CO_2 is demonstrated in the next 5–6 years.

Finding S.12 (see Finding 4.3 in Chapter 4)

Indirect liquefaction of combined coal and biomass to transportation fuels is close to being commercially deployable today. Coal can be combined with biomass at a ratio of 60:40 (on an energy basis) to produce liquid fuels that have greenhouse gas life-cycle emissions comparable with those of petroleum-based fuels if CCS is not implemented. With CCS, production of fuels from coal and biomass would have a carbon balance of about zero to slightly negative.

Recommendation S.10 (see Recommendation 4.4 in Chapter 4)

A program of aggressive support for first-mover commercial plants that produce coal-to-liquid transportation fuels and coal-and-biomass-to-liquid transportation fuels with integrated geologic storage of CO_2 should be undertaken immediately to address U.S. energy security and to provide fuels with greenhouse gas emissions similar to or less than those of petroleum-based fuels. The demonstration and deployment of "first-mover" coal or coal-and-biomass plants should be encouraged on the basis of the primary technologies, including CCS to demonstrate the technological viability of CTL and CBTL fuels and to reduce the technical and investment risks associated with funding of future plants. If decisions to proceed with commercial demonstrations are made soon so that the plants could start up in 4–5 years and if CCS is demonstrated to be safe and viable, those technologies would be commercially deployable by 2020.

Finding S.13 (see Finding 4.8 in Chapter 4)

The technology for producing liquid transportation fuels from biomass or from combined biomass and coal via thermochemical conversion has been demonstrated but requires additional development to be ready for commercial deployment.

Recommendation S.11 (see Recommendation 4.6 in Chapter 4)

Key technologies should be demonstrated for biomass gasification on an intermediate scale, alone and in combination with coal, to obtain the engineering and operating data required to design commercial-scale synthesis gas-production units.

Finding S.14 (see Finding 4.4 in Chapter 4)

Geologic storage of CO_2 on a commercial scale is critical for producing liquid transportation fuels from coal without a large adverse greenhouse gas impact. This is similar to the situation for producing power from coal.

Recommendation S.12 (see Recommendation 4.2 in Chapter 4)

The federal government should continue to partner with industry and independent researchers in an aggressive program to determine the operational procedures, monitoring, safety, and effectiveness of commercial-scale technology for geologic storage of CO_2. Three to five commercial-scale demonstrations (each with about 1 million tonnes of CO_2 per year and operated for several years) should be set up within the next 3–5 years in areas of several geologic types.

The demonstrations should focus on site choice, permitting, monitoring, operation, closure, and legal procedures needed to support the broad-scale application of geologic storage of CO_2. The development of needed engineering data and determination of the full costs of geologic storage of CO_2—including engineering, monitoring, and other costs based on data developed from continuing demonstration projects—should have high priority.

COSTS, GREENHOUSE GAS EMISSIONS, AND SUPPLY

This section compares the life-cycle costs, CO_2 life-cycle emissions,[2] and potential supply of the alternative-fuel technologies deployable by 2020 by analyzing the supply chain that begins with the biomass and coal feedstocks and ends with

---

[2]This section assesses only CO_2 life-cycle emission because the panel was not able to determine changes in emission of other greenhouse gases throughout the life cycle of fuel production. Changes in emission of greenhouse gases other than CO_2 are likely to be small or none.

alternative liquid fuels. The result of the analysis is a supply curve of potential alternative liquid fuels that use biomass, coal, or combined biomass and coal as feedstocks on the basis of technologies deployable by 2020. The supply curve does not represent the amounts of fuels that would be commercially deployed. The actual supply in 2020 could well be smaller than the potential supply because there are important lags in decisions to construct new conversion plants and in construction, as discussed in the section "Deployment of Alternative Liquid Transportation Fuels" below. In addition, some of the coal and biomass supplies that appear to be economical might not be made available for conversion to alternative fuels because of logistical, infrastructure, and agricultural organization issues or because they would have already been committed to power plants. The analysis shows how the potential supply curve might change with alternative CO_2 prices and alternative capital costs. As mentioned earlier, the panel worked with several research groups to develop the costs and CO_2 life-cycle emission for the individual conversion technologies and the cost of biomass.

To examine the potential supply of liquid transportation fuels from nonpetroleum sources, the panel developed estimates of the unit costs and quantities of various biomass sources that could be made available. The panel's analysis was based on land that is currently not used for growing foods, although the panel cannot ensure that none of that land will be used for food production in the future. The estimates of biomass supply were combined with estimates of supply of corn grain to satisfy the current legislative requirement to produce 15 billion gallons of ethanol. The analysis allowed the estimation of a supply curve for biomass that shows the quantities of biomass feedstocks that would potentially be available at the various unit costs. Coal was assumed to be available in sufficient quantities at a constant unit cost if used with biomass in thermochemical conversion processes. Quantitative analyses were developed to compare alternative pathways to convert biomass, coal, or combined coal and biomass to liquid transportation fuels using thermochemical technologies. Biochemical technology that produced ethanol from biomass was also evaluated quantitatively on as consistent a basis as possible. Various combinations of biomass feedstocks could, in principle, be converted with either thermochemical or biochemical conversion processes.[3] However, rather than examining all possible combinations, the panel first examined the cost of and CO_2 emission associated with each of the various

---

[3]In addition, the panel included a biochemical conversion of corn grain to ethanol but did not focus the quantitative analysis on this process.

thermochemical and biochemical conversion processes by using a generic biomass feedstock with approximately a median cost and biochemical composition (the panel used *Miscanthus* in the analysis) and then examined the costs, supplies, and CO_2 emissions associated with one thermochemical conversion process and one biochemical conversion process that would use each of the different biomass feedstocks.

The analyses involved a set of assumptions, changes in which would likely change the estimated supply curve:

- The panel's analyses assume that all available CRP land discussed earlier will be made available for growing biomass for liquid fuels. Conversion plants that use biomass as feedstock by itself and in combination with coal (60 percent coal and 40 percent biomass on an energy basis) would have the capacity of using about 4,000 dry tons of biomass per day.

- All product prices are assumed to be without government subsidies. The costs of CO_2 avoided—which include the cost of drying, compression, pipelining, and geologic storage of CO_2—are estimates of engineering costs to implement geologic storage and are in the range of $10–15/tonne of CO_2 avoided.

- If a carbon price is imposed, the assumption is that it applies to the entire life-cycle CO_2 net emission, which is the balance of CO_2 removal from the atmosphere by plants, CO_2 released in the production of biomass (for example, CO_2 released from machinery used in the production), and emissions from conversion of feedstock to fuels and from combustion of the liquid fuels. A fuel that removes more CO_2 from the atmosphere than it produces over its life cycle would receive a net payment for CO_2.

- The panel's analyses assume that no indirect greenhouse gas emissions result from land-use changes in the growing and harvesting of biomass. All biomass volumes in Chapter 2 were estimated under the constraint that they could be grown and harvested without creating indirect greenhouse gas emissions.

- The price of subbituminous Illinois no. 6 coal is assumed to be $42 per dry ton.

- Electricity produced as a coproduct is assumed to be valued at $80/MWh in the absence of any price placed on CO_2.

Costs and Carbon Dioxide Emissions

The estimated 2020 supply function for biomass costs versus availability is shown in Figure S.3. The costs of two feedstocks—corn grain and hay—are based on recent market prices. In particular, it is assumed that corn price will have dropped sharply from the 2008 high of $7.88/bushel to $3.17/bushel in 2020, corresponding to $130 per dry ton, a price more consistent with its historical levels. The price of dryland or field-run hay is assumed to be $110/ton, similar to historical prices. Finally, the cost of using wastes is based on a rough estimate of the costs of gathering, transporting, and storing municipal waste. Such costs can be expected to be highly variable, but the panel assumes that gathering, transporting, and storing add up to $51 per dry ton.

The costs of producing alternative liquid fuels through the various pathways were estimated on the basis of feedstock, capital, and operating costs; conversion efficiencies; and the assumptions outlined above. Figure S.4 shows the esti-

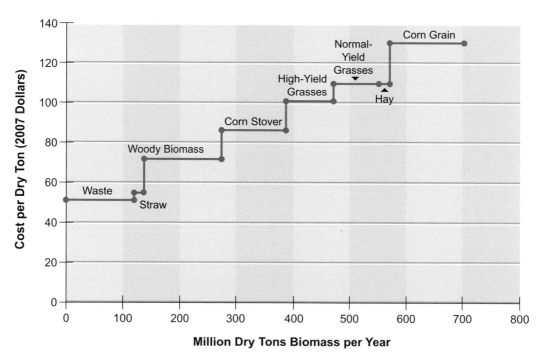

FIGURE S.3 *Supply function for biomass feedstocks in 2020. High-yield grasses include* Miscanthus *and normal-yield grasses include switchgrass and prairie grasses.*

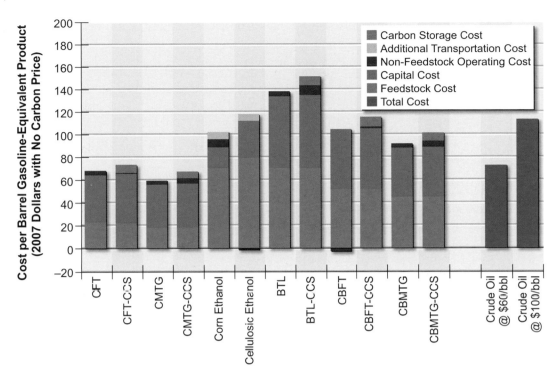

FIGURE S.4 *Costs of alternative liquid fuels produced from coal, biomass, or coal and biomass with zero carbon price.*
Note: BTL = biomass-to-liquid fuel; CBFT = coal-and-biomass-to-liquid fuel, Fischer-Tropsch; CBMTG = coal-and-biomass-to-liquid fuel, methanol-to-gasoline; CCS = carbon capture and storage; CFT = coal-to-liquid fuel, Fischer-Tropsch; CMTG = coal-to-liquid fuel, methanol-to-gasoline.

mated gasoline-equivalent[4] costs of alternative liquid fuels, without a CO_2 price, produced from biomass (B), coal (C), or combined coal and biomass (CB). As indicated above, liquid fuels would be produced with biochemical conversion to produce ethanol from a generic biomass (cellulosic ethanol), thermochemical conversion via FT, or MTG. For thermochemical conversion, FT and MTG are shown both with and without CCS. The supply of ethanol produced from corn grain is also included in the figure. For comparison, costs of gasoline are shown for two crude-oil prices: $60/bbl and $100/bbl (that is, $73/bbl and $113/bbl of gasoline equivalent). Results are also shown in Table S.3.

---

[4]Costs per barrel of ethanol are divided by 0.67 to put ethanol costs on an energy-equivalent basis with gasoline. For FT liquids, the conversion factor is 1.0.

TABLE S.3 Estimated Costs of Fuel Products With and Without a CO_2 Equivalent Price of $50/tonne

Fuel Product	Cost Without CO_2 Equivalent Price ($/bbl of gasoline equivalent)	Cost With CO_2 Equivalent Price of $50/tonne ($/bbl of gasoline equivalent)
Gasoline at crude-oil price of $60/bbl	75	95
Gasoline at crude-oil price of $100/bbl	115	135
Cellulosic ethanol	115	110
Biomass-to-liquid fuels without carbon capture and storage	140	130
Biomass-to-liquid fuels with carbon capture and storage	150	115
Coal-to-liquid fuels without carbon capture and storage	65	120
Coal-to-liquid fuels with carbon capture and storage	70	90
Coal-and-biomass-to-liquid fuels without carbon capture and storage	95	120
Coal-and-biomass-to-liquid fuels with carbon capture and storage	110	100

Note: Numbers are rounded to the nearest $5. Estimated costs of fuel products for coal-to-liquids conversion represent the mean costs of fuels produced via FT and MTG.

Figure S.5 shows the net CO_2 emissions per gasoline-equivalent barrel produced by various production pathways. The CO_2 released on combustion of the fuel is similar among the various pathways, with ethanol releasing less CO_2 than either gasoline or synthetic diesel and gasoline produced from coal or combined coal and biomass (that is, CFT, CMTG, CBFT, and CBMTG). The large variation in net releases is the result of the large variations in the CO_2 taken from the atmosphere in growing biomass and the large variations in the release of CO_2 into the atmosphere in the conversion process. CO_2 emission from corn-grain ethanol is slightly lower than that from conventional gasoline. In contrast, CO_2 emission from cellulosic ethanol without CCS is close to zero.

Figure S.4 shows that CTL fuel products with and without geologic CO_2 storage are cost-competitive at gasoline-equivalent prices of about $70/bbl and $65/bbl, respectively (this represents equivalent crude-oil prices of around $50–55/bbl). Gasoline prices from MTG are similar. Figure S.5 shows that without CCS the process vents a large amount of CO_2, and the CO_2 life-cycle emission

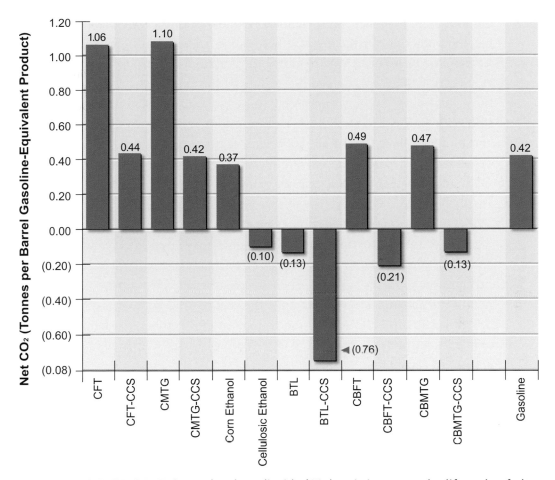

FIGURE S.5 *Estimated carbon dioxide (CO₂) emissions over the life cycle of alternative fuels, from the mining and harvesting of resources to the conversion process to the consumption of fuels.*
Note: BTL = biomass-to-liquid fuel; CBFT = coal-and-biomass-to-liquid fuel, Fischer-Tropsch; CBMTG = coal-and-biomass-to-liquid fuel, methanol-to-gasoline; CCS = carbon capture and storage; CFT = coal-to-liquid fuel, Fischer-Tropsch; CMTG = coal-to-liquid fuel, methanol-to-gasoline.

is about twice that from petroleum-based gasoline. With CCS, the CO_2 life-cycle emission is about the same as that from petroleum-based gasoline.

The biochemical conversion of biomass produces fuels that are more expensive than CTL fuels because the conversion plants are smaller and the feedstock more expensive—biomass costs almost 4 times as much as coal on an energy-equivalent basis. The production cost of cellulosic ethanol is around $115/bbl on a gasoline-equivalent basis. The cost of thermochemical conversion of biomass, without coal, is higher than cellulosic ethanol on an energy-equivalent basis and has the potential of large negative net releases of CO_2 with geologic storage; that is, the process involves a net removal of CO_2 from the atmosphere. For BTL and venting of CO_2, the estimated fuel cost is $140/bbl if electricity is sold back to the grid at $80/MWh; with geologic storage of CO_2, it is $150/bbl if electricity is sold back to the grid at $80/MWh. The results of the relatively small cofed coal and biomass plant (total feed, 8000 tons/day) are particularly interesting. Fuels produced by that plant cost about $95/bbl on a gasoline-equivalent basis without CCS, and CO_2 atmospheric releases from plants with CCS are negative. Those results point to the importance of that option in the U.S. energy strategy.

The important influence of a carbon price on fuel price is shown in Figure S.6. It is important to note that Figure S.6 shows the breakdown of all costs, including negative costs such as credit from electricity generation or from carbon uptake. The negative costs must be subtracted from the positive costs to obtain the actual costs. For example, the cost of BTL with CCS is $151/bbl–$37/bbl = $114/bbl. CO_2 emission from corn-grain ethanol is slightly lower than that from gasoline. In contrast, CO_2 emission from cellulosic ethanol without CCS is close to zero.

Figure S.6 shows that a CO_2 price of $50/tonne significantly increases the costs of the fossil-fuel options, including the costs of petroleum-based gasoline. The carbon price brings the cost of biochemical conversion options down to about $110/bbl (crude price, about $95/bbl). The large amount of CO_2 vented in the CTL process without CO_2 storage almost doubles the cost of product once the carbon price of $50/tonne of CO_2 is imposed.

Inclusion of a carbon price does not increase the total costs of all pathways. For example, although thermochemical conversion of biomass costs about $140/bbl of gasoline equivalent without CCS, the produced fuels become competitive with petroleum-based fuels at $115/bbl of gasoline equivalent with the carbon price and CCS. In general, if any pathway takes more CO_2 from the atmosphere than it releases in other parts of its life cycle, the inclusion of a carbon price reduces the total cost of producing liquid fuel by that pathway. Those estimates are all based on costs of small gasification units operating at a feed rate of 4,000 tons/day. Each of those units is capital-intensive. Therefore, larger units can be expected to be deployed in regions where potential biomass availability is large—for example, 10,000 tons/day. Such units could result in much lower costs.

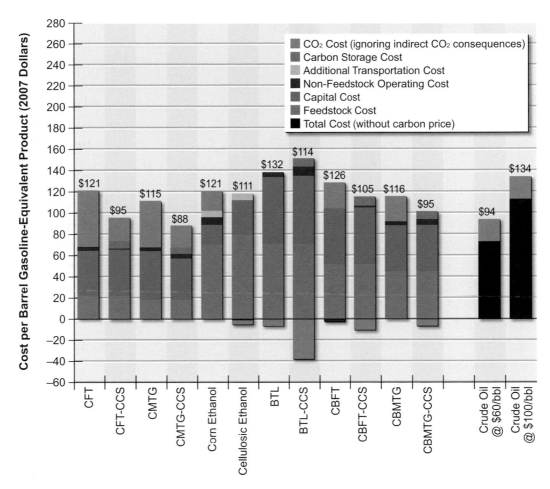

FIGURE S.6 *Cost of alternative liquid fuels produced from coal, biomass, or coal and biomass with a CO$_2$-equivalent price of $50/tonne. Negative cost elements must be subtracted from the positive elements; the number at the top of each bar indicates the net costs.*
Note: BTL = biomass-to-liquid fuel; CBFT = coal-and-biomass-to-liquid fuel, Fischer-Tropsch; CBMTG = coal-and-biomass-to-liquid fuel, methanol-to-gasoline; CCS = carbon capture and storage; CFT = coal-to-liquid fuel, Fischer-Tropsch; CMTG = coal-to-liquid fuel, methanol-to-gasoline.

Costs and Supply

As noted previously, the cost estimates for biochemical conversion and thermo-chemical conversion are based on one generic biomass source. Figures S.4 and S.6 do not show how much fuel could be produced at the estimated costs. To provide a complete supply function for alternative liquid fuels, the supply function from Figure S.3 for all biomass feedstocks has been combined with the conversion cost estimates. (The potential supply of gasoline and diesel from CTL technology is discussed below in the section "Deployment of Alternative Liquid Transportation Fuels.") The results are shown in Figures S.7 and S.8. Figure S.7 shows the potential gasoline-equivalent supply of ethanol from biochemical conversion of lignocellulosic biomass and corn grain with 2020-deployable technology. The supply of grain ethanol satisfies the current legislative requirement to produce 15 billion gallons of ethanol in 2022. Figure S.7 shows potential supply, not the panel's projected penetration of cellulosic ethanol in 2020; it does not incorporate lags in implementation of the technology that result from the need to permit and build the infrastructure to produce and transport the alternative liquid fuels. The estimated supply of synthetic gasoline and diesel (G/D) derived from coal and biomass is shown in Figure S.8. Two supply functions are shown: one with CCS and

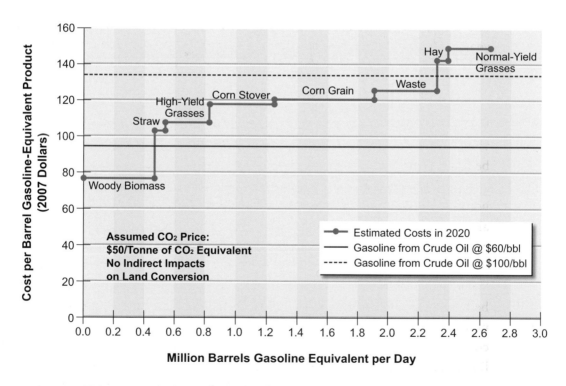

FIGURE S.7 *Estimated supply of cellulosic ethanol plus corn-grain ethanol at different price points in 2020. The red solid and dotted lines show the supply of crude oil at $60/bbl and $100/bbl for comparison.*

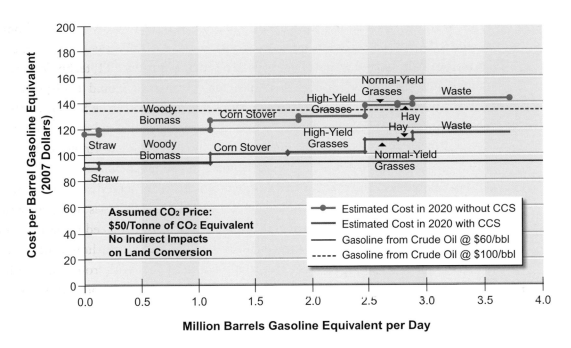

FIGURE S.8 *Estimated supply of gasoline and diesel produced by thermochemical conversion of combined coal and biomass via Ficher-Tropsch with or without carbon capture and storage at different price points in 2020.*

the other without CCS. The comparison shows that if the CCS technologies are viable and a CO_2-eq price of $50/tonne is implemented, for each feedstock it will be less expensive to use CCS than to release the CO_2 into the atmosphere.

Either of the production processes underlying Figures S.7 and S.8 would use the same supplies of biomass. Therefore, the quantities cannot be added. If all the production (in addition to ethanol produced from corn grain) is based on cellulosic conversion, Figure S.7 would be potentially applicable. If all production is based on thermochemical conversion cofed with biomass and coal, Figure S.8 would be potentially applicable. Most likely, some of the production would be based on cellulosic processes and some on thermochemical processes, so the potential supply function would lie between the two supply functions shown. If corn-grain ethanol has not been phased out by then, it would add about 0.67 million bbl/day of gasoline-equivalent production to the supply.

To put the results in perspective, the light-duty vehicle gasoline and diesel use in the United States in 2008 is estimated to be about 9 million barrels of oil equivalent per day (1 bbl of crude oil produces about 0.85 bbl of gasoline equivalent).

Total liquid fuel used in the United States in 2008 was 21 million barrels per day, of which 14 million was used for transportation and 12 million was imported. Thus, 2 million barrels of gasoline-equivalent ethanol produced from cellulosic biomass and the 0.7 million barrels of gasoline-equivalent ethanol produced from corn grain have the potential to replace about 30 percent of the petroleum-based fuel consumed in the United States by light-duty vehicles.

The potential supply of gasoline or diesel fuel from thermochemical CBTL with CCS is greater than that from biochemical or thermochemical conversion of cellulosic biomass. The costs of thermochemical CBTL are lower than those of either biochemical or thermochemical conversion of biomass. The cost difference occurs because coal is a lower-cost feedstock than biomass. In addition, cofeeding coal and biomass allows a larger plant to be built and reduces capital costs per unit volume of product. Thus, the combination of coal with biomass allows a larger amount of alternative fuels to be produced than would be possible with biomass alone because the quantity of biomass limits overall production. The addition of coal increases the total amount of liquids that could be produced from a fixed quantity of biomass. Using coal and biomass at 60 and 40 percent, respectively, on an energy basis, almost 4 million barrels per day of gasoline equivalent can potentially be displaced from transportation (60 billion gallons of gasoline equivalent per year, or 45 percent of gasoline and diesel used by light-duty vehicles in 2008). That assumes that all of the 550 million dry tons of cellulosic biomass sustainably grown for fuel will be used for CBTL fuel production, so the estimates represent the maximum potential supply.

Finding S.15 (see Finding 6.1 in Chapter 6)

Alternative liquid transportation fuels from coal and biomass have the potential to play an important role in helping the United States to address issues of energy security, supply diversification, and greenhouse gas emissions with technologies that are commercially deployable by 2020.

- **With CO_2 emissions similar to those from petroleum-based fuels,** a substantial supply of alternative liquid transportation fuels can be produced with thermochemical conversion of coal with geologic storage of CO_2 at a gasoline-equivalent cost of $70/bbl.
- **With CO_2 emissions substantially lower than those from petroleum-based fuels,** up to 2 million barrels per day of gasoline-equivalent

fuel can technically be produced with biochemical or thermochemical conversion of the estimated 550 million dry tons of biomass available in 2020 at a gasoline-equivalent cost of about $115–140/bbl. Up to 4 million barrels per day of gasoline-equivalent fuel can be technically produced if the same amount of biomass is combined with coal (60 percent coal and 40 percent biomass on an energy basis) at a gasoline-equivalent cost of about $95–110/bbl. However, the technically feasible supply does not equal the actual supply inasmuch as many factors influence the market penetration of fuels.

DEPLOYMENT OF ALTERNATIVE LIQUID TRANSPORTATION FUELS

The discussion above has focused on the potential supply of alternative fuels from technologies ready to be deployed commercially by 2020, but the potential supply does not translate to the alternative supply that could be available by 2020. Apart from technological readiness, the penetration rates of alternative liquid fuels into the market will depend on many factors, including oil price, carbon taxes, construction environment, and labor availability. The panel developed a few plausible scenarios to illustrate the lag between when technology becomes commercially deployable, and when substantial market penetration will be seen.

Deployment of Cellulosic-Ethanol Plants

For biochemical conversion to cellulosic ethanol, the panel developed two scenarios on the basis of the current activities of demonstration plants, the announced commercial plants, the U.S. Department of Energy roadmap, and the rate of construction of grain-ethanol plants. The two scenarios assume that the cellulosic-ethanol capacity by 2015 will be 1 billion gallons per year, resulting from overall commercial development and demonstration activities, and that capacity-building beyond 2015 tracks one of two scenarios based on the capacity-building experienced by grain ethanol. One scenario assumes the maximum capacity-building experienced for grain ethanol (about a 25 percent yearly increase in capacity over a 6-year period); the second is a scenario of aggressive capacity-building of about twice that achieved for grain ethanol. The two scenarios project 7–12 billion gallons of cellulosic ethanol per year by 2020. Continued aggressive capacity-building could achieve the Renewable Fuel Standard[5] mandate capacity of 16 billion gallons of cellulosic ethanol per year by 2022, but it would be a stretch. Continued aggressive capacity-building could yield 30 billion gallons of cellulosic ethanol per year by 2030 and up to 40 billion gallons per year by 2035, consuming about 440 million dry tons of biomass per year and replacing 1.7 million barrels of petroleum-based fuels per day.

Deployment of Alternative Liquid Fuels from Coal-to-Liquids Plants with Carbon Capture and Storage

If commercial demonstrations of CTL with CCS are started immediately (as discussed in Recommendations S.10 and S.12) and CCS is proved viable and safe by 2015, commercially viable plants could be starting up before 2020. The growth rate after that could be about two or three plants per year. That would reduce dependence on imported oil but would increase CO_2 emission from transportation. At a build-out rate of two plants (at 50,000 bbl/d of fuel) per year, liquid fuel would be produced at 2 million barrels per day from 390 tons of coal per year by 2035 at a total cost of about $200 billion for all the plants built. At a build-out rate of three plants per year, liquid fuels would be produced at 3 million barrels per day from about 580 million tons of coal per year. The latter case would replace about one-third of the current U.S. oil use in light-duty transportation and increase U.S. coal production by 50 percent. At a build out of three plants starting up per year, five or six plants would be under construction at any time.

Deployment of Alternative Fuels from Coal-and-Biomass-to-Liquids Plants

For cofed biomass and coal plants, the technology is close to being developed, and several commercial plants without CCS have started cofeeding biomass. However, gaining operational experience in the plants with CCS is critical; CCS will probably be required, and plants are going through early commercialization to gain operating experience and to reduce costs. Because coal-and-biomass plants are much smaller than CTL plants (plant size, one-fifth the size of CTL plants,

---

[5]The Renewable Fuel Standard (RFS) was created by the 2005 U.S. Energy Policy, and the 2007 U.S. Energy Independence Act (EISA) amended the RFS to set forth "a phase-in for renewable fuel volumes beginning with 9 billion gallons in 2008 and ending at 36 billion gallons in 2022." The 36 billion gallons would include 16 billion gallons of cellulosic ethanol.

or fuel production at 10,000 bbl/d) and biomass feed rates are similar to those in cellulosic biochemical conversion plants, penetration rates should follow the cellulosic-plant build out more closely. But most likely, the coal-and-biomass build out will be much slower than the aggressive cellulosic-plant buildout presented above because of issues of siting the plants near both biomass and coal production and because plant design is more complex. The panel assumed that penetration rates for the coal-and-biomass plants would be slightly less than the rate for the cellulosic-ethanol build-out case that follows the experience of grain ethanol discussed above (which has experienced a 25 percent growth rate). At a 20 percent growth rate until 2035 with 280 plants in place, 2.5 million barrels of gasoline equivalent would be produced per day. That would consume about 300 million dry tons of biomass and about 250 million tons of coal per year—less than the projected biomass availability. Siting to have access to both biomass and coal is probably the limiting factor for CBTL plants. This analysis shows that the rates of capacity growth would have to exceed historical rates considerably if 550 million dry tons of biomass per year is to be converted to liquid fuels by 2035.

Finding S.16 (see Finding 6.2 in Chapter 6)

If commercial demonstration of cellulosic-ethanol plants is successful and commercial deployment begins in 2015 and if it is assumed that capacity will grow by 50 percent each year, cellulosic ethanol with low CO_2 life-cycle emissions can replace up to 0.5 million barrels of gasoline equivalent per day by 2020 and 1.7 million barrels per day by 2035.

Finding S.17 (see Finding 6.3 in Chapter 6)

If commercial demonstration of coal-and-biomass-to-liquid plants with carbon capture and storage is successful and the first commercial plants start up in 2020 and if it is assumed that capacity will grow by 20 percent each year, coal-and-biomass-to-liquid fuels with low CO_2 life-cycle emissions can replace up to 2.5 million barrels of gasoline equivalent per day by 2035.

Finding S.18 (see Finding 6.4 in Chapter 6)

If commercial demonstration of coal-to-liquid plants with carbon capture and storage is successful and the first commercial plants start up in 2020 and if it is

assumed that capacity will grow by two to three plants each year, coal-to-liquid fuels with CO_2 life-cycle emissions similar to those of petroleum-based fuels can replace up to 3 million barrels of gasoline equivalent per day by 2035. That option would require an increase in U.S. coal production by 50 percent.

Finding S.19 (see Finding 7.2 in Chapter 7)

The deployment of alternative liquid transportation fuels aimed at diversifying the energy portfolio, improving energy security, and reducing the environmental footprint by 2035 would require aggressive large-scale demonstration in the next few years and strategic planning to optimize the use of coal and biomass to produce fuels and to integrate them into the transportation system. Given the magnitude of U.S. liquid-fuel consumption (14 million barrels of crude oil per day in the transportation sector) and the scale of current petroleum imports (about 56 percent of the petroleum used in the United States is imported), a business-as-usual approach is insufficient to address the need to find alternative liquid transportation fuels, particularly because development and demonstration of technology, construction of plants, and implementation of infrastructure require 10–20 years per cycle.

Recommendation S.13 (see Recommendation 7.8 in Chapter 7)

The U.S. Department of Energy should partner with industry in the aggressive development and demonstration of cellulosic-biofuel and thermochemical-conversion technologies with carbon capture and storage to advance technology and to address challenges identified in the commercial demonstration programs. The current government and industry programs should be evaluated to determine their adequacy to meet the commercialization timeline required to reduce U.S. oil use and CO_2 emissions over the next decade.

Recommendation S.14 (see Recommendation 6.1 in Chapter 6)

Detailed scenarios of market penetration rates of biofuels, coal-to-liquid fuels, and associated biomass and coal supply options should be developed to clarify hurdles and challenges to achieving substantial effects on U.S. oil use and CO_2 emissions. The analysis will provide policy makers and business leaders with the information needed to establish enduring policies and investment plans for accelerating the development and penetration of alternative-fuels technologies.

Finding S.20 (see Finding 7.1 in Chapter 7)

A potential optimal strategy for producing biofuels in the United States could be to locate thermochemical conversion plants that use coal and biomass as a combined feedstock in regions where biomass is abundant and locate biochemical conversion plants in regions where biomass is less concentrated. Thermochemical plants require larger capital investment per barrel of product than do biochemical conversion plants and thus benefit to a greater extent from economies of scale. This strategy could maximize the use of cellulosic biomass and minimize the costs of fuel products.

Recommendation S.15 (see Recommendation 7.6 in Chapter 7)

The U.S. Department of Energy and the U.S. Department of Agriculture should determine the spatial distribution of potential U.S. biomass supply to provide better information on the potential size, location, and costs of conversion plants. The information would allow determination of the optimal size of conversion plants for particular locations in relation to the road network and the costs and greenhouse gas effects of feedstock transport. The information should also be combined with the logistics of coal delivery to such plants to develop an optimal strategy for using U.S. biomass and coal resources for producing sustainable biofuels.

ENVIRONMENTAL EFFECTS OTHER THAN GREENHOUSE GAS EMISSIONS

Biomass Supply

Although greenhouse gas emissions have been the central focus of research concerning the environmental effects of biomass production for liquid fuels, other key effects must be considered. On the whole, lignocellulosic-biomass feedstocks present distinct advantages over food-crop feedstocks with respect to water-use efficiency, nutrient and sediment loading into waterways, enhancement of soil fertility, emissions of criteria pollutants that affect air quality, and habitat for wildlife, pollinators, and species that provide biocontrol services for crop production. But dedicated fuel crops have the potential to become invasive, and many of the ideal traits of biomass crops have been shown to contribute to invasiveness.

Biochemical Conversion

The biochemical conversion of cellulosic biomass to ethanol requires process water for mixing with fermentation substrates and for cooling, heating, and making reagents that are associated with hydrolysis and fermentation. The amount of water required for processing biomass into ethanol or other biofuels is estimated to be 2–6 gallons per gallon of ethanol produced. The lower levels would be approached if a plant were designed to recycle process water. The processing of cellulosics to ethanol will result in a residual water stream that would need to undergo treatment. However, an efficient process, by definition, will ferment most of the sugars to ethanol and leave only small amounts of organic residue.

Air emissions resulting from bioprocessing include CO_2, water vapor, and possibly sulfur and nitrogen. Fermentation processes release CO_2 as a result of microbial metabolism. Water vapor is released particularly if the lignin coproduct is dried before being shipped from the plant for use as boiler fuel at an off-site power-generation facility. The sulfur and nitrogen content of fermentation residues would be expected to be low unless chemicals are used in the pretreatment of the biomass materials. The chemicals used in pretreatment can be recovered.

Thermochemical Conversion

CTL plants can be configured to minimize their effects on the environment. Clean-coal technologies have been developed for the electric-power industry but can be used in CTL applications. CTL plants need to produce clean synthesis gas from coal by using gasification and gas-cleaning technologies. As a result, concerns over emission of criteria pollutants and toxicants—such as sulfur oxides, nitrogen oxides, particulates, and mercury—would be minimal because CTL plants will use clean-coal technologies.

The sulfur compounds in coal are converted into elemental sulfur, which can be sold as a by-product. The ammonia in synthesis gas can be recovered and sold as fertilizer or sent to wastewater treatment, where it is absorbed by bacteria. All the mercury, arsenic, and other heavy metals in the syngas are adsorbed on activated charcoal. The mineral matter (or ash) in the coal has been exposed to extremely high temperatures during gasification and has become vitrified into slag; the slag is nonleachable and finds use in cement or concrete for buildings, bridges, and roads. Nitrogen oxide emissions are reduced to about 3 parts per million (ppm) by using existing conversion technologies.

Water use in thermochemical conversion plants depends primarily on the water-use approach used in designing the plants. For the conversion of coal and combined coal and biomass to transportation fuels with all water streams recycled or reused, the major consumptive uses of water are for cooling, producing hydrogen, and handling solids. If water availability is unlimited because of access to rivers, conventional forced- or natural-draft cooling towers would be used. In arid areas, air cooling would be used as much as possible. Depending on the magnitude of air cooling, water consumption could range from about 1 to 8 bbl per barrel of product. CTL plants will have environmental effects associated with the mining of additional coal, as discussed in the National Research Council reports *Evolutionary and Revolutionary Technologies for Mining* (NRC, 2002) and *Coal Research and Development to Support National Energy Policy* (NRC, 2007).

BARRIERS TO DEPLOYMENT

The development of a biomass-supply industry for the production of cellulosic biofuels faces substantial challenges. The technological and sociological issues are not trivial, but they can be successfully overcome. The challenges are as follows.

Challenge 1

Issues related to cellulosic-feedstock production include:

- Developing a systems approach through which farmers, biomass integrators, and those operating biofuel-conversion facilities can develop a well-organized and sustainable cellulosic-ethanol industry that will address multiple environmental concerns (for example, biofuel; soil, water, and air quality; carbon sequestration; wildlife habitat; rural development; and rural infrastructure) without creating unintended consequences through piecemeal development efforts.
- Determining the full-life-cycle greenhouse gas emissions of various biofuel crops.
- Certifying the greenhouse gas benefits for different potential biofuel scenarios.

Those issues, although formidable, can be overcome by developing a systems approach with multiple end points that collectively can provide a variety of credits or incentives (for example, carbon sequestration, water quality, soil quality, wildlife habitat, rural development) and thus contribute to a stronger U.S. agricultural industry. Failure to link the various critical environmental, economic, and social needs and to address them as an integrated system could reduce the availability of biomass for conversion to levels far below the 550 million tons technically deployable by 2020.

Challenge 2

For thermochemical conversion of coal or combined coal and biomass to have any substantial effect on U.S. reliance on crude oil and CO_2 emissions in the next 20–30 years, CCS will have to be shown to be safe and economically and politically viable. The capture of CO_2 is proven, but commercial-scale demonstration plants are needed now to both quantify and improve cost and performance. Separate large-scale programs will be required to resolve storage and regulatory issues associated with geologic CO_2 storage approaching a scale of gigatonnes per year. In the analyses presented in this report, the viability of CCS was assumed to be demonstrated by 2015 so that integrated CTL plants could start up by 2020. In that scenario, the first coal or coal-and-biomass gasification plant would not be in operation until 2020. That assumption is ambitious and will require focused and aggressive government action to realize. Uncertainty about the regulatory environment arising from concerns of the general public and policy makers have the potential to raise storage costs above the costs assumed in this study. Ultimate requirements for selection, design, monitoring, carbon-accounting procedures, liability, and associated regulatory frameworks have yet to be developed, so there is a potential for unanticipated delay in initiating demonstration projects and, later, in licensing individual commercial-scale projects. Large-scale demonstrations and establishment of procedures for operation and long-term monitoring of CCS projects have to be pursued aggressively in the next few years if thermochemical conversion of biomass and coal with CCS is to be ready for commercial deployment by 2020.

Challenge 3

Cellulosic ethanol is in the early stages of commercial development, and a few commercial demonstration plants are expected to begin operations in the next several years. Over the next decade, process improvements are expected to come from evolutionary developments and learning gained through commercial experience and increases in scale of operation. Incremental improvements in biochemical conversion technologies can be expected to reduce nonfeedstock process costs by

25 percent by 2020 and 40 percent by 2035. It will take focused and sustained industry and government action to achieve those cost reductions. The key technical barriers to achieving cost reduction are as follows:

- More efficient pretreatment to free up celluloses and hemicelluloses and to enable more efficient downstream conversion. Improved pretreatment is unlikely to reduce product cost substantially because pretreatment cost is small relative to other costs.
- Better enzymes that are not subject to end-product inhibition to improve the efficiency of the conversion process.
- Maximizing of solids loading in the reactors.
- Engineering organisms capable of fermenting the sugars in a toxic biomass hydrolysate and producing high concentrations of the final toxic product biofuel; improving microbial tolerance of toxicity is a key issue.

Challenge 4

If ethanol is to be used in large quantities in light-duty vehicles, an expanded ethanol transportation and distribution infrastructure will be required. Ethanol cannot be transported in pipelines used for petroleum transport. Ethanol is currently transported by rail or barges and not by pipelines, because it is corrosive in the existing infrastructure and can damage the seals, gaskets, and other equipment and induce stress-corrosion cracking in high-stress areas. If ethanol is to be used in fuel at concentrations higher than 20 percent (for example, E85, which is a blend of 85 percent ethanol and 15 percent gasoline), the number of refueling stations offering it will have to be increased. The distribution challenges have to be addressed to enable widespread availability of ethanol in the fuel system. However, if cellulosic biomass were dedicated to thermochemical conversion with FT or MTG, the resulting fuels would be chemically equivalent to conventional gasoline and diesel, and the infrastructure challenge associated with ethanol would be minimized.

Challenge 5

The panel's analyses provide a snapshot of the potential costs of liquid fuels derived from biomass with biochemical or thermochemical conversion and from biomass and coal with thermochemical conversion. Costs of fuels are dynamic and fluctuate as a result of other externalities, such as the costs of feedstock, labor, and construction; the economic environment; and government policies. Given the wide variation in most commodity prices, especially oil prices, investors will have to have confidence that such policies as carbon caps, a carbon price, and tariffs on imported oil will ensure that alternative liquid transportation fuels can compete with fuels derived from crude oil. The price of carbon emissions or the existence of fuel standards that require specified reductions in greenhouse gas life-cycle emission will affect the economic choices.

OTHER TRANSPORTATION FUELS

Technologies for producing transportation fuels from natural gas are ready for deployment by 2020. Compressed natural gas already fuels vehicles. Other liquid fuels can be produced from syngas, including gas-to-liquid diesel, dimethyl ether, and methanol. Only if large supplies of natural gas are available—for example, from natural-gas hydrates—will the United States be likely to use natural gas as the feedstock for transportation-fuel production.

Hydrogen has the potential to reduce U.S. CO_2 emissions and oil use, as discussed in two recent National Research Council reports, *Transitions to Alternative Transportation Technologies—A Focus on Hydrogen* (NRC, 2008) and *The Hydrogen Economy: Opportunities, Costs, Barriers, and R&D Needs* (NRC, 2004). Hydrogen fuel-cell vehicles can yield large and sustained reductions in U.S. oil consumption and greenhouse gas emissions, but several decades will be needed to realize these potential long-term benefits.

REFERENCES

NRC (National Research Council). 2002. Evolutionary and Revolutionary Technologies for Mining. Washington, D.C.: National Academy Press.

NRC. 2004. The Hydrogen Economy: Opportunities, Costs, Barriers, and R&D Needs. Washington, D.C.: The National Academies Press.

NRC. 2007. Coal: Research and Development to Support National Energy Policy. Washington, D.C.: The National Academies Press.

NRC. 2008. Transitions to Alternative Transportation Technologies—A Focus on Hydrogen. Washington: D.C.: The National Academies Press.

1

Liquid Fuels for Transportation

Worldwide demand for energy has been increasing as a result of continued population increases and economic growth, particularly in developing countries. Because fossil fuels continue to dominate the global energy market, rising energy use results in increased greenhouse gas emissions from that sector. In fact, emissions from the use of fossil fuels and emissions of carbon from plants and soil as a result of changes in land use have been identified as two primary sources of carbon dioxide[1] (CO_2) emission (Solomon et al., 2007). Increasing energy supply to support population growth and economic growth while reducing CO_2 emission from the energy sector certainly poses a serious challenge to the current generation and future generations because our very way of life is at stake. An option for the energy sector to secure supply and to reduce its greenhouse gas emissions is to diversify its energy sources and invest in technological change to provide energy with low or zero CO_2 emission.

The National Academies initiated a series of studies, "America's Energy Future," in 2007 to provide authoritative estimates of the current contributions and future potential of existing and new energy supply technologies, their effects, and their projected costs. Because of considerable uncertainty and disagreements about the prospective costs and performance of alternative liquid transportation fuels (biofuels and coal-to-liquid fuels in particular), the National Research Council appointed an independently operating panel to examine those issues in

---

[1]CO_2 is one of the most important greenhouse gases. Other greenhouse gases include water vapor, methane (CH_4), nitrous oxide (N_2O), halocarbons, and ozone (O_3).

depth. This is the report of the Panel on Alternative Liquid Transportation Fuels. (See Appendix C for information on the panel members.)

DEMAND FOR LIQUID TRANSPORTATION FUELS

Transport activity is one of the key components of continued economic growth and social stability in industrialized societies. Demand for transportation fuels increases around the world as economies grow. Oil has been the primary source of liquid transportation fuels since the early 1900s largely because of its favorable energy density, ease of distribution, low cost, and abundance. The world demand for oil has increased from 11 million barrels per day of oil equivalent (MBDOE) in 1950 to 57 MBDOE in 1970 to about 85 MBDOE in 2009 and is projected to be 116 MBDOE in 2030 (ExxonMobil, 2008; IEA, 2009). From 1985 to 2005, global energy demand for transportation increased by an average of 2.2 percent per year (ExxonMobil, 2007), and it is expected to increase by an average of 1.4 percent per year from 2005 to 2030 (Figure 1.1) (ExxonMobil, 2008). As seen in late 2008 and early 2009, oil demand dropped rapidly as the global economy

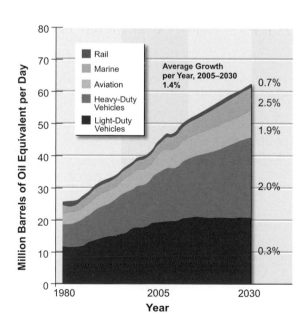

FIGURE 1.1 *Worldwide demand for energy for transportation, 1980–2030. Reprinted from ExxonMobil, 2008. Copyright 2008, with permission from ExxonMobil.*

slowed down into a recession. When the global economy recovers in the long term and as the economies of developing countries grow with their populations, the demand for oil will grow again.

The majority of current demand for liquid transportation fuel is met by the Organization of the Petroleum Exporting Countries (OPEC) crude and non-OPEC crude and condensate (ExxonMobil, 2007). Other energy sources that contribute a small fraction of transportation fuels include oil sands, natural gas, and biofuels. Whether and when global petroleum production will reach its peak (beyond which it will decline) and be unable to meet global crude-oil demand is uncertain. By the end of 2006, the proven worldwide reserves of oil—that is, resources that are discovered, recoverable with current technology, commercially feasible, and remaining in the ground—was reported to be 1,372 billion barrels (BP, 2007).

The primary source of energy for transportation in the United States and elsewhere is oil. Between 2007 and 2008, when this report was written, the crude-oil price fluctuated from about $70/bbl when the committee convened its first meeting in November 2007 to a record high of $147/bbl in July 2008 and then dropped to about $35/bbl at the end of 2008. Volatile oil prices, oil importation in large quantities and its associated tremendous shift of U.S. wealth overseas, a tight worldwide supply-demand balance, and fears that oil production would peak in the next 10–20 years all motivate a search for domestic sources of alternative fuels. The United States uses 25 percent of global oil production for 4.5 percent (U.S. Census Bureau, 2008) of the global population. The United States imports about 56 percent of its oil and in 2008 spent $10–38 billion each month (depending on oil price and demand) overseas for oil.

U.S. oil demand stems from four main sectors: transportation, industry, electricity generation, and residential and commercial use. Transportation is by far the largest consumer, at nearly 70 percent (EIA, 2009) (Table 1.1). Domestic demand

TABLE 1.1 Consumption of Liquid Fuel in United States in 2008, by Sector

Sector	Liquid Fuel Consumption (millions of barrels per day)
Residential and commercial use	1.10
Industry	4.94
Transportation	13.66
Electricity generation	0.22
Total	19.54

Source: EIA, 2009.

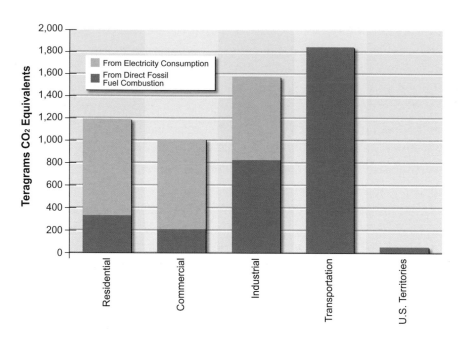

FIGURE 1.2 *CO$_2$ emissions from fossil-fuel combustion by different end-use sectors. Source: EPA, 2008.*

for oil steadily increased by 1.4 percent from 1980 to 2005 (NPC, 2007), but domestic oil production has been decreasing (Zittel and Schindler, 2007). Proven oil reserves in the United States at the end of 2006 were 29.9 billion barrels (2.5 percent of the total world reserves); that is in stark contrast with the 35.1 billion barrels at the end of 1986 (BP, 2007). Options for reducing reliance of U.S. transportation on oil are few. The nation could reduce the amount of oil that it uses by reducing driving and improving vehicle fuel efficiency; energy efficiency of the transportation sector is discussed in detail in another report in the America's Energy Future series, *Real Prospects for Energy Efficiency in the United States* (NAS-NAE-NRC, 2009c). Or it could diversify its portfolio of fuels.

The U.S. transportation sector contributes the most greenhouse gas emissions among all end users of domestic fossil fuel. Transportation activities accounted for one-third of total greenhouse gas emissions in the United States in 2006 (Figure 1.2) (EPA, 2008). The strategies for reducing greenhouse gas emissions from the transportation sector are similar to those for reducing oil dependence: reducing driving, improving vehicle fuel efficiency, and using fuels that have low greenhouse gas emissions over the life cycle of their production and use.

ALTERNATIVE TRANSPORTATION FUELS

The use of alternative transportation fuels constitutes one of the few options for reducing U.S. reliance on oil. Alternative fuels include liquid fuels produced from unconventional oil (such as that from oil sands, heavy oils, and oil shale), natural gas, hydrogen, biomass-based fuels, and fuels produced from coal. Successful development and commercialization of those fuels would depend on their cost competitiveness compared with that of conventional gasoline and on the amount of fuel that could be supplied annually. In addition to economics, technological status, and potential supply, sustainable development of alternative transportation fuels would take environmental and social concerns into consideration. Use of fuels that contribute less greenhouse gas emission than gasoline needs to be part of the strategy to reduce greenhouse gas emissions while improving energy security. Interest in domestic alternative fuels that contribute less greenhouse gas than gasoline has led to large increases in U.S. ethanol production—primarily from corn— and increased use of biodiesel fuel from soybean and other vegetable oils. Because corn and vegetable oils are sources of food for humans and feed for animals, their use to increase biofuel production has sparked the debate of "food versus fuel." Furthermore, actual greenhouse gas emission reductions resulting from the substitution of grain ethanol for gasoline or biodiesel for diesel are small.

In short, for alternative transportation fuels to take hold in the United States, they have to be price-competitive, environmentally sustainable, and socially acceptable. A comparison of the economies of various alternative transportation fuels requires estimation of their total costs of production, from the cost of raw materials for fuel production to the resources used in the process of distributing the fuel to vehicles. A life-cycle assessment of greenhouse gas emissions that takes into account the uptake and release of greenhouse gases as a result of feedstock production and materials and energy used in production and consumption of each fuel type would have to be conducted to assess the environmental effects. An assessment of the potential supply, life-cycle costs, and environmental effects of different alternative liquid fuels can help guide policy to improve America's energy security and to reduce the greenhouse gas emissions from the transportation sector.

As history and many studies—for example, NRC (2004, 2005, 2008)—have shown, it will take decades to transform the U.S. transportation and fuel system to one that uses primarily domestic sources, has lower CO_2 emission, and meets the nation's transportation energy demand during the transition. A potential conflict between the need for more domestic fuel supply and the need to reduce carbon emission from transportation and transportation fuel can be avoided if high priority is placed on improving fuel-consumption efficiency and on developing and implementing alternative, low-carbon, new fuel technologies. The joint challenge of providing transportation fuel and reducing greenhouse gas emissions drives the U.S. Department of Energy toward the vision of producing transportation fuels with low greenhouse gas life-cycle emissions from domestic sources.

PURPOSE OF THIS STUDY

The purpose of this study was to examine the technical potential for reducing reliance on oil for transportation, principally in automobiles and trucks, through the use of alternatives fuels. (See Appendix B for the complete statement of task.) Hydrogen and natural gas as sources of energy for transportation have been discussed extensively in the published literature (Ingersoll, 1996; Di Pascoli et al., 2001; NRC, 2004, 2005, 2008) and are discussed briefly in this report. There is no substantial production of oil from tar sands and no production of oil from shale in the United States. The potential of those sources is discussed in the report *America's Energy Future: Technology and Transformation* (NAS-NAE-NRC, 2009a). The panel recognizes that biomass can be used for power generation and that the electricity generated could be used to power electric vehicles or plug-in hybrid electric vehicles. However, those topics are discussed extensively in two other reports in the America's Energy Future series. Biomass for electricity is discussed in *Electricity from Renewable Resources: Status, Prospects, and Impediments* (NAS-NAE-NRC, 2009b), and electric and plug-in hybrid electric vehicles are discussed in *Real Prospects for Energy Efficiency in the United States* (NAS-NAE-NRC, 2009c). The focus of this panel's study was limited to liquid fuels that can be derived from biomass and coal feedstocks.

Coal and biomass are abundant in the United States, but as feedstocks for transportation fuels they have different constraints and environmental effects. Although the United States has at least 20 years' worth of coal reserves in active mines and probably has enough to meet the nation's needs for more than 100 years at current rates of consumption (NRC, 2007), coal is a nonrenewable source of energy. Thus, coal-to-liquid fuels would not be a sustainable solution to the problem of oil dependence. Combustion of coal also releases the highest greenhouse gas emission per unit energy of all fossil fuels. Although technologies for producing liquid fuels from coal are well developed and are being used commercially, they have not been integrated with technologies that would capture the CO_2 stream released from the coal facility and store it geologically. In contrast, biomass is a renewable resource that can also offer net CO_2 benefits because living plants take up CO_2 through the process of photosynthesis. Although biomass can be produced continuously over a long term, the amount that can be produced at a given time is limited by the availability of the natural resources that support biomass production. Most arable land in the United States is already being used for food, feed, and fiber production. Although the technologies for producing fuels from plant sugar and starch are known and used commercially, the technologies for producing fuels from lignocellulosic feedstock have yet to be demonstrated on a commercial scale.

To address the statement of task, the panel focused on technologies for converting biomass and coal to alternative liquid fuels that are commercially deployable by 2020. Technologies deployable after 2020 were also evaluated, but in less depth. For the purpose of this study, *commercially deployable* technologies are ones that have been scaled up from research to development to pilot plant and then have gone through several commercial-size demonstrations. Thus, the capital and operating costs of a plant using commercially deployable technologies have been optimized so that the technology can compete with other options. *Commercial deployment* of a technology is the rate at which it penetrates the market. Deployment depends on market forces, capital and human availability, other competitive technologies, public policies, and other factors. To be consistent with the other studies in the America's Energy Future series, the panel:

- Evaluated the state of technology development on the basis of estimated times to initial commercial deployment,
- Evaluated key research, development, and demonstration challenges for technologies to be ready for commercial deployment,
- Developed current and projected costs and CO_2 emissions of technologies deployable by 2020,
- Evaluated environmental, economic, policy, and social factors that would enhance or impede development and deployment,
- Estimated the potential supply curve for liquid fuels produced from coal or biomass with the technologies that could be deployed by 2020, and
- Reviewed other alternative fuels that would compete with coal-based and biomass-based fuels over the next 15 years.

The panel was asked not to include recommendations on policy choices.

CONTEXT OF REPORT

The panel's work began when prices of fossil fuels and of raw materials and capital for infrastructure were rising rapidly (November 2007). As the study progressed, those prices reached a peak (for example, crude oil reached $147/bbl on July 11, 2008) and then fell steeply. Although this report makes no attempt to forecast fuel prices, it is clear to the panel that the incentives for businesses and individuals to invest in and deploy technologies for alternative transportation fuels will depend largely on fossil-fuel and raw-material prices and on public policies and regulations that govern fuel production, distribution, and use.

The oil crises of the 1970s sparked a number of energy-policy changes at the federal, state, and local levels. Price controls and rationing were instituted nationally with a reduced speed limit to save gasoline. The Energy Policy and Conservation Act of 1975 created the strategic petroleum reserve (SPR) and mandated the doubling of fuel efficiency in automobiles from 13 to 27.5 miles/gal through the corporate average fuel economy (CAFE) standards. Alternative fuels have been promoted in several government incentives and mandates.

- *Synthetic Liquid Fuels Act of 1944*—authorized the construction and operation of demonstration plants to produce synthetic liquid fuels from coal, oil shale, agriculture, and other substances.
- *Energy Security Act of 1980*—provided insured loans to small ethanol plants that produced less than 1 million gallons per year and established the Synthetic Fuels Corporation.
- *Alternative Motor Fuels Act of 1988*—encouraged auto manufacturers to produce vehicles that operate on E85 (ethanol and gasoline blend that contains 85 percent ethanol) or other alternative fuels.
- *Energy Policy Act of 1992*—set a number of alternative-fueled vehicle (AFV) requirements for government and state motor fleets. It also extended the fuel tax exemption and the blender's income tax credit to two blend rates of 5.7 percent and 7.7 percent in addition to the blend rate of 10 percent. The federal government never met the mandated use of alternative fuels in its own fleet.
- *Energy Policy Act of 2005*—established a national renewable fuel standard (RFS) that mandates an increase use of renewable fuels from 4.0 billion gallons per year in 2006 to 7.5 billion gallons per year in 2012.

- *Energy Independence and Security Act of 2007*—amends RFS to set forth a phase-in for renewable fuel volumes beginning with 9 billion gallons in 2008 and ending at 36 billion gallons in 2022.

In addition to energy policies, the American Jobs Creation Act of 2004 also encourages the production of biofuels by providing a $0.51 tax credit per gallon of ethanol blended to companies that blend gasoline with ethanol and a $0.50–1.00 tax credit to biodiesel producers. Many U.S. state programs are designed to encourage the growth in alternative transportation fuel use (NASEO, 2008). Even though many public policies have addressed transportation energy supply and use over the past 60 years, the use of alternative transportation fuels in the U.S. market is still small as of 2008. Although many factors contribute to the low market penetration of alternative fuels (for example, low oil prices), the fact that many of the policies have not been durable and sustainable over time has played a significant role.

There are many choices of biomass feedstocks and technologies for converting biomass and coal to liquid fuels. In the time available, the panel could not provide detailed assessments of every potential biomass feedstock or conversion technology. Thus, the panel focused on biomass feedstock and technologies that could potentially (1) be commercially deployable over the next 10–15 years, (2) be cost competitive with petroleum fuels, and (3) result in significant reductions in U.S. oil use and greenhouse gas emissions.

The panel identified what it judged to be "aggressive but achievable" deployment opportunities for the alternative fuels. Over the course of this study, it became clear that given the costs of alternative fuels compared to petroleum-based fuels, significant deployment for alternative fuels into the market will likely not be achieved without some realignment of public policies, regulations, and other incentives and by substantial investments by both the public and private sectors.

There continues to be a great deal of uncertainty about some of the factors that will directly influence the rate of deployment of new transportation fuel supplies. Because of these uncertainties, the transportation fuel supply and cost estimates provided in this report should be considered as important first-step assessments rather than forecasts.

STRUCTURE OF THIS REPORT

The panel approached the statement of task on three parallel tracks. First, it estimated the biomass resources that would be available for fuel production without affecting the cost and supply of food and feed or incurring adverse environmental effects (Chapter 2). Second, it assessed the cost, energy use, and environmental effects of the conversion of biomass to liquid transportation fuels by biological processes (Chapter 3). Third, it assessed the same characteristics for the conversion of biomass, coal, or combined biomass and coal by thermochemical processes (Chapter 4). The panel discusses the distribution of ethanol, which is not compatible with the existing petroleum-distribution infrastructure (Chapter 5). The three sets of assessments were then integrated to provide a life-cycle assessment of the costs and CO_2 emissions of various liquid fuels produced from biomass and coal with different conversion processes. The cost and CO_2 life-cycle emissions estimates of biofuels and coal-to-liquid fuels produced biochemically or thermochemically were set on a consistent basis for comparison, and the supply of liquid transportation fuels produced from biomass and coal was estimated at different price points (Chapter 6). The overarching findings of the study and the panel's recommendations for research and development (Chapter 7) and the key challenges to commercial deployment (Chapter 8) are then presented. Other alternative transportation fuels are also discussed (Chapter 9).

The chapters on biomass supply, biochemical conversion, and thermochemical conversion (Chapters 2, 3, and 4, respectively) are structured to address the following issues in order: the feasibility of biomass supply or the commercial readiness of each technology, research and development needs, modeling to estimate costs, estimated CO_2 emissions and other environmental effects, and challenges for each technical subject. Although Chapter 2 provides estimates of costs of various biomass feedstocks, one assumed feedstock cost was used in the model simulations in Chapter 3 to assess whether biorefinery size or feedstock composition has any effect on the costs of biochemical conversion. Those trends would not be as apparent if variations in feedstock costs had been included in the simulations. Similarly, Chapter 4 used an assumed biomass cost and coal cost. In Chapter 6, the estimated costs of biochemical and thermochemical conversion are put on a consistent basis and are integrated with the different feedstock costs to provide life-cycle costs of biomass-based and coal-based fuels. Likewise, the CO_2 uptake and emission estimates in Chapters 2, 3, and 4 are integrated and put on a consistent basis in Chapter 6 to provide estimates of life-cycle emissions that can be compared. Quantities are expressed in standard units that are commonly used in the United States. Greenhouse gas emissions, however, are expressed in tonnes of CO_2 equivalent (CO_2 eq), the common metric used by the Intergovernmental Panel on Climate Change.

The Panel on Alternative Liquid Transportation Fuels provides estimates of total costs of fuel products that include the feedstock, technical, engineering, construction, and production costs that were put on a consistent basis and at one time. However, the price of fuel products is dynamic because the costs of feedstock, labor, and construction fluctuate and are influenced by multiple factors, including shortages in labor or construction material, government policies, and the economic environment; and the cost estimates are sensitive to debt-to-equity ratios, interest rates, the discount rate, and specific corporate goals (such as return on capital and risks). Therefore, the cost estimates in this report are not predictions of fuel costs in 2020; rather, they provide a comparison of technologies on a level playing field. Cost estimates in this report do not include taxes or subsidies. Gasoline and other potential taxes and carbon prices could change the relative competitiveness of alternative fuel choices.

REFERENCES

BP (British Petroleum). 2007. BP Statistical Review of World Energy: June 2007. Available at www.bp.com/statisticalreview. Accessed March 2, 2009.

Di Pascoli, S., A. Femia, and T. Luzzati. 2001. Natural gas, cars and the environment: A (relatively) "clean" and cheap fuel looking for users. Ecological Economics 38:179-189.

EIA (Energy Information Administration). 2009. Annual Energy Outlook 2008 with Projections to 2030. U.S. Department of Energy. Available at http://www.eia.doe.gov/oiaf/aeo/aeoref_tab.html. Accessed April 2, 2009.

EPA (U.S. Environmental Protection Agency). 2008. Inventory of U.S. Greenhouse Gas Emission and Sinks: 1990-2006. EPA 430-R-08-005. Washington, D.C.: U.S. Environmental Protection Agency.

ExxonMobil. 2007. The Outlook for Energy: A View to 2030. Irving, Tex.: ExxonMobil.

ExxonMobil. 2008. The Outlook for Energy: A View to 2030. Irving, Tex.: ExxonMobil.

IEA (International Energy Administration). 2009. Oil Market Report. Available at http://omrpublic.iea.org/currentissues/full.pdf. Accessed January 30, 2009.

Ingersoll, J.G. 1996. Natural Gas Vehicles, Lilburn, Ga.: Fairmont Press, Inc.

NASEO (National Association of State Energy Officials). 2008. State Alternative Fuels. Available at http://www.naseo.org/committees/energyenvironment/archive/alt_fuels. htm. Accessed January 30, 2009.

NAS-NAE-NRC (National Academy of Sciences-National Academy of Engineering-National Research Council). 2009a. America's Energy Future: Technology and Transformation. Washington, D.C.: The National Academies Press.

NAS-NAE-NRC. 2009b. Electricity from Renewable Resources: Status, Prospects, and Impediments. Washington, D.C.: The National Academies Press.

NAS-NAE-NRC. 2009c. Real Prospects for Energy Efficiency in the United States. Washington, D.C.: The National Academies Press.

NPC (National Petroleum Council). 2007. Hard Truths—Facing the Hard Truths About Energy: A Comprehensive View to 2030 of Global Oil and Natural Gas. Washington, D.C.: NPC.

NRC (National Research Council). 2004. Review of the Research Program of the FreedomCAR and Fuel Partnership: First Report. Washington, D.C.: The National Academies Press.

NRC. 2005. Assessment of Resource Needs for Development of Fuel Cell and Hydrogen Technology. Washington, D.C.: The National Academies Press.

NRC. 2007. Coal Research and Development to Support National Energy Policy. Washington, D.C.: The National Academies Press.

NRC. 2008. The Hydrogen Economy: Opportunities, Costs, Barriers, and R&D Needs. Washington, D.C.: The National Academies Press.

Solomon, S., D. Qin, R.B. Alley, T. Berntsen, N.L. Bindoff, Z. Chen, A. Chidthaisong, J.M. Gregory, G.C. Hegerl, M. Heimann, B. Hewitson, B.J. Hoskins, F. Joos, J. Jouzel, V. Kattsov, U. Lohmann, T. Matsuno, M. Molina, N. Nicholls, J. Overpeck, G. Raga, V. Ramaswamy, J. Ren, M. Rusticucci, R. Somerville, T.F. Stocker, P. Whetton, R.A. Wood, and D. Wratt. 2007. Technical summary. In Climate Change 2007: The Physical Science Basis. Contribution of Working Group I to the Fourth Assessment Report of the Intergovermental Panel on Climate Change, S. Solomon, D. Qin, M. Manning, Z. Chen, M. Marquis, K.B. Averyt, M. Tignor, and H.L. Miller, eds. Cambridge, Mass.: Cambridge University Press.

U.S. Census Bureau. 2008. Population Clocks. Available at http://www.census.gov/. Accessed July 9, 2008.

Zittel, W., and J. Schindler. 2007. Crude Oil: The Supply Outlook. EWG Series 3/2007. Ottobrunn, Germany: The Energy Watch Group.

2

Biomass Resources for Liquid Transportation Fuels

America's transportation systems will undergo major and multifaceted transformations as the nation addresses human-driven climate change, the availability and cost of liquid transportation fuels, and the need for energy security. Plant biomass has the potential to play an important role in America's energy future. Plants convert solar energy to chemical energy naturally for their growth and development through the process of photosynthesis. Plant biomass can be produced sustainably and converted into liquid transportation fuels via biochemical conversion (Chapter 3) or thermochemical conversion (Chapter 4). Liquid transportation fuels derived from biomass feedstock are often referred to as biofuels. The amount of biomass that can be produced in an area depends on the local availability of sunlight, water, and other resources. In principle, biofuels are attractive alternatives to gasoline because they are made from renewable feedstocks and can decrease the net release of greenhouse gases by the transportation sector. Although those benefits are important, they must be viewed in the context of other societal needs that are also met by the nation's land base, especially needs for food, feed, fiber, potable water, carbon storage in ecosystems, and preservation of native habitats and biodiversity. Responsible development of feedstocks for biofuels and expansion of biofuel use in the transportation sector would be economically, environmentally, and socially sustainable. This chapter addresses the questions raised in the statement of task regarding the following:

- The quantities of biomass that could potentially be produced and collected in a sustainable manner for use as feedstocks for liquid transportation fuels.

- The input and costs involved in growing and harvesting the crops or in collecting the feedstock and delivering it to a biorefinery for production of liquid transportation fuels.
- The land-use, agricultural, price, greenhouse gas, and other environmental implications of biomass production for liquid fuels.
- Research and development (R&D) needed to advance production of biomass feedstock for transportation fuels.

The chapter examines the quantities of different types of biomass that can be harvested or produced while minimizing competition between food and fuel and minimizing adverse environmental effects. It also assesses the total costs of various feedstocks that will be delivered to a processing plant for conversion to biofuel. The panel considered societal needs on the basis of recent analyses that have explored tradeoffs between using land for biofuel production and using it for food, feed, fiber, and other ecological services that land resources provide.

CURRENT BIOMASS PRODUCTION FOR BIOFUELS

Biofuel produced in the United States is overwhelmingly dominated by ethanol made from corn grain; biodiesel derived from soybean oil makes up most of the remainder. In the 2007 crop year (from September 2, 2007, to August 31, 2008), 3.0 billion bushels of corn, or 23 percent of the year's harvest, was used to produce 8.2 billion gallons of ethanol (NCGA, 2008). Around 450 million gallons of biodiesel were also produced, about 90 percent of which was derived from the oil extracted from 275 million bushels of soybean, 17 percent of the year's harvest (USDA-NASS, 2008a; NBB, 2008). On an energy-equivalent basis (in British thermal units), corn grain ethanol and soybean biodiesel together made up 2.1 percent of the liquid transportation fuel used in the United States in 2007 (EIA, 2008).

The social, economic, and environmental effects of domestic biofuels have been mixed. Diverting corn, soybean oil, or other food crops to biofuel production could induce competition among food, feed, and fuel, but increases in crop price have helped to revive rural economies. From the perspective of farmers and small rural communities, development of ethanol plants has created greater local demand and higher prices for corn grain (and for soybean through parallel efforts associated with production of biodiesel). Local investment in and control of these plants have also provided well-paying employment opportunities that reinvigorated many small midwestern communities, but some argue that the number of jobs added to the local economy is overestimated (Low and Isserman, 2009). For farmers, the increase in corn grain prices, which averaged $2.36 ± 0.40 per bushel of grain (25 kg) in 1973–2005 but $3.04 and $4.00 per bushel in 2006 and 2007 (USDA-NASS, 2008a), was of great importance. The increased prices were results of an increased global demand for corn as animal feed and for grain ethanol production. Higher commodity prices have also led to markedly higher values of fertile farmland, and have adversely affected low-income consumers in the United States and abroad and the drawing of land out of the U.S. Conservation Reserve Program (CRP). On a global scale, high commodity prices are expected to accelerate clearing of rain forest and savanna. There is growing concern about the use of grain for fuel instead of food. Other environmental concerns, especially the loss of nitrogen by leaching (Donner and Kucharik, 2008), have also been pointed out. Corn and soybean are renewable biofuel feedstocks, but large amounts of fertilizer and pesticide are often needed to grow them (Hill et al., 2006). The resulting greenhouse gas and other pollutant effects of those practices can be harmful to human health and the environment.

Corn grain ethanol and soybean biodiesel are viewed by some as intermediate fuels in the transition from oil to advanced biofuels made from cellulosic biomass. As a biofuel feedstock, cellulosic biomass has numerous advantages over food and feed crops, including its availability from sources that do not compete with food and feed production. Biomass can be reclaimed from municipal solid waste streams and from residual products of some forestry and farming operations. It can also be grown on idle or abandoned cropland, on which food or feed production is already minimal. Growing cellulosic biomass can require less fossil fuel, fertilizer, and pesticide inputs than growing corn and soybean (Tilman et al., 2006), especially if legumes (nitrogen-fixing plants) are included in the mix (NRC, 1989). In addition, cellulosic biomass can serve not only as a feedstock for biofuel production but also as a source of the heat and power required for biorefineries and thus displace fossil fuels and fossil-fuel-derived electric power (Morey et al., 2005). Therefore, this chapter focuses on the biomass resources available for cellulosic biofuel production.

Sustainable Production of Biomass for Conversion to Biofuels

Globally, about 12 billion acres of land are used for agriculture, about 4 billion of which are cultivated and the remainder used for grazing. Any substantial expansion of agriculture to accommodate dedicated biofuel crops via the direct conversion of natural ecosystems—such as native rain forests, savannas, and grasslands—into cropland could threaten those ecosystems and reduce their biodiversity. Biofuels can also indirectly cause land to be cleared when fertile agricultural soils or food crops are used for biofuels. Such indirect land clearing provides land used to grow "replacement" food crops. Moreover, intact ecosystems are major storehouses of carbon: terrestrial vegetation stores as much carbon as the atmosphere does, and terrestrial soils store twice as much (Schlesinger, 1997). Dry biomass—whether wood of trees, hay, or corn stover—contains about 45 percent carbon. On combustion or decomposition, every ton of dry biomass contributes about 1.5 tons of carbon dioxide (CO_2) to the atmosphere. In many cases, conversion of intact ecosystems to grain or fuel-crop production could incur losses of biomass and soil carbon to the atmosphere as CO_2 that greatly exceeds the greenhouse gas savings associated with biofuel production on such lands for many years (Box 2.1) (Fargione et al., 2008; Searchinger et al., 2008).

Biofuels offer opportunities for greenhouse gas reductions, but large amounts of cellulosic biomass will be needed. Sustainably produced biomass would be derived from various agricultural or forestry residues, from current waste streams, or from dedicated fuel crops grown on agricultural reserve land or on land so degraded that it is no longer cost-effective for commodity production (Tilman et al., 2006). The United Nations Environment Programme and other sources estimate that globally there are about 400–500 million hectares of such land (Campbell et al., 2008; Field et al., 2008).

Collecting agricultural residues and producing biofuel crops both have environmental benefits and costs. Removing biomass and crop residues, such as corn stover, could increase soil erosion by wind and water and deplete soil carbon reserves, ultimately affecting water entry, retention, runoff, nutrient cycling, productivity, and other critical functions. Depending on the crop, soil type, and terrain, various amounts of biomass or crop residues need to be left on a field to mitigate soil erosion and sustain soil carbon and nitrogen (Wilhelm et al., 2007). The proportion of biomass that has to be left on the soil surface to prevent erosion is higher for annual crops than for perennial crops because of the tillage generally used to establish a new crop each year. Perennial crops, especially grasses, have dense long-lived root systems that can maintain soil resources.

When ecosystems are cleared of perennial vegetation and converted to annual row crops, soil carbon stores tend to decline by 30–50 percent until a new equilibrium is reached (Davidson and Ackerman, 1993). Removal of plant residues, such as corn stover or wheat or rice straw, without such offsetting practices as growing cover crops or decreasing tillage intensity, could reduce soil carbon to a lower equilibrium. A portion of the crop residue is needed for erosion and nutrient management and to sustain soil organic carbon, which is the carbon fraction associated with all types of organic matter, including plant and animal litter, microbial biomass, water-soluble organic compounds, and stabilized or recalcitrant organic matter (Stevenson, 1994; Johnson et al., 2006a). Removing plant residues for any purpose would decrease the annual carbon input, gradually diminish soil organic carbon (Figure 2.1), and threaten the soil's production capacity (Johnson et al., 2006a). Therefore, a "systems" approach[1] is required for sustainable biomass production to ensure that its production has a low impact on global food, feed, and fiber production and that addressing the biofuel problem does not aggravate other critical challenges, including soil, water, and air quality; carbon sequestration; greenhouse gas emissions; rural development; and wildlife habitat.

A Landscape Vision of Feedstock Production

The rapidly emerging technologies to develop and use lignocellulosic materials for production of bioenergy and bioproducts might offer an opportunity to reduce the environmental footprint of the transportation sector and improve the environmental sustainability of agriculture. For example, periodically mown perennial biomass crops could be used to reduce some of the agricultural production "externalities" if they are planted as buffer strips and in locations that would help to reduce soil erosion, improve water quality, sequester carbon, and provide wildlife habitat (Tilman et al., 2006; Doornbosch and Steenblik, 2007; Ernsting and Boswell, 2007; Fargione et al., 2008; Searchinger et al., 2008).

Implementation of a landscape approach for producing biofuel feedstocks while addressing some of the externalities associated with agriculture could be made more feasible by precision agriculture (Giles and Slaughter, 1997; Tian et al., 1999; Ferguson et al., 2002; Khosla et al., 2002; Robert, 2002) and other changes (Zhang et al., 2002; Berry et al., 2003; Dinnes, 2004). Examples of how watershed-scale or landscape-scale management could potentially address those multiple

---

[1]A "systems approach" to agriculture is a holistic or integrated framework that recognizes the connectivity of multiple processes that occur on the farm and in the ecosystem and that reach across spatial, temporal, and trophic dimensions and scales. The systems approach examines the connections and interactions between the different components that make up a system so that the relative effects of change on each component can be understood.

BOX 2.1 **Effects of Land-Use Change on Greenhouse Gas Emissions**

Recent studies have focused on the life-cycle greenhouse gas emissions from different biofuels compared with gasoline or diesel. Despite some important disagreement, the prevailing view is that corn grain ethanol emitted less greenhouse gas than did gasoline, that biofuels from sugarcane provided an even greater benefit, and that cellulosic ethanol, once commercialized, would further increase the benefit (Farrell et al., 2006; Hill et al., 2006; Wang et al., 2007).

Some earlier studies recognized that carbon sequestration achieved by changing practices to reduce carbon on a landscape could be offset by increased carbon releases on other landscapes, which would result in a smaller net decrease or even a net increase in total carbon emission (Murray et al., 2004; IPCC, 2006). However, emissions from change in land use were not explicitly included in the comparative analyses of different biofuels in the life-cycle assessment. If land is cleared and used to grow plants for biofuels, much of the carbon stored in the biomass and some of that in the soil is released as CO_2. A more complete life-cycle analysis than the earlier biofuel analyses would deduct the carbon lost into the atmosphere from land-clearing and no longer being stored in an ecosystem. That approach is being implemented by the U.S. Environmental Protection Agency under the 2007 renewable fuels standard mandated in the Energy Independence and Security Act (EISA).

Fargione et al. (2008) determined the greenhouse gas carbon released in converting forest or grassland to biofuel production, which they called the "carbon debt," and the years of biofuel production required to "pay back" the debt. They argued that the carbon debt would arise from land where upland and lowland forest in Southeast Asia was converted to produce palm oil, where various forms of cerrado forest in Brazil were converted for production of biodiesel fuel, and where CRP grasslands were converted to corn for ethanol—scenarios with payback periods of 48 years (the CRP) to more than 400 years (lowland palm oil).

If land converted to biofuel production had been sequestering carbon, as would occur with regrowing forests and conservation grasslands, it is also necessary to consider the greenhouse gas effects of forgoing the benefits that would have occurred on the same land if it had not been used for biofuel production. Similarly, in a global agricultural system, if land used for food production is converted to biofuel production, some portion of the decrease in food production will be replaced by cultivation elsewhere and to a substantial extent through the conversion of noncropland to cropland (Searchinger et al., 2008). This indirect cause of converting forest and grassland to cropland also creates a carbon debt that needs to be accounted for in order to evaluate effects on greenhouse gases fully.

Searchinger et al. (2008) used the international model developed by the Center for Agricultural and Rural Development at Iowa State University and the Food and Agricultural Policy Research Institute at Iowa State University and the University of Missouri–Columbia to estimate emissions from such indirect land-use changes. They found that each acre of corn diverted to ethanol in the United States would result in roughly 0.8 acre of new cropland worldwide. They concluded that U.S. corn grain ethanol increased greenhouse gas emissions over 30 years by 93 percent relative to gasoline and that it would take 167 years to pay back the carbon debt. Ethanol made from switchgrass grown on former corn land increased emissions over 30 years by 50 percent relative to gasoline.

Estimating effects of land-use change on greenhouse gas emissions requires a worldwide agricultural land-use model, a basis of estimating which ecosystems will contribute to new cropland, and a basis of estimating carbon release per hectare. Each step contains uncertainties. The path through which land conversion takes place is complicated, requiring consideration of animal-feed by-products of biofuel production, crop-switching, reduction in food demand as a result of higher prices, likely regions of expansion, different yields in different countries, and alternative means of increasing production (for example, increased fertilizer use, drainage, or irrigation). All those factors interact and require at least a partial-equilibrium model for analysis. Because of the complexity, the exact magnitude of indirect carbon debt is difficult to determine with great certainty.

Nevertheless, the analyses of Searchinger et al. (2008) show that conversion of fertile farmlands to biofuel production is likely to have caused substantial indirect greenhouse gas release via land conversion to pasture or row crops and that indirect effects cannot be ignored in determining the full life-cycle greenhouse gas effects of biofuels. The best way to minimize such indirect effects is to avoid using for biofuel production those fertile lands that are well suited for food and feed production.

One way to deal with indirect land-use conversion, followed by the EISA, is to require calculation of indirect land-use change for each source or type of biofuel and to mandate only the biofuels that achieve specified reductions in greenhouse gas emissions relative to those from gasoline. The requirement has the obvious benefit of encouraging only biofuels that, on balance, reduce greenhouse gas emissions. One limitation of the approach is the failure to consider effects on food production and prices or other environmental effects of agricultural expansion, including loss of biodiversity and other ecosystem services. An alternative approach that takes into account both food and carbon limitations would mandate or provide incentives only for biofuels that present little risk of substantial emissions from land-use change. Such a policy would emphasize biofuel production from waste products or from feedstocks grown on marginal land (that is, areas that sequester little carbon or produce little food but can produce biomass for biofuels). Such a policy would be designed to avoid greenhouse gases, biofuel-food competition, and other potential environmental effects of agricultural expansion on water quality or quantity and biodiversity. It would also avoid the difficulty of estimating greenhouse gas emissions from indirect land-use conversions accurately.

In summary, the greenhouse gas benefit of biofuels compared with petroleum-based fuels depends not only on direct greenhouse gas emissions from biofuels during their life cycle (that is, from the growth of biomass to the production and burning of the biofuels) but also on any indirect emissions that might be incurred by changes in land use. The appropriate quantification of indirect greenhouse gas emissions is being debated. Policies could play an important role in ensuring that the biofuels produced provide environmental benefits.

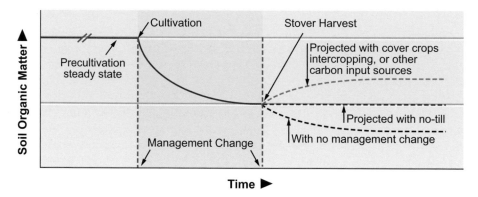

FIGURE 2.1 *Conceptual diagram of how agriculture has affected soil organic matter and what may occur after various strategies for crop-residue removal.*

concerns while supplying the necessary volume of biofuel feedstocks are presented in Appendix E.

BIOMASS RESOURCES

The following is an assessment of biomass resources for liquid fuel production using technologies available and management practices known in 2008 and projected to be available in 2020. It is predicated on two fundamental principles: (1) that biomass production for liquid fuels should not compete for land on which an existing crop is produced for food, feed, or fiber or compete for pasture land that will be needed to feed a growing and increasingly affluent population, even with yield increases, and (2) that the environmental impact on land used for biomass production should be no worse than that of its previous use and provide greater benefits wherever possible (for example, reducing fuel loads in fire-prone areas, managing volumes of urban waste, and increasing soil carbon sequestration in restoration of native grassland ecosystems). Although many other possible visions of biomass availability that do not hold as closely to those two principles are possible, the panel chose to conduct its assessment with those two in mind.[2]

---

[2]These criteria are consistent with those of Johnson et al. (2006a,b), who concluded that (1) biomass feedstocks should come first from wastes that would otherwise go to landfills, (2) agricultural residues should be harvested only when the needs for protecting soil from wind and water erosion and loss of soil organic carbon have been met, (3) dedicated fuel crops should

Among the biomass sources considered are corn stover, straw from wheat and seed grasses (for example, bluegrass and fescue), traditional hay crops (for example, alfalfa and clover), normal and high-yielding fuel crops, woody residues, animal manure, and municipal solid waste. Advantages of and concerns about each of these feedstocks are described below. The resource amounts that could be made available by using technologies and management practices of 2008 and the resource amounts projected to be available by 2020 are also described.

Corn Stover

In 2007, 13.1 billion bushels of corn grain was harvested in the United States from 86.5 million acres of cropland. Assuming a 1:1 ratio of dry weight of corn grain to stover (Johnson et al., 2006a), the amount of stover produced was estimated to be 370 million tons. Not all the corn stover can be used to produce biofuel, however, because this crop residue is also a "resource" that farmers use to mitigate wind and water erosion and to maintain soil organic matter. The amount of stover that needs to be left on the land for those purposes depends on the tillage practice being used as soil is being prepared for planting by plowing, ripping, or turning (Johnson et al., 2006a; Wilhelm et al., 2007). Perlack et al. (2005) estimated that no-tillage requires 0.35–0.5 ton of stover per acre to protect against wind and water erosion, but that amount of crop residue is not sufficient to control soil erosion if more aggressive tillage is used (Figure 2.2) and is not sufficient to sustain soil organic matter (soil carbon). To maintain soil organic matter, 2.3–5.6 tons/acre needs to be left in the field, depending on crop rotation and tillage practice (Wilhelm et al., 2007). Maintaining soil organic matter is crucial for sustaining soil structure, water entry and retention, nutrient cycling, biological activity, and other critical soil processes.

Erosion control and maintenance of soil organic matter are critical factors to be considered in the estimation of the sustainable amount of corn stover that could be harvested to produce biofuel. The national average corn-grain production in 2008 was 151 bushels/acre (USDA-NASS, 2008b). If erosion is to be controlled and soil organic matter maintained, the potential harvestable corn stover even with no-tillage practices is reduced from 3.58 tons/acre to 0.06–1.25 tons/acre depending on the crop rotation (Figure 2.2). If more intensive tillage

---

be developed regionally to meet local needs, and (4) management strategies must ensure that soils do not lose their ability to provide food, feed, fiber, and fuel.

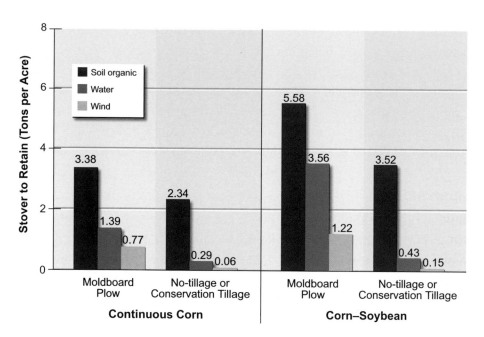

FIGURE 2.2 *Tillage and crop-rotation effects on the amount of corn stover required to protect soil resources against wind or water erosion and to sustain soil carbon (organic matter) levels.*
Source: Adapted from Wilhelm et al., 2007.

equivalent to moldboard plowing is used, as is the case for about 17 percent of U.S. corn cropland, all the corn stover in the corn–soybean rotation is required for maintenance of soil organic matter, and only 0.2 ton/acre would be available as a feedstock if corn is grown continuously. Using the higher harvestable value[3] of 1.25 tons/acre and recognizing that only 70 percent of the available cropland area at most would be planted continuously in corn because of disease, insects, and other factors, the maximum sustainable amount of corn stover available as biofuel feedstock in 2007 would have been 75.7 million tons. That value, rounded to the nearest million tons, was used for the panel's baseline estimate for the amount of corn stover that could be harvested sustainably. The panel's projection of 112 million tons available by 2020 was calculated in a similar manner and allowed for increased yield as a result of genetic improvement and improved management. (See Appendix F for details of those estimates.)

---

[3]The panel used the high harvestable value because it took a conservative approach to estimating the amount of stover that has to be left in the field to maintain soil.

Wheat Straw and Seed-Grass Straw

Wheat straw and grass straw can be biofuel feedstocks as the technology to convert them to liquid fuels develops. Banowetz et al. (2008) estimate that the Pacific Northwest states of Idaho, Oregon, and Washington could provide at least 6.5 million tons of crop residues with the wheat straw and grass straw yields in 2007 after the appropriate amount of wheat straw and grass straw are left on the field to protect soil resources. However, those straws are often distributed across the landscape at an average available density of about 1 ton/acre, so transporting them to centralized biomass-processing plants would probably increase transportation costs and transportation-dependent greenhouse gas releases. One approach for overcoming those limitations would be to establish localized preprocessing and densification centers. Similar estimates by Nelson (2002) for Kansas, Texas, Ohio, Illinois, and Missouri projected the availability of another 8.8 million tons of wheat straw for harvest from the Great Plains each year. The panel estimated that 15 million dry tons of wheat and grass straws per year could be available for fuel production on the basis of earlier studies. It assumed a 20 percent increase in available wheat and grass straws by 2020.

Hay

U.S. hay production ranged from about 50 million tons in 1999 to about 142 million tons in 2006 (USDA-NASS, 2008b). The average yield in 2007 was 2.4 tons/acre. Most hay is consumed as animal feed, but as with corn grain, average yields are often lower than many good producers achieve. On the basis of the 30-year record of increases in hay yields, the panel estimated that 10 percent of the average production for 2003–2007 (15 million tons) could be available for biofuel production without substantially affecting the hay price and supply. The portion of the hay crop used as biofuel feedstock was assumed to have very low nutritional quality for animal production because of excessive weathering. The low-quality hay would be marketed only in areas where biofuel plants provided an alternative marketing option to local farmers. The assumed supply of hay for use as a biofuel feedstock is small because hay production is dispersed, bulky, and expensive to transport. As with wheat and grass straws, the panel's projection of available hay for biofuels in 2020 was based on a 20 percent yield increase as a result of better genetics and management practices.

Dedicated Fuel Crops

When this report was being written, most agricultural land in the United States was being used for food, feed, hay, livestock, and forestry production or was enrolled in the CRP. This section considers the potential for producing biofuels on CRP lands to avoid potential conflicts over land requirements for existing or future food, feed, and fiber needs. Other lands—such as power-line rights of way, road rights of way, land classified by the U.S. Department of Agriculture (USDA) as "idle" land, or lands abandoned by agriculture sufficiently long ago as not to be classified—merit further study for their potential to produce biomass crops. Although they are not formally considered in this report, such lands might be used for production of dedicated fuel crops in lieu of or in addition to the CRP lands discussed below. The potential yields from those lands, however, have not been assessed, because few side-by-side studies of dedicated fuel crops grown on lands of different fertility and climate have been performed.

The CRP compensates farmers for removing land from crop production for environmental reasons (such as erosion control, water-quality improvement, and provision of wildlife habitat by planting appropriate perennials) and economic reasons (such as curbing production of surplus commodities and providing income support for land owners) (USDA-FSA, 2008a). If the land has not been severely eroded or depleted of essential nutrients and if expected rainfall patterns are not disrupted by increasing climate variability, a portion of it could be used for dedicated perennial fuel-crop production with appropriate site-specific agricultural practices. Planting an appropriate species or mixture of perennials and harvesting them late in the growing season could produce biofuel feedstock while potentially providing many of the same environmental benefits envisioned for CRP land. Because some land was enrolled in the CRP because of low yields of annual crops, the panel focuses on using such lands for perennials, which generally are more efficient in using nutrients in resource-poor soil than are annuals.

As of early 2008, about 35 million acres was enrolled in the CRP (USDA-FSA, 2008b). However, not all types of CRP land can be used for dedicated fuel-crop production without losing their current environmental benefits. The different types of conservation practices used on CRP land and those considered potentially compatible with biofuel-crop production by this panel are listed in Table 2.1. The categories of practices considered by the panel to be unavailable for biofuel-crop production included those already in wooded areas, in wetland restorations, or containing particular wildlife habitat. Using that classification, about 24 million

TABLE 2.1 CRP Acreage by Conservation Practice as of June 2008

Category[a]	Practice	Total Acres	Acres Potentially Suitable for Biomass
CP1	New introduced grasses and legumes	3,066,914	3,066,914
CP2	New native grasses	6,953,918	6,953,918
CP3	New softwood trees (not longleaf)	367,203	0
CP3A	New longleaf pines	213,011	0
CP3A	New hardwood trees	473,552	0
CP4	Permanent wildlife habitat	2,518,289	0
CP4B	Wildlife-habitat corridors	10,609	0
CP5	Field windbreaks	90,643	0
CP6	Diversions	540	0
CP7	Erosion-control structures	406	0
CP8	Grass waterways	129,655	0
CP9	Shallow-water areas for wildlife	52,685	0
CP10	Existing grasses and legumes	13,848,334	13,848,334
CP11	Existing trees	1,056,369	0
CP12	Wildlife-food plots	85,998	0
CP15	Contour grass strips	82,430	0
CP16	Shelterbelts	35,713	0
CP17	Living snow fences	5,826	0
CP18	Salinity-reducing vegetation	256,442	0
CP20	Alternative perennials	13	0
CP21	Filter strips (grass)	1,056,700	0
CP22	Riparian buffers	848,533	0
CP23	Wetland restoration	1,491,794	0
CP23	Wetland restoration (floodplain)	115,883	0
CP23A	Wetland restoration (nonfloodplain)	43,879	0
CP24	Cross-wind trap strips	725	0
CP25	Rare and declining habitat	1,221,521	0
CP26	Sediment retention	12	0
CP27	Farmable wetland pilot (wetland)	53,767	0
CP28	Farmable wetland pilot (upland)	127,609	0
CP29	Wildlife-habitat buffer (marginal pastureland)	97,489	0
CP30	Wetland buffer (marginal pastureland)	24,843	0
CP31	Bottomland hardwood initiative	41,976	0
CP32	Hardwood trees (previously expired)	8,563	0
CP33	Upland bird-habitat buffer initiative	197,036	0
CP36	Longleaf pine initiative	57,915	0
CP37	Duck nesting-habitat initiative	37,088	0
CP38	State acres for wildlife enhancement	37,041	0
Unknown		401	0
Total		34,711,325	23,869,166

[a]CP, conservation practice.
Source: USDA-FSA, 2008b.

acres of CRP land could potentially be converted to appropriate dedicated fuel-crop production. Landowners, however, might choose to leave land in the CRP for various reasons or to return it to food, feed, and fiber production, an option that becomes more profitable as crop prices rise (Secchi and Babcock, 2007).

Biomass yields depend on a host of factors, including location, choice of crop, cultivation practices, fertility status, and seasonal weather patterns. Switchgrass (*Panicum virgatum*) is the most immediately implementable and has been the focus of the Department of Energy's Bioenergy Feedstock Development Program for more than a decade. Although more is known about switchgrass yields than about the yields of any other proposed biofuel crop, the available data cannot yet adequately address the yields likely to be achieved on potentially usable, typical CRP lands. A recent review of published switchgrass yield trials across the United States showed an average annual yield of 4.6 tons/acre (Heaton et al., 2004a). Farmers are more likely to plant the cultivated varieties (cultivars) that had the highest yields in those trials, and a separate tally of the two highest-yielding switchgrass cultivars in independent trials across the United States showed an average of 6.1 tons/acre (McLaughlin and Kszos, 2005). Such trials are generally used as the basis of models for predicting yields. Two studies predicted an average yield of 5.4 tons/acre on existing cropland across the United States with the use of best management practices (Graham and Walsh, 1999; McLaughlin et al., 2002). Their predicted yield might not be achievable on CRP land if, as might often be the case, its soil has degraded physical, chemical, and biological conditions or if it is isolated in small fields or the terrain is not suitable for efficient mechanical harvesting. In general, the land most likely to be put into switchgrass production (for example, CRP acreage) tends to be of lower quality than test plots that are typically situated on fertile ground. For example, McLaughlin et al. (2002) estimated switchgrass yields on previously idled land to be 85 percent of those on land most recently in food production.

Trials like those described above are typically conducted on small plots, and although they are useful for evaluating ranking of cultivars best adapted to local environmental conditions, the results are not necessarily indicative of what can be expected of farm-scale production (Monti et al., 2009). Schmer et al. (2008) noted that most biofuel-crop data are derived from small plots of less than 6 yd^2 each, and they assisted farmers in establishing farm-scale switchgrass trials in Nebraska, South Dakota, and North Dakota. Average postestablishment yields in 2003–2005 were 2.7, 3.6, and 3.2 tons/acre in Nebraska, South Dakota, and North Dakota, respectively. In contrast, values predicted on the basis of small-plot trials were

5.4, 5.1, and 4.4 tons/acre (Graham and Walsh, 1999). Moreover, small-plot trials conducted concurrently in Nebraska with the cultivars represented in the farm-scale trials yielded an average of 6.4 tons/acre, or over twice the average yield of the larger plots (Schmer et al., 2006; Vogel, 2007). Thus, actual farm-scale fuel-crop production results in harvested yields about 35–50 percent lower than those of small-scale plots. Lower yields in large-scale production might be a result of farmers' inexperience with the cropping system or differences in cropland quality. But in the experiments of Schmer et al. (2008), farmers worked closely with the researchers, and the land that was used had been in active annual crop production until it was converted to switchgrass production.

An alternative biomass source is diverse mixtures of native prairie species—about equal initial densities of legume species and warm-season grass species—and seems likely to fare better in drier areas and on soils that are nitrogen-limited. In the only side-by-side comparison done to date, a high-diversity mixture of perennial grasses, legumes, and forbs had biomass yields about 200 percent greater than those of switchgrass monocultures (Tilman et al., 2006). That one study, however, was done without fertilization, with unimproved cultivars, and on a highly degraded soil of much lower fertility than the land used in the studies of switchgrass mentioned earlier and *Miscanthus*. Further field trials are necessary to assess the yield of switchgrass and mixtures of perennial grasses. To provide a preliminary estimate of potential yield of perennial grasses, the panel assumed that their yield is about 4 tons/acre, for two reasons: many studies report yields of 2–6 tons/acre (Heaton et al., 2004a; Fike et al., 2006; Perrin et al., 2008; Vadas et al., 2008), and producers are likely to use species or cultivars that have high yields, and 4 tons/acre is about 60 percent of the high yield reported.

Another dedicated perennial fuel crop being evaluated and developed is *Miscanthus*. *Miscanthus* is an exotic and potentially invasive grass species (unless sterile hybrids that reproduce only vegetatively are used) from Asia that has high yield potential. Recent European trials have resulted in average biomass yields of 10 tons/acre (Heaton et al., 2004b). Yield trials in the United States have been limited to Illinois, where the average yield was 13.2 tons/acre (Heaton et al., 2008), close to the 14.7 tons/acre predicted for that state on the basis of European data (Heaton et al., 2004b). *Miscanthus* has higher water requirements than switchgrass does and therefore would have a more restricted production range.

Although the initial result suggests that high-diversity mixtures rich in warm-season grasses and cool-season legumes have the potential to be a viable source of biomass on highly degraded land, further field trials are needed to test that possibility; to determine the regions, soil types, and other conditions for which such mixtures, switchgrass, *Miscanthus,* or other feedstock species would be superior biomass sources; and to assess the effects of dedicated fuel crops on other ecosystem services. The panel emphasizes that much work is needed to achieve greater confidence in any projections of perennial grassland biomass production for biofuels.

Short-rotation woody crops, such as hybrid poplar and willow, could also provide biomass while maintaining environmental benefits of the CRP (Johnson et al., 2007). Woody-crop yield might be greater than average in New England and the northern regions of the Great Lakes states (Graham and Walsh, 1999). (See also figures in Milbrandt, 2005.) Additional research to identify appropriate woody species on various land types is needed.

The panel's estimates of biomass that could be produced from dedicated fuel crops are presented in Table 2.2. The yields are the amounts potentially achievable with current technology if production of biomass feedstocks had high priority. CRP land would also have to be made available for dedicated fuel-crop production. In reality, the amounts would not be achievable for at least a few years.

TABLE 2.2 Estimated Biomass Supply That Could Be Available from Dedicated Fuel Crops with 2008 Technologies and Management Practices and in 2020

	Yield (tons/acre)	Area[a] (millions of acres)	Total (millions of tons)
2008			
Normal yielding[b]	4	12	24
High yielding[c]	9	6	54
Total			102
2020			
Normal yielding	5	16	40
High yielding	10.5	8	84
Total			164

[a]CRP land has not been used for dedicated fuel-crop production as of 2008. The panel assumed that two-thirds of the CRP land would be used for dedicated fuel production as an illustration.

[b]Normal-yielding candidate crops include high-diversity perennial grass-legume mixtures, such perennial grasses as switchgrass, and short-rotation woody plants grown on upland degraded soils.

[c]High-yielding biomass crops include varieties of *Miscanthus* grown on low-lying, moist, and fertile soils.

Woody Residues

Woody biomass is available from four sources other than dedicated fuel crops: forestry-industry residues, fuel-treatment residues,[4] forest-product residues, and urban wood residues. Perlack et al. (2005) estimated that as much as 41 million tons of forestry-industry residues could be collected after adjusting for recovery losses of 35–50 percent. That estimate is consistent with that of Milbrandt (2005), which is 62 million tons before adjusting for recovery losses. The panel supports the Perlack et al. (2005) recommendation that the nutrient-rich fraction of harvestable residues—which includes leaves, needles, and fine branches—be uncollected to maintain soil fertility.

The dead material on the forest floor provides readily available fuel for forest fires. The U.S. Forest Service has estimated that the amount of dead material on the forest floor could be as great as 60 million tons per year or nearly 2 billion tons over 30 years. If economically viable methods to thin overstocked forests mechanically can be developed, much of that material could be removed from forests in the western states (Perlack et al., 2005). Forest thinning could have the additional beneficial effect of reducing the amount of high-quality timber lost to forest fires each year (Fight and Barbour, 2005).

Most residue from forest-products industries are already used, but Perlack et al. (2005) estimated that an additional 8 million tons per year is available, which is higher than Milbrandt's estimate of 5 million tons (Milbrandt, 2005). Urban wood residues include wood from tree trimmings by utilities and private companies, construction and demolition, and municipal solid waste. Perlack et al. (2005) estimated that urban wood residues collectively could provide 36 million tons of woody biomass. That estimate is comparable with Milbrandt's (2005) estimate of 34 million tons. After subtraction of the currently used 8 million tons (Perlack et al., 2005) and 13 million tons of municipal solid waste wood (accounted for later in this chapter), it is assumed that 15 million tons would be available as a feedstock for biofuel production.

Overall, the panel considers that, with proper forethought and planning and demand, the infrastructure necessary to produce about 124 million tons of woody biomass could potentially be developed by 2020.

---

[4]Residues (for example, limbs and brush) from the manipulation or reduction of natural fuels or activity-caused fuels (generated by a management activity, such as slash left from logging) to reduce fire hazard.

Animal Manure

The USDA Natural Resources Conservation Service (USDA-NRCS, 2003) calculated the amount of recoverable livestock manure on a national scale on the basis of the USDA National Agricultural Statistics Service (USDA-NASS) 1997 Census of Agriculture data to determine the costs associated with establishing national comprehensive nutrient-management plans for animal-feeding operations. The calculations were based on a minimum number of on-site animal units and related characteristics. All farm livestock operations that produced less than 200 lb/yr of recoverable manure nitrogen were excluded. With those criteria and data, USDA-NRCS estimated that 60.6 million tons of dry manure could be recovered each year. The optimal use of the manure material would be as fertilizer, but many concentrated animal-feeding operations produce more manure than can be effectively used locally as fertilizer. Thus, the panel estimated that 10 percent could currently be diverted to biofuel production. The panel's estimate for 2020 assumes a 20 percent increase in the supply of manure that could be diverted to biofuel production by that year.

Municipal Solid Waste

In 2006, U.S. residents, businesses, and institutions produced more than 251 million tons of municipal solid waste, which is about 4.6 pounds of waste per person per day (Table 2.3) (EPA, 2007). Residential waste (including waste from apartment houses) accounted for 55–65 percent of the total waste generated. Waste from schools and commercial locations, such as hospitals and businesses, amounted to 35–45 percent. The largest component of municipal solid waste is organic material. Of the municipal solid waste generated in 2006, paper and paperboard products accounted for 34 percent; yard trimmings and food scraps 25 percent; plastics 12 percent; metals 8 percent; rubber, leather, and textiles 7 percent; wood 6 percent; glass at 5 percent; and other miscellaneous wastes about 3 percent.

Several municipal solid-waste management practices—such as source reduction, recycling, and composting—divert materials from the waste stream. As of 2008, 32.5 percent is recovered and recycled or composted, 12.5 percent is burned at combustion facilities, and the remaining 55 percent is disposed of in landfills. The panel agrees with Perlack et al. (2005) and Milbrandt (2005) that more municipal solid waste could be used as a biofuel feedstock. The panel assumed

TABLE 2.3 Estimated Municipal Solid Waste Available Each Year for Production of Liquid Transportation Fuels

Municipal Solid-Waste Component	Millions of Tons			
	Generated	Currently Recovered	Currently Unrecovered	Potentially Usable for Bioenergy
Paper and paperboard	85.3	44.0	41.3	41.3
Glass	13.2	2.9	10.3	0.0
Steel	14.2	5.1	9.1	0.0
Aluminum	3.3	0.7	2.6	0.0
Other nonferrous metals	1.7	1.2	0.5	0.0
Plastics	29.5	2.0	27.5	27.5
Rubber and leather	6.5	0.9	5.7	5.7
Textiles	11.8	1.8	10.0	10.0
Wood	13.9	1.3	12.6	12.6
Other materials	4.6	1.1	3.4	0.0
Food	31.3	0.7	30.6	30.6
Yard trimmings	32.4	20.1	12.3	12.3
Miscellaneous inorganic wastes	3.7	0.0	3.7	0.0
Total	251.3	81.8	169.6	139.9
Fraction recoverable for bioenergy				About 2/3
Total amount recoverable for bioenergy				About 90.0

that 90 million tons of the unrecovered organic and plastic fractions of municipal solid waste are available for bioenergy and that about two-thirds of that could be collected (Table 2.3). In comparison, San Francisco recycles about 70 percent of all urban waste, and city administrators have set a target of 75 percent. By 2020, the panel estimates that 100 million tons of municipal solid waste will be available for production of liquid transportation fuels. That estimate is based on the assumption that per capita municipal solid-waste generation will remain constant at 4.6 lb/person per day, as it has since 1990 (when it was 4.5 lb/person per day), and that additional municipal solid-waste generation is a result of population growth of 12 percent (from 304 million in 2008 to 341 million in 2020) (U.S. Census Bureau, 2008).

Summary of Lignocellulosic Feedstocks

The panel's baseline and 2020 projections for potential biofuel-feedstock supplies are summarized in Table 2.4. The estimated supplies are much lower than previous estimates (Milbrandt, 2005; Perlack et al., 2005; Biomass Research and Development Board, 2008), but the estimates are justified because of the emphasis on the amounts that could be collected in a sustainable manner without unintended consequences for soil, water, and air resources or for society as a whole and because they take into account the effects of climatic variation (including drought

TABLE 2.4 Estimated Lignocellulosic Feedstock That Could Potentially Be Produced for Biofuel with 2008 Technologies and Agricultural Practices and in 2020

| | Millions of Tons | |
Feedstock Type	2008	2020
Corn stover	76	112
Wheat and grass straw	15	18
Hay	15	18
Dedicated fuel crops	104	164
Woody residues	110	124
Animal manure	6	12
Municipal solid waste	90	100
Total	416	548

Key assumptions:

Corn stover—For continuous corn, 2.3 tons/acre must be left on fields to sustain soil carbon and control erosion. Anything above that can be harvested for biofuel production. Corn rotated with soybean requires that 3.5 tons/acre be left to meet those needs. The panel assumes that no more than 70 percent of the corn will be grown continuously and that future yield increases will mirror those achieved during the last 30 years.

Wheat straw and grass straw—Estimates are based on those of Banowetz et al. (2008) and Nelson (2002). Future increases are based on historical rates of increase in crop yields.

Hay—The panel assumes that price increases for biomass will encourage higher yields in hay, creating a 10 percent yield increase that can be dedicated to biofuel production.

Dedicated biofuel-biomass crops—The panel assumes that 18 million acres of CRP land could be planted currently as an illustration and that 24 million acres of CRP or similar land would be planted with perennial plants (switchgrass, mixed prairie species, *Miscanthus*, and so on). The field-scale yields are assumed to be 60 percent of those reported for small-scale test plots.

Woody biomass—Estimates are based on the Milbrandt (2005) and Perlack et al. (2005) reports but exclude all currently marketed woody biomass residues and municipal solid-waste wood (about 13 million tons).

and extreme weather patterns) on yields. Despite the low estimates, the panel reaffirms that research and development directed toward sustainable lignocellulosic-biofuel production are important for the nation's energy security.

Alternative Scenarios

The panel presented a scenario in which 550 million dry tons of cellulosic feedstock can be harvested or produced sustainably in 2020. Its estimates are not predictions of what would be available for fuel production in 2020. The actual supplies of biomass could exceed the panel's estimates if existing croplands are used more efficiently (Heggenstaller et al., 2008) or if genetic improvement of dedicated fuel crops exceeds the panel's estimate. In contrast, the panel's estimates could be lower if producers decide not to harvest agricultural residues or not to grow dedicated fuel crops on their CRP land.

Genetic and genomic advances could result in improved species and cultivars of dedicated fuel crops that have much higher yields than estimated by the panel. The agricultural industry aims to achieve a 40–50 percent increase in yield of commodity crops per acre in the next 10–20 years (Associated Press, 2008; York News-Times, 2009). If its goal is achieved, less acreage might be needed for food and feed production, and some agricultural land could be freed up for fuel-crop production. However, if historical trends in U.S. corn yields continue (Cassman and Liska, 2007), the predicted increases in yields would be 12 percent in 10 years and 24 percent in 20 years (see Figure 3 in Cassman and Liska, 2007). The historical yield increases have been achieved through major advances in corn-production technology, including new breeding methods, expansion of irrigated area, soil testing and balanced fertilization, conservation tillage, integrated pest management, and transgenic hybrids (Cassman and Liska, 2007). Moreover, at least part of any increase in commodity yields would be used to meet the increasing demand for food for a growing population and the increasing demand for feed due to an increasing preference for meat-based diets (Myers and Kent, 2003).

Geographic Distribution

The geographic distribution of biomass of dedicated fuel crops was estimated from the locations of CRP lands suitable for growth of switchgrass and mixed high-diversity prairie or *Miscanthus* (assuming equal land dedicated to each perennial crop and using published farm-scale yields), residue from agriculture and forestry (modified from Milbrandt, 2005), and municipal solid waste (estimated county by county on the basis of the population of each county and the national average per capita rate of municipal solid-waste production). The geographic distribution of biofuel feedstocks allows an estimation of the current amounts of biomass that could be grown within a given distance of a biorefinery. For illustrative purposes, the amounts for areas within a radius of 40 miles (equivalent to a driving distance of about 50 miles) are shown in Figure 2.3. Biomass transportation costs are high because of the low density of biomass. With the exception of woody material (primarily pulpwood), 40–50 miles has historically been the maximum distance considered economically feasible for biomass transport. On the basis of the projected biomass supply in Table 2.4, the number of sites within a 40-mile radius of where biorefineries could be established is shown in Figure 2.4. The wide variation in potential biomass supply means that the size of biorefineries might vary widely. For example, there are 290 sites where 1,500–10,000 dry tons of biomass per day could be supplied to a biorefinery within a 40-mile radius (about a 50-mile driving distance).

Barriers and Challenges to Deployment

One potential rate-limiting step in achieving large-scale lignocellulosic biofuel production might be the gathering of farmers, biomass integrators, and bio-fuel-conversion facilities into a well-organized and sustainable cellulosic-ethanol industry. Several factors need to be addressed to bring the three groups together: the efficient delivery of geographically distributed, low-density, logistically difficult materials to biorefineries in a timely manner without harming the "normal" farming operations associated with modern agriculture; determination of the full life-cycle greenhouse gas signatures of various cellulosic feedstocks when grown in a particular region with locally prescribed "best practices"; and the certification of greenhouse gas benefits so that resulting biofuels can qualify for subsidies or carbon credits associated with greenhouse gas standards, such as the standard in the 2007 EISA. A fourth critical factor might be the perception by biofuel conversion facilities that crop residues and other similar materials are literally "trash" or waste products and thus have low or no value for farmers.

Crop residues are often perceived as trash because farmers sometimes use the same term when they speak of crop residues with regard to later tillage or planting operations. Those residues can create a nuisance if they are not managed properly. Referring to crop residues as trash also reflects in part traditional American perceptions regarding the beauty of clean, weed-free fields, straight rows, and other

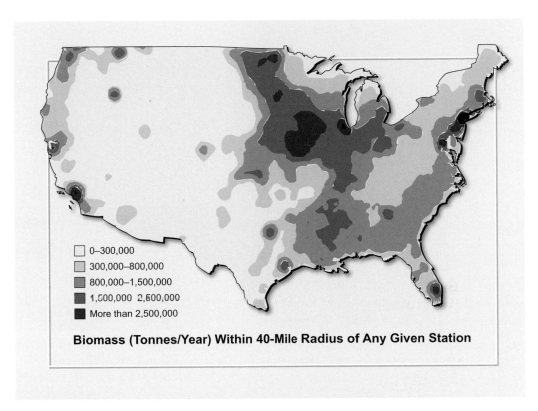

Biomass (Tonnes/Year) Within 40-Mile Radius of Any Given Station

Legend:
- ☐ 0–300,000
- 300,000–800,000
- 800,000–1,500,000
- 1,500,000–2,500,000
- More than 2,500,000

FIGURE 2.3 *Geographic distribution of potential biomass supply for biofuel production. Shading shows the annual supply of all potential biomass feedstocks within a 40-mile radius of any point in the lower 48 states. Potential biomass supplies considered were municipal solid wastes, dedicated perennial crops on degraded lands, and environmentally appropriate proportions of crop and forestry residues.*
Source: Modified from Milbrandt, 2005. County-by-county data provided by A. Milbrandt, National Renewable Energy Laboratory.

visual characteristics that Coughenour and Chamala (2000) referred to as the culture of agriculture. However, the same farmers who call crop residues trash also recognize the importance of crop residues for protecting soil resources from wind and water erosion, for cycling essential plant nutrients, for building and sustaining soil organic matter and soil fertility, and for sustaining the biological life in the soil. The environmental concerns of removing crop residues from fields have been well documented (Johnson et al., 2006a,b; Blanco-Canqui and Lal, 2007; Lal, 2007; Wilhelm et al., 2007). Farmers will insist on being adequately compensated not only for the time, labor, and other expenses incurred in harvesting, storing, and delivering crop residues but also for the nutrients, carbon content, and erosion control that will have to be replaced with increased fertilization and

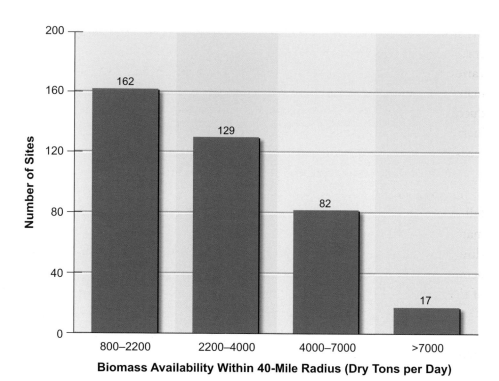

FIGURE 2.4 *The number of sites in the United States with a potential to supply the indicated daily amounts of biomass within a 40-mile radius of each site. Note that 17 sites can provide more than 7000 dry tons of biomass per day, which is equivalent to more than 2.5 million dry tons per year.*

other management strategies if the materials are harvested and sold as a biofuel feedstock. From the farmers' perspective, the rationale for those actions is that economic growth that destroys ecological support systems is neither sustainable nor true progress.

The wide geographic distribution of lignocellulosic materials is an issue that affects the supply of all potential feedstock sources (including crop residues, dedicated perennials, cover crops, and woody species). Farmers recognize the spatial variability in their fields and know that some areas (such as sideslopes and hilltops) can tolerate no crop-residue removal, whereas crops in other areas (such as depressions and toeslopes) might show a positive yield response to residue removal because seedbed conditions are more favorable. To address that concern on a field scale, a single-pass harvesting system is being developed to collect both grain and stover from some areas but to collect only grain in areas where the crop residue is needed to sustain biological, chemical, and physical properties and processes in the soil. Spatial variability will also affect production of perennial grasses on land enrolled in the CRP and similar land that has been taken out of production for various reasons. CRP and similar land is highly erodible and often is dissected by many gullies or ditches, is encumbered by rock outcrops, or has thin and nonproductive soils. Collectively, all those factors will hinder mechanical harvesting and increase the cost and logistical problems associated with delivering lignocellulosic feedstocks to biorefineries.

Another important concern, especially for midwestern farmers, is the amount of time available to handle stover when grain harvesting, trucking, storage, and fall tillage already occupy almost all their available time and labor between crop maturity and the onset of winter weather. Any new operation that threatens to slow grain harvesting will be viewed with skepticism because of current time demands and the unpredictable vagaries of fall weather. That concern is also one of the reasons for developing a one-pass harvesting system, but there is also substantial concern that additional wheel traffic will increase soil compaction and that on-farm storage will require space or even facilities that are not available on many parcels of land. To address the latter concern, several research programs in the Department of Energy, universities, and the USDA Agricultural Research Service are examining approaches for increasing the weight and fuel density of cellulosic feedstocks through various pretreatment and storage techniques (for example, Kaliyan and Morey, 2009). The low density of the materials and the high cost of transporting them to centralized refineries is also being addressed by developing more distributed networks for preprocessing and densification. Current ideas range from developing local farmer cooperatives to developing mobile processing equipment that can be assembled on a site, used, disassembled, and moved to another site. The latter appears most probable with respect to thermochemical conversion with mobile pyrolyzers—a system that has been commercialized by the Dynamotive Energy Systems Corporation in Canada.

Because of increasing global concerns about atmospheric greenhouse gases, future U.S. legislation will probably require certification of the full life-cycle greenhouse gas effects of crop production and of conversion to biofuels for each method of growing a crop and each method of converting the crop to a biofuel. It will be impractical to do such certification farm by farm, so it will be necessary to establish certifiable best practices for a given crop in a given region and then to establish ways to certify that individual farmers follow the practices. The determination of the greenhouse gas signatures of a crop and associated best practices will require a partnership between agricultural and environmental researchers and farmers.

Another approach to bringing all parties together for successful lignocellulosic-ethanol production is to implement the landscape approach described earlier in this chapter. That will require not only cooperation among biomass producers and purchasers but also policies that address energy, water quality, air quality, soil quality, wildlife habitat, carbon sequestration, community development, and other land-use issues in a coordinated manner. With the current patterns of land ownership and high rental rates (62 percent in Iowa), incentives have to be provided to farmers to address various environmental concerns simultaneously and to encourage optimal use of all types of land. Incentives will be required to manage fields that are near streams or that have depressions that help to recharge groundwater resources. In those fields, biomass crops that can tolerate wet soil conditions during spring can be used to mitigate the effect of the drainage of nutrient-rich waters to streams, rivers, and ultimately the Gulf of Mexico. Those fields would no longer be tilled annually, so there would be greater opportunities for sequestering carbon and thus helping to mitigate the increasing atmospheric CO_2 concentrations. Crop-production practices that improve yield could be developed and adopted and would help to alleviate global concerns that food and fuel production are not compatible. This land-management approach could also lead to what has been referred to as "sustainable cropping systems" that would have increased value not only for the commodities to be sold but also for their environmental and social benefits.

In summary, if managed properly, the production of lignocellulosic biofuels could contribute to U.S. energy security. It could also have benefits for some sectors of American agriculture, for the environment, and for rural communities. There are challenges to the development of the biomass-supply industry for the production of lignocellulosic biofuels: organizing farmers, biomass integrators, and biofuel conversion facilities into a well-organized and sustainable cellulosic-ethanol industry; determining the full life-cycle greenhouse gas signatures of various biofuel crops; certifying the greenhouse gas benefits; and addressing the perception that crop residues and other similar materials are trash or waste products and thus have low or no value for farmers. The challenges are formidable, but they can be overcome with an incentive-based, organized, and planned U.S. agricultural industry. As outlined in the 2007 EISA, grain ethanol is expected to provide 15 billion of the 36 billion gallons of annual biofuel production that is to be available by 2020. The incentives for achieving the EISA's goal of biofuel production are in place to accelerate the production of cellulosic or other future biofuels and the benefits that they can provide.

RESEARCH AND DEVELOPMENT

Genetic Improvement of Feedstocks

Most plants used by humans have been adapted for agriculture and human food, feed, or fiber preferences by genetic selection. In some cases, such as with corn and sugarbeet, the gains in yield of product have been enormous. In addition, the advances in understanding the genetic bases of biological processes and mechanisms during the last 25 years, accompanied by the development of methods for genetic modification of most species (NRC, 2008a), have led to optimism that future advances in plant improvement can be engineered. Indeed, Robert Fraley, the chief technology officer for Monsanto Company, was quoted in the *Financial Times* (Cameron, 2007) as saying in 2007 that "we think we can double yields over the next 25 years." He went on to note that new "traits" in soybean would lead to increases in yield similar to the increase seen in corn in recent years. The breeding of corn and soybean is now carried out largely by the commercial sector, so predictions from leading companies might be relevant. In contrast, corn yields have increased in a markedly linear fashion in the past (1.8 bushels/acre per year), even with such major advances as the use of double-cross hybrids, improved cultivation practices, and transgenic resistance to insect pests with *Bacillus thuringiensis* (Cassman and Liska, 2007).

Because relatively little effort has gone into improving the productivity of dedicated herbaceous energy crops (grasses and legumes in particular), it seems likely that substantial gains in total biomass accumulation can be realized during the next several decades. The most rapid gains in both herbaceous and woody species will almost certainly be obtained through selection of superior genotypes accompanied by conventional breeding. The application of modern genomic technologies to conventional breeding could greatly accelerate progress by providing measurements of natural genetic diversity and by allowing unambiguous identification of genotypes in segregating populations (Bouton, 2007; NRC, 2008a). Moreover, recent advances in analytical instrumentation would facilitate characterization of the chemical composition of biomass and the selection of varieties that are optimized for processing into fuels. For instance, it would be advantageous to identify varieties that are low in compounds that are toxic to microorganisms used in fuel production. Similarly, it would be advantageous to select for altered lignin composition or for changes in polysaccharide composition that facilitate decomposition of biomass to sugars, although such traits might actually reduce the overall fitness of the varieties expressing them.

Because there is still substantial uncertainty about which species of plants (besides sugarcane and its relatives) will ultimately be used as dedicated energy crops, it is challenging to estimate how much biomass yield can be improved in energy crops. However, in general terms, it seems unlikely that yield will increase to an extent similar to that observed in some food crops, such as maize, wheat, and rice, in which some of the yield increase has been obtained by increasing harvest index (that is, the ratio of grain to stover) rather than by increasing total biomass accumulation. In contrast, some of the important factors, such as resistance to disease or tolerance of abiotic stress, are likely to be as important for energy crops as for conventional crops, so some gains are to be expected by breeding for such traits. It also seems likely that, for much of the land that will be available to produce energy crops, the first species selected will be those already adapted to the regional water and temperature conditions and having reasonable biomass production in the designated habitat. Because the amount of water used by plants is determined by physical principles, it will not be possible to develop plants with substantially reduced needs for water. The most that could be achieved is to develop plants that can withstand periods of water deficit without serious physiological damage or loss of yield.

In addition to conventional breeding and selection, directed genetic modification could play a role in maximizing biomass production in the next 15–20 years. There has been substantial progress in identifying the genes that control tolerance of drought, cold, salt, insects, nematodes, and pathogens in crop species (Meksem et al., 2001; Brueggeman et al., 2002; Skinner et al., 2005; Rice Chromosomes 11 and 12 Sequencing Consortia, 2005). In some cases, it might be possible to extend the range of highly productive tropical species into temperate regions by enhancing their cold tolerance. For instance, there has recently been progress in developing *Eucalyptus* varieties that survive freezing and thrive in regions far north of the current limit for the species. The use of such a species for energy in addition to fiber may increase biomass production in the southeastern United States greatly. The results of research on crop species will translate directly to applications in many energy crops that are closely related to some important crops.

It might be possible to develop plants that have fundamental changes in chemical composition or architecture that are useful in the downstream processing to fuels. For instance, many of the molecular details of lignin synthesis and deposition are known, and it is theoretically possible to develop plants with novel lignin structure that are easier to process (Liang et al., 2008). It might also be possible to make useful changes in the structure of some of the polysaccharides, such as cellulose and hemicellulose, which would allow less energetically expensive preprocessing of biomass during conversion to fuels. In that respect, it has been proposed that it would be useful to develop transgenic plants in which the enzymes required to hydrolyze the biomass to sugars are produced by the plants themselves and stored in the plant until they are activated during preprocessing. That concept might, in principle, decrease the cost of the conversion of biomass to fuel substantially.

More speculatively, some scientists are interested in the idea of developing plants in which liquid fuels similar to gasoline and diesel accumulate in the tissues and can be harvested directly in the field by cold-pressing. That idea is attractive because it might allow a higher capture of solar energy, would greatly decrease processing costs, and would leave all mineral nutrients and much biomass in the field. Such a modification of plants would draw as much on knowledge of developmental biology as on knowledge of metabolism and photosynthesis. Plant improvement for energy or for food, feed, and fiber depends on the development of comprehensive knowledge about plants.

Systems Research for Lignocellulosic-Biomass Production

In addition to research on feedstock improvement to increase yield, studies need to be conducted to understand the favorable and unfavorable effects of lignocellulosic-biomass production on different landscapes. To ensure that potential energy security and environmental benefits of biofuels are achieved while minimizing effects on food and feed production, feedstock and commodity-crop production would have to be addressed together. The landscape vision discussed earlier in this chapter and in Appendix E is needed to balance the production of commodity crops and biomass for fuel against the externalities that can result in adverse environmental effects, such as the hypoxia in the Gulf of Mexico, unintended soil carbon release, and increasing food prices. Systems research that recognizes the connections between processes that occur on the farm and in the ecosystem and that reach across spatial, temporal, and trophic dimensions and scales is needed to develop and implement the landscape vision. Such research is necessary to determine the regions, soil types, and conditions in which different dedicated fuel crops would be appropriate and to assess the environmental, ecological, and social effects of such crop production. That complex topic and the research needs associated with it are addressed in another in-process National Research Council study on 21st-century systems agriculture.

COSTS OF SUPPLYING BIOFUEL FEEDSTOCKS

Cellulosic biomass has come to the fore as a potential source of biofuels, but there have been few attempts to provide a comprehensive accounting of the economic costs of supplying crop residue and dedicated fuel-crop feedstocks. In this section, the panel presents a simple but comprehensive economic model to provide a break-even cost to a farmer or forester for supplying the marginal or last unit of cellulosic biomass to a biofuel-processing plant. Six cellulosic biomass-feedstock sources are considered: corn stover, wheat straw, switchgrass, native prairie plants, *Miscanthus*, and woody biomass.

The biomass suppliers' willingness-to-accept (WTA) price for the marginal or last dry ton of delivered cellulosic material is assumed to be equal to the marginal cost of producing the last ton. The WTA price or marginal cost of the last ton is assumed to include land-rental cost or other forgone net returns from not selling or using the cellulosic material for feed or bedding and to include all other costs incurred in sustainably producing, harvesting and storing the biomass, and transporting it to the processing plant. The cost or feedstock price is the long-run equilibrium price that would induce suppliers to deliver biomass to the processing plant. The WTA price or marginal cost curve (or supply curve) slopes upward to the right, which implies that as the biomass processor seeks larger supplies to operate the plant on a continuous basis, the processor not only will have to pay more for each successive ton but also will have to pay the same price for all tons delivered in a competitive market environment. The biomass-feedstock costs might appear higher than anticipated by previous studies, but the panel assumed a price that will be sufficient to induce biomass suppliers to deliver 1000–4000 dry tons per day 350 days per year to sustain production of a single plant and assumed a national industry that will use more than 500 million tons per year by 2020. Because an established market for cellulosic biomass does not exist, the panel did not have long-run marginal-cost estimates to determine feedstock-supply costs in this assessment. Instead, the analysis relied on published parameter values and cost estimates that were updated with 2007 prices. Those WTA prices are analogous to prices used by economists in nonmarket valuation analyses and experiments. On the basis of published research, estimates were developed for low-cost, baseline, and high-cost alternatives. Those alternatives provide a representative range of research values and an indication of how sensitive the WTA price is to the range of parameter values found in the literature. Particular parameter values could change as a result of technological improvements, changes in energy prices and input costs, and alternative assumptions used in the literature.

The supplier's WTA price for the last dry ton of delivered cellulosic material is equal to the total costs that the supplier incurs at market equilibrium. Costs include establishment and seeding, harvesting and maintenance, transportation, storage, nutrient replacement, and land and biomass opportunity costs. For woody biomass, additional costs include chipping and stumpage fees. The discussion that follows indicates the nature and range of cost estimates that appear in the literature, differences in assumptions and reported costs, and inclusion in the comprehensive accounting of all the economic costs of supplying cellulosic biomass.

Biomass Input Costs

Harvested biomass contains essential plant nutrients (Appendix G) that need to be resupplied to the soil if the availability of the biomass is to be sustainable. Nutrient replacement cost varies with feedstock and harvesting technique. After adjustment for 2007 costs,[5] estimated costs of nutrient replacement range from $4/ton to $21/ton of harvested biomass (Aden et al., 2002; Perlack and Turhollow, 2003; Edwards, 2007; Hoskinson et al., 2007; Khanna and Dhungana, 2007; Khanna et al., 2008; Petrolia, 2008; Karlen and Birrell, Unpublished). Details of those estimates are provided in Appendix H. A baseline nutrient-replacement cost of $15/ton was used for corn stover, with low and high costs of $10/ton and $20/ton. For switchgrass and *Miscanthus*, a baseline cost of $10/ton and low and high cost estimates of $5/ton and $15/ton are used. For wheat straw, the baseline cost is $5/ton, and the low and high costs are $0/ton and $10/ton. Presumably, no nutrient replacement is necessary for woody biomass if leaves or needles and fine stems are left on the landscape. For mixed-species prairie biomass, which uses nitrogen fixed by legumes and is harvested in fall after senescence and translocation

---

[5]Nutrient-replacement costs were updated by using USDA-NASS agricultural-fertilizer prices from 1999–2007 (USDA-NASS, 2007a,b).

of the macronutrients to perennial roots, nutrient-replacement cost is estimated at $0–$5/ton. A baseline cost of $5/ton and low and high costs of $0/ton and $10/ton were used for this material because if unpublished research trials evaluating nitrogen applications to prairie grasses are successful, both yield and nutrient-replacement costs would increase compared with costs of current prairie grass production and nutrient removal by those species.

Several reports have estimated costs of harvesting and maintenance of cellulosic material (McAloon et al., 2000; Aden et al., 2002; Sokhansanj and Turhollow, 2002; Suzuki, 2006; Duffy, 2007; Edwards, 2007; Hess et al., 2007; Khanna and Dhungana, 2007; Kumar and Sokhansanj, 2007; Mapemba et al., 2007; Khanna et al., 2008; Mapemba et al., 2008; Perrin et al., 2008). Harvesting costs include costs of labor, equipment, and fuel. Maintenance costs include costs of general equipment and storage. Key points of the studies are summarized in Appendix H. Given their results, the baseline harvesting and maintenance cost is $40/ton for switchgrass, *Miscanthus*, prairie grasses, and wheat straw. The panel was not aware of any published estimates of the harvesting and maintenance cost for woody biomass and assumed that it was about $40/ton. The low and high costs are $35/ton and $45/ton. For corn stover, the baseline cost is $45/ton and the low and high costs are $40/ton and $50/ton. It is important to note that those costs include the extra labor required during a relatively narrow timeframe for harvesting corn stover and are based on sustainably harvesting 2 tons of corn stover per acre. Sustainable harvesting incurs higher costs per ton than harvesting all stover.

Transportation and storage costs will play a critical role in the development of the cellulosic-ethanol industry. The low density of grass biomass complicates the logistical dimensions of transportation and storage and contributes to the marginal cost of delivered biomass. The panel's model captures transportation and storage costs separately; transportation cost is determined according to a variable cost per dry ton per mile times the miles to the refinery. Estimates for corn stover transportation range from $11/ton to $31/ton (Aden et al., 2002; Perlack and Turhollow, 2002; English et al., 2006; Hess et al., 2007; Mapemba et al., 2008; Vadas et al., 2008). Switchgrass transportation costs have been estimated to cost around $14–36 per ton (Duffy, 2007; Kumar and Sokhansanj, 2007; Mapemba et al., 2007; Khanna et al., 2008; Mapemba et al., 2008; Perrin et al., 2008; Vadas et al., 2008), and *Miscanthus* transportation costs have been estimated to cost around $14–36 per ton (Mapemba et al., 2007; Khanna et al., 2008; Mapemba et al., 2008), all adjusted to 2007 costs.[6] Woody biomass transportation costs are expected to range from $11 to $22 per dry ton (Summit Ridge Investments, 2007).

Other research has separated the cost of transportation into distance variable cost (DVC) and distance fixed cost (DFC). DVC estimates range from $0.09 to $0.63/ton per mile (Kaylen et al., 2000; Kumar et al., 2003, 2005; Searcy et al., 2007; Petrolia, 2008). DFC estimates, mainly for biomass loading and unloading, range from $7.30 to $9.80/ton (Kumar et al., 2003, 2005; Searcy et al., 2007). Expected one-way transportation distances range from 22 to 67 miles (Perlack and Turhollow, 2002, 2003; English et al., 2006; Mapemba et al., 2007; Khanna et al., 2008; Vadas et al., 2008).

On the basis of those values, the panel's baseline transportation cost for corn stover, switchgrass, *Miscanthus*, prairie grasses, and wheat straw is $0.35/mile per ton, with a baseline distance of 30 miles. The low- and high-cost estimates are $0.25/mile per ton for 20 miles and $0.45/mile per ton for 40 miles. For woody biomass, variable transportation costs and a chipping fee were used. The low, baseline, and high transportation costs are $0.40/ton for 40 miles, $0.50/ton for 50 miles, and $0.60/ton for 70 miles. The chipping fee is $8/ton, $10/ton, and $12/ton for the low-cost, baseline, and high-cost scenarios. The panel validated its estimates with a model developed by French (1960) that included both fixed and variable distance costs and biomass-density estimates (McCarl et al., 2000).

Biomass-storage estimates were found to range from $2/ton to $17/ton (Duffy, 2007; Hess et al., 2007; Khanna et al., 2008; Mapemba et al., 2008; Petrolia, 2008) after adjustment for 2007 costs.[7] Given those estimates and information from the industry, a baseline storage cost for corn stover, switchgrass, *Miscanthus*, prairie grasses, and wheat straw of $15/ton was used. The assumed low and high costs are $10/ton and $20/ton. The baseline storage cost for woody biomass is assumed to be $10/ton and the low and high costs $0/ton and $20/ton.

Presumably, corn stover, wheat straw, and woody biomass suppliers will not incur costs of establishment and seeding. But because switchgrass, mixed prairie grasses, and *Miscanthus* do not produce another cash crop, sellers need to be compensated for their establishment and seeding, which were assumed to recur every 10 years in the case of switchgrass and every 20 years in the case of *Miscanthus*.

---

[6]Transportation costs were updated by using USDA-NASS agricultural-fuel prices from 1999 to 2007 (USDA-NASS, 2007a,b).

[7]Storage costs were updated by using USDA-NASS agricultural-building material prices from 1999 to 2007 (USDA-NASS, 2007a,b).

Cost estimates for switchgrass establishment and seeding, adjusted to 2007 costs,[8] are between $30–200/acre (Duffy, 2007; Khanna et al., 2008; Perrin et al., 2008; Vadas et al., 2008). *Miscanthus* establishment and seeding cost adjusted to 2007 costs was estimated to be around $43–350/acre (Lewandowski et al., 2003; Khanna et al., 2008). The panel's baseline value for switchgrass establishment and seeding cost is $100/acre and the low and high costs $75/acre and $125/acre. For *Miscanthus* and mixed prairie grasses, the baseline establishment cost is $225/acre and the low and high costs $175/acre and $275/acre. In the future, *Miscanthus* establishment costs could be similar to those of switchgrass as new seeded cultivars are developed (Christian et al., 2005) and commercialized, but current cost estimates are based on rhizome propagation.

To provide a complete accounting of economic costs incurred by the producer of biomass on a long-run basis, cropland rental costs (or the forgone net returns from using biomass in its next-best use) are included. Economists refer to these costs as opportunity costs because they represent the net returns forgone by the producer for not using cropland to produce the next-best crop or product. For example, land-rental rates typically reflect the net returns from producing the most profitable crop in the region, such as corn and soybean in the Corn Belt. The net returns from those crops determine how much a farmer can pay to rent an additional acre of cropland. When farmers plant perennial grasses instead of corn and soybeans, they need at least as high or higher net returns to compete for that cropland. In addition, it is argued that the farmer might require a premium beyond the WTA price to cover the risk of growing a perennial crop for 10 years or longer. Likewise, cellulosic biomass might incur an opportunity cost if there are alternative uses of the biomass, such as animal feed. If the biomass is sold for ethanol production rather than used for feed, the farmer incurs an opportunity cost equal to the net returns of using or selling biomass for livestock feed or bedding.

The panel categorized cropland-rental cost and alternative biomass-use cost in a single opportunity-cost category. Corn stover opportunity costs range from $22/acre to $143/acre (Edwards, 2007; Khanna and Dhungana, 2007). Given those research estimates, a baseline opportunity cost of $50/acre was assumed for corn stover and low and high costs of $0/acre and $100/acre. The opportunity costs of switchgrass and *Miscanthus* are much higher if they are grown on land of sufficient fertility to be suitable for corn, soybean, or other higher-value crops,

---

[8]Establishment and seeding costs were updated by using USDA-NASS agricultural-fuel and seed prices from 1999 to 2007 (USDA-NASS, 2007a,b).

with estimates ranging from $76/acre to $230/acre (Khanna and Dhungana, 2007; Khanna et al., 2008). Estimates of opportunity cost of nonspecific biomass range from $10/acre to $76/acre (Khanna et al., 2008; Mapemba et al., 2008). The opportunity cost of woody biomass is estimated to range from $0/ton to $25/ton (Summit Ridge Investments, 2007). *Miscanthus* grows best on sandy or silty loam soils with high water-holding capacity and organic-matter content and in regions with good rainfall during the growing season, much like productive corn land. Likewise, switchgrass might do well on lands that are less productive than those for corn but in a region with a longer growing season and more rainfall. Thus, the opportunity cost of a given biomass crop will depend on the type of land on which it is produced and on alternative uses for the biomass. For switchgrass and *Miscanthus*, a baseline cost of $200/acre was assumed with low and high costs of $150/acre and $250/acre.

Although it could be argued that switchgrass and mixed prairie grasses will perform well on similar land and thus have similar opportunity costs, mixed prairie grasses are reported to perform well on abandoned land (Tilman et al., 2006) that is likely to have low agricultural value and thus earn lower cash returns similar to CRP payments and have a lower opportunity cost. Thus, mixed prairie grasses were assumed to be planted on land with lower opportunity cost, such as CRP land or cropland pasture. For illustrative purposes, the baseline and low opportunity costs of mixed prairie grasses assume that the supplier still receives a CRP payment and therefore has an opportunity cost of $0/acre. The high opportunity cost of mixed prairie grasses was assumed to be $85/acre. The wheat straw baseline opportunity cost was assumed to be $0/ton and the low and high costs $10/ton and $30/ton. The negative and zero opportunity costs of wheat straw are based on the nuisance cost of seeding a new crop of small grain. Occasionally, straw is burned at harvest to avoid grain-planting problems in the following crop year. Instead of an opportunity cost for woody biomass, a stumpage fee was included with an assumed baseline of $0/ton, and low and high costs of $0/ton and $5/ton. The $5/ton cost assumes that a portion of the stumpage fee charged at timber harvest is attributed to the slash or woody biomass by-product.

The final variable in the model is biomass yield per acre of land. Biomass yield is a parameter that has the potential to be variable in both the near and the distant future. Corn stover yield per acre will vary with soil quality and other topographical characteristics, and the current yield is estimated to be 2–3 tons/acre (Edwards, 2007; Khanna and Dhungana, 2007; Vadas et al., 2008). Estimates of potential switchgrass yield range from 0.89 to 9.8 tons/acre (McLaughlin et al.,

2002; Vogel et al., 2002; Lewandowski et al., 2003; Heaton et al., 2004a; Berdahl et al., 2005; McLaughlin and Kszos, 2005; Fike et al., 2006; Shinners et al., 2006; Duffy, 2007; Khanna and Dhungana, 2007; Khanna et al., 2008; Perrin et al., 2008; Vadas et al., 2008). Grass-yield estimates range from 2.2 to 6.2 tons/acre (Banowetz et al., 2008). *Miscanthus* has much higher yield estimates, from 3.4 to 17.8 tons/acre (Lewandowski et al., 2003; Heaton et al., 2004a,b; Khanna and Dhungana, 2007; Christian et al., 2008; Khanna et al., 2008). Therefore, a baseline yield value for corn stover of 2 tons/acre was assumed with low and high yields of 1.5 and 2.5 tons/acre. For switchgrass and mixed prairie grass, the low, baseline, and high yields are assumed to be 2, 4, and 6 tons/acre, respectively. *Miscanthus* yields, are assumed to be 6, 9, and 12 tons/acre. Note that high yields are associated with low costs and vice versa.

Baseline and Sensitivity Results

The biomass supplier's WTA price or marginal cost is described by the equation below, which is generalized for nonspecific biomass. Depending on the type of biomass, some parameter values will equal zero.

$$P_{Sbiomass} = C_{NR} + C_{HM} + C_{T} \cdot D + CF + SF + C_{S} + (C_{ES} + C_{Opp})(1/Y_{B}),$$

in which $P_{Sbiomass}$ is the WTA price or marginal cost that the biomass producer would require to produce, store, and deliver 1 dry ton of biomass feedstock to the processing plant; C_{NR} is the nutrient replacement cost per dry ton of biomass; C_{HM} is the harvesting and maintenance cost per dry ton of biomass; C_{T} is the transportation cost per dry ton of biomass per mile; D is distance (in miles) to the biorefinery; CF is the chipping fee per dry ton of biomass at the roadside; SF is the stumpage fee per dry ton of biomass; C_{S} is the storage cost per dry ton of biomass; C_{ES} is the annualized biomass establishment and seeding costs per acre; C_{Opp} is the land and biomass opportunity cost in the best alternative use of biomass delivered per acre; and Y_{B} is the biomass yield of the biomass crop per acre.

Given the baseline values specified, the biomass supplier's baseline WTA per dry ton of biomass is $110 for corn stover, $151 for switchgrass, $123 for *Miscanthus*, $127 for prairie grasses, $85 for woody biomass, and $70 for wheat straw. Table 2.5 lists the WTAs for the six feedstocks in the low-cost, baseline, and high-cost scenarios.

The baseline costs (and yield) of different feedstocks were estimated on the

TABLE 2.5 Willingness-to-Accept Price for Biomass in Low-Cost, Baseline, and High-Cost Scenarios in 2020

Biomass	Willingness-to-Accept Price (dollars per ton)				
	Low Cost	50% Low[a] (2020)	Baseline[a] (2008)	50% High	High Cost
Corn stover	65	**86**	**110**	140	175
Switchgrass	93	**118**	**151**	199	286
Miscanthus	82	**101**	**123**	150	186
Prairie grasses	79	**101**	**127**	179	273
Woody biomass	59	**72**	**85**	104	124
Wheat straw	40	**55**	**70**	97	123

[a]Bolded numbers represent the panel's estimated willingness-to-accept prices in 2008 and 2020.

basis of all baseline values established above. The low-cost scenario uses all low cost estimates coupled with high yield to estimate costs, and the high-cost scenario uses all high cost estimates coupled with low yield to estimate costs for different biomass feedstocks. Next, the panel assumed that the low- and high-cost scenarios occurred with a probability of 0.5, or half the cost deviation between the baseline scenario and the high- and low-cost scenario estimates (comparable with Monte Carlo values).

The estimates listed in Table 2.5 can be interpreted in two ways. First, the low- and high-cost scenarios can be viewed as best and worst cases, assuming that everything worked or everything went wrong. More realistically, the 50 percent low- and 50 percent high-cost scenarios constitute a more reasonable range of outcomes around the baseline estimates for WTA or supply price per dry ton of cellulosic biomass feedstocks delivered to biofuel-processing plants when all costs incurred by suppliers of dry biomass delivered to the plant are considered. Second, the estimates reported in Table 2.5 as baseline (2008) can be viewed as estimates of today's costs, and the 50 percent low estimates can be viewed as estimates of biomass costs in 2020. On the basis of the research information available, the 2008 or baseline cost estimates in Table 2.5 are the best estimates of current biomass supply costs today. The 50 percent low estimates are the panel's projections of what has a high probability of happening in going from 2008 technologies to

2020 technologies. For example, the panel assumes that biomass crop yields will increase because of plant breeding and improvements in seed and plant physiology for propagation, germination, and disease, weed, and insect control. The management of biomass crops will improve, including adaptation to more suitable land types and climatic regions, production practices, and improved harvesting practices. Finally, substantial advances in the logistical dimensions of biomass handling, storage, and transportation will be needed to make biomass feedstock more economically competitive with fossil fuels. Many of the studies referred to in deriving the cost estimates assume yields and crop management reflecting future potential and goals, which are the basis of the low-biomass-cost scenario.

ENVIRONMENTAL EFFECTS

Potential Carbon Reductions from the Use of Biofuels

The panel was tasked to consider not only the economic costs of supplying various crop, forest, and dedicated biomass feedstocks for biofuel production but also the environmental costs and benefits. In general, the production of lignocellulosic biomass feedstocks for liquid fuels would probably have smaller adverse environmental effects than crop production for current biofuels and in some cases might provide additional benefits, such as lowering forest-fire risk (by removal of fuel-treatment biomass), increasing crop yield (by harvesting corn stover in specific areas), reducing the need for landfills (by collecting municipal solid waste), and restoring wildlife habitat (by re-establishing diverse prairies for growing dedicated fuel crops). Numerous environmental effects of biomass production, both favorable and unfavorable, have been described throughout this chapter, and the following brief sections are intended largely to highlight additional supporting studies in a broader discussion of next-generation biofuel sustainability.

Effects on Greenhouse Gas Emissions

Greenhouse gas emissions from biofuel-feedstock production have two primary sources: fossil fuels and the land itself. Biofuels produced from lignocellulosic feedstocks would have distinct advantages over current biofuels in both respects. Crop and forest residues, animal manure, and municipal waste—all parts of existing product streams—can be collected with minimal use of fossil fuels and, if collected in accordance with the sensitivities described earlier in this chapter, with little or no effect on soil greenhouse gas flux. Compared with corn, dedicated fuel crops can be grown with the use of less diesel for running farm equipment and less natural gas for producing nitrogenous fertilizers (Farrell et al., 2006; Hill et al., 2006). With respect to greenhouse gas flux from the land itself, perennial biomass crops may increase soil carbon (Liebig et al., 2008; Anderson et al., 2009; Pineiro et al., 2009), and corn by and large does not, even with conservation tillage practices (Baker et al., 2007). Lower nitrogen-fertilizer requirements for dedicated fuel crops also lead to lower nitrous oxide emission directly from the soil and indirectly from nitrogen runoff into waterways. The net effect is best evaluated in the context of a fuel's full life cycle of production and use.

A number of recent studies have consistently shown that biofuels produced from lignocellulosic biomass are likely to emit less greenhouse gas than is emitted by petroleum-based fuels (Farrell et al., 2006; Tilman et al., 2006; Adler et al., 2007; Wang et al., 2007) if there is no indirect greenhouse gas emissions by the conversion of native ecosystems to plant displaced crops (Fargione et al., 2008; Searchinger et al., 2008).

Schmer et al. (2008) evaluated perennial herbaceous plants, including switchgrass (*Panicum virgatum* L.), as cellulosic bioenergy crops. They addressed two major concerns: net energy efficiency and economic feasibility of switchgrass and similar crops. Prior energy analyses were based on data on smaller research plots (less than 5 m^2), but Schmer et al. (2008) managed switchgrass as a biomass-energy crop in field trials of 3–9 ha (1 ha = 10,000 m^2) on marginal cropland on 10 farms across a wide precipitation and temperature gradient in the midcontinental United States. Agricultural-energy input costs, biomass yield, estimated ethanol output, greenhouse gas emissions, and net energy results were reported. Annual biomass yields of established fields averaged 5.2–11.1 tonnes/ha with a resulting average estimated net energy yield of 60 GJ/ha per year. Average greenhouse gas emissions from cellulosic ethanol derived from switchgrass were 94 percent lower than the estimated emissions from gasoline. That study is a baseline that represents the genetic material and agronomic technology available for switchgrass production in 2000 and 2001, when the fields were planted. Improved genetics and agronomics may improve energy sustainability and biofuel yield of switchgrass further.

The panel was tasked to develop a comprehensive accounting of greenhouse gas emissions associated with the different cellulosic feedstocks. The panel estimated the greenhouse gases released during different phases of production, harvesting, transportation, and storage by using the coefficients presented by Farrell et al. (2006). Details for the calculations are provided in Appendix H; although some of the values might underestimate engineering assumptions regarding carbon input, the panel did not have any more appropriate data from which to derive a more accurate set of carbon-input numbers. The estimates were then integrated with the estimates of greenhouse gas emissions from the biochemical conversion or thermochemical conversion and from the combustion of fuel products to provide an estimate of life-cycle emissions from different fuel products in Chapter 6.

Environmental Effects Beyond Greenhouse Gas Emissions

Although greenhouse gas emissions have been the central focus of research concerning the environmental effects of biomass production for liquid fuels, other key effects must also be considered. These, not surprisingly, tend to be in the suite of effects that have long been considered for agriculture and forest management. On the whole, lignocellulosic-biomass feedstocks present distinct advantages over food-crop feedstocks in efficiency of water use (NRC, 2008b), nutrient and sediment loading into waterways (Schilling et al., 2008; Broussard and Turner, 2009), enhancement of soil fertility (Fornara and Tilman, 2008), emissions of criteria pollutants that affect air quality (Wu et al., 2006; Hill et al., 2009), and habitat for wildlife, pollinators, and species that provide biocontrol services for crop production (Landis et al., 2008). In contrast, dedicated fuel crops might pose problems not typically associated with the first-generation biofuel-crop feedstocks, including the potential to become invasive. Indeed, many of the ideal traits of biomass crops have been shown to contribute to invasiveness, including C_4 photosynthesis, long canopy duration, rapid spring growth, translocation of nutrients underground in fall, and high efficiency of water use (Raghu et al., 2006).

To guide the development of lignocellulosic biomass sources that have high overall environmental benefit, a host of recent reports have outlined sustainability criteria in ways that can aid policy decisions (Reijnders, 2004; Groom et al., 2007; Firbank, 2008; Robertson et al., 2008). The approaches that they advocate support the panel's landscape vision of biomass production in which a wide array of wastes, residues, and dedicated fuel-crop feedstocks are considered for producing liquid fuels; this not only protects purchasers of biomass against fluctuations in availability but also provides additional ecosystem services engendered by landscape diversity.

FINDINGS AND RECOMMENDATIONS

Cellulosic biomass can be produced domestically and used to produce liquid transportation fuels to improve U.S. energy security and to reduce greenhouse gas emissions from the transportation sector.

Finding 2.1

An estimated annual supply of 400 million dry tons of cellulosic biomass could be produced sustainably with technologies and management practices already available in 2008. The amount of biomass deliverable to conversion facilities could probably be increased to about 550 million dry tons by 2020. The panel judges that this quantity of biomass can be produced from dedicated energy crops, agricultural and forestry residues, and municipal solid wastes with minimal effects on U.S. food, feed, and fiber production and minimal adverse environmental effects.

The 2020 cost of biomass feedstocks, when produced in sufficient quantities to support biofuel-production facilities, is estimated to range from $55 to $118 per dry ton delivered to either biochemical-conversion or thermochemical-conversion plants.

Finding 2.2

Improvements in agricultural practices and in plant species and cultivars will be required to increase the sustainable production of cellulosic biomass and to achieve the full potential of biomass-based fuels. A sustained research and development (R&D) effort in increasing productivity, improving stress tolerance, managing diseases and weeds, and improving the efficiency of nutrient use will help to improve biomass yields.

Recommendation 2.1

The federal government should support focused research and development programs to provide the technical bases for improving agricultural practices and biomass growth to achieve the desired increase in sustainable production of cellulosic biomass. Focused attention should be directed toward plant breeding, agronomy, ecology, weed and pest science, disease management, hydrology, soil physics, agricultural engineering, economics, regional planning, field-to-wheel biofuel systems analysis, and related public policy.

Crop residues; residues from pulp, timber, and other forestry operations; forest thinnings; and some cover crops can be used to produce fuels that have much lower CO_2 emission than fossil fuels do if the biomass sources are harvested so as to preserve soil carbon and nutrients and to minimize erosion. Some components of municipal solid waste can also be used as cellulosic feedstock to reduce and reuse waste. Using biomass as a resource for energy in a sustainable manner requires holistic assessment of the effects of biomass production or harvesting on soil, water, and air quality; food, feed, and fiber production; carbon sequestration; wildlife habitat and biodiversity; rural development and related issues; and the resulting supply of energy so that multiple concerns are addressed simultaneously. If food crops or lands used for food production are diverted to produce biofuel rather than food, additional land will probably be cleared elsewhere in the world and drawn into food production. The greenhouse gas emissions caused by such clearing of land, especially forests, will decrease or even negate the greenhouse gas benefits of the resulting biofuels.

Finding 2.3

Incentives and best agricultural practices will probably be needed to encourage sustainable production of biomass for production of biofuels. Producers need to grow biofuel feedstocks on degraded agricultural land to avoid direct and indirect competition with the food supply and also need to minimize land-use practices that result in substantial net greenhouse gas emissions. For example, continuation of CRP payments for CRP lands when they are used to produce perennial grass and wood crops for biomass feedstock in an environmentally sustainable manner might be an incentive.

Finding 2.4

Depending on the locations in which it is grown and the management practices used to produce it, the production of cellulosic biomass for fuels has the potential to improve agricultural sustainability. Research that emphasizes the relationship between cellulosic-biomass production and its surrounding landscape as a system is needed to improve knowledge and understanding of the environmental effects of harvesting crop or woody residues or growing the fuel crops and the potential ecosystem services that they provide. Such research would require expertise in a wide array of topics.

Recommendation 2.2

A framework should be developed to assess the effects of cellulosic-feedstock production on various environmental characteristics and natural resources. Such an assessment framework should be developed with input from agronomists, ecologists, soil scientists, environmental scientists, and producers and should include, at a minimum, effects on greenhouse gas emissions and on water and soil resources. The framework would provide guidance to farmers on sustainable production of cellulosic feedstock and contribute to improvements in energy security and in the environmental sustainability of agriculture.

Large regions of the United States could produce sufficient biomass to provide about 300,000 tons of biomass per year within a 40-mile radius of strategically located biomass-conversion facilities. Biomass is also available in other regions but at lower densities. The major U.S. regions that can deliver large quantities of biomass include portions of the Northwest, the upper Midwest, and the East.

Finding 2.5

Biomass availability could limit the size of a conversion facility and thereby influence the cost of fuel products from any facility that uses biomass irrespective of the conversion approach. Biomass is bulky and difficult to transport. The density of biomass growth will vary considerably from region to region in the United States, and the biomass supply available within 40 miles of a conversion plant will vary from less than 1,000 tons/day to 10,000 tons/day. Longer transportation distances could increase supply but would increase transportation costs and could magnify other logistical issues.

Recommendation 2.3

Technologies that increase the density of biomass in the field to decrease transportation cost and logistical issues should be developed. The densification of available biomass enabled by a technology such as field-scale pyrolysis could facilitate transportation of biomass to larger-scale regional conversion facilities.

REFERENCES

Aden, A., M. Ruth, K. Ibsen, J. Jechura, K. Neeves, J. Sheehan, B. Wallace, L. Montague, A. Slayton, and J. Lukas. 2002. Lignocellulosic Biomass to Ethanol Process Design and Economics Utilizing Co-Current Dilute Acid Prehydrolysis and Enzymatic Hydrolysis for Corn Stover. Golden, Colo.: National Renewable Energy Laboratory.

Adler, P.R., S.J. Del Grosso, and W.J. Parton. 2007. Life-cycle assessment of net greenhouse gas flux for bioenergy cropping systems. Journal of Applied Ecology 17:675-691.

Anderson, S.H., R.P. Udawatta, T. Seobi, and H.E. Garrett. 2009. Soil water content and infiltration in agroforestry buffer strips. Agroforestry Systems 75(1):5-16.

Associated Press. 2008. Monsanto says its seeds will double yield of corn, soybeans and cotton by 2030. Available at http://www.iht.com/articles/ap/2008/06/04/business/NA-FIN-COM-USMonsanto-Future.php. Accessed March 4, 2009.

Baker, T., I. Bashmakov, L. Bernstein, J.E. Bogner, P.R. Bosch, R. Dave, O.R. Davidson, B.S. Fisher, S. Gupta, K. Halsnaes, G.J. Heij, S. Kahn Ribeiro, S. Kobayashi, M.D. Levine, D.L. Martino, O. Masera, B. Metz, L.A. Meyer, G-J. Nabuurs, A. Najam, N. Nakicenovic, H.H. Rogner, J. Roy, J. Sathaye, R. Schock, P. Shukla, R.E.H. Sims, P. Smith, D.A. Tirpak, D. Urge-Vorsatz, and D. Zhou. 2007. Technical summary. In Climate Change 2007: Mitigation. Contribution of Working Group III to the Fourth Assessment Report of the Intergovernmental Panel on Climate Change, B. Metz, O.R. Davidson, P.R. Bosch, R. Dave, and L.A. Meyer, eds. New York: Cambridge University Press.

Banowetz, G.M., A. Boatang, J.J. Steiner, S.M. Griffith, V. Sethi, and H. El-Nashaar. 2008. Assessment of straw biomass feedstock resources in the Pacific Northwest. Biomass and Bioenergy 32(7):629-634.

Berdahl, J., A. Frank, J. Krupinsky, P. Carr, J. Hanson, and H. Johnson. 2005. Biomass yield, phenology, and survival of diverse switchgrass cultivars and experimental strains in western North Dakota. Agronomy Journal 97:549-555.

Berry, J.K., J.A. Delgado, R. Khosla, and F.J. Pierce. 2003. Precision conservation for environmental sustainability. Soil and Water Conservation 58:332-339.

Biomass Research and Development Board. 2008. Increasing Feedstock Production for Biofuels. Economic Drivers, Environmental Implications, and the Role of Research. Available at http://www.brdisolutions.com/Site%20Docs/Increasing%20Feedstock_revised.pdf. Accessed February 10, 2008.

Blanco-Canqui, H., and R. Lal. 2007. Soil and crop response to harvesting corn residues for biofuel production. Geoderma 141:355-362.

Bouton, J.H. 2007. Molecular breeding of switchgrass for use as a biofuel crop. Current Opinion in Genetic Development 17:553-558.

Broussard, Whitney, and R. Eugene Turner. 2009. A century of changing land-use and water-quality relationships in the continental US. Frontiers in Ecology and the Environment 7(6):302-307.

Brueggeman, R., N. Rostoks, D. Kudrna, A. Kilian, F. Han, J. Chen, A. Druka, B. Steffenson, and A. Kleinhofs. 2002. The barley stem rust-resistance gene Rpg1 is a novel disease-resistance gene with homology to receptor kinases. Proceedings of the National Academy of Sciences USA 99:9328-9333.

Cameron, D. 2007. Rivals give Monsanto food for thought. Financial Times, September 4.

Campbell, J.E., D.B. Lobell, R.C. Genova, and C.B. Field. 2008. The global potential of bioenergy on abandoned agriculture lands. Environmental Science and Technology 42:5791-5794.

Cassman, K.G., and A.J. Liska. 2007. Food and fuel for all: Realistic or foolish? Biofuels, Bioproducts and Biorefining 1:18-23.

Christian, D., A. Riche, and N. Yates. 2008. Growth, yield and mineral content of *Miscanthus x Giganteus* grown as a biofuel for 14 successful harvests. Industrial Crops and Products 28(3):320-327.

Christian, D.G., N.E. Yates, and A. Riche. 2005. Establishing Miscanthus sinensis from seed using conventional sowing methods. Industrial Crops and Products 21(1):109-111.

Coughenour, C.M., and S. Chamala. 2000. Conservation Tillage and Cropping Innovation: Constructing the New Culture of Agriculture. Ames: Iowa State University Press.

Davidson, E.A., and I.L. Ackerman. 1993. Changes in soil carbon inventories following cultivation of previously untilled soils. Biogeochemistry 20:161-193.

Dinnes, D.L. 2004. Assessments of Practices to Reduce Nitrogen and Phosphorus Nonpoint Source Pollution of Iowa's Surface Waters. Iowa Department of Natural Resources and USDA-ARS National Soil Tilth Laboratory. Available at ftp://ftp.nstl.gov/pub/NPS/NPS%20Nutrient%20Pollution%20Assessments%20of%20Conservation%20Practices.pdf. Accessed April 25, 2008.

Donner, S.D., and C.J. Kucharik. 2008. Corn-based ethanol production compromises goal of reducing nitrogen export by the Mississippi River. Proceedings of the National Academy of Sciences USA 105(11):4513-4518.

Doornbosch, R., and R. Steenblik. 2007. Biofuels: Is the cure worse than the disease? In Round Table on Sustainable Development. Paris: Organisation for Economic Cooperation and Development.

Duffy, M. 2007. Estimated Costs for Production, Storage, and Transportation of Switchgrass. Available at http://www.extension.iastate.edu/agdm/crops/pdf/a1-22.pdf. Accessed April 25, 2008.

Edwards, William. 2007. Estimating a Value for Corn Stover. Available at http://www.extension.iastate.edu/agdm/crops/pdf/a1-70.pdf. Accessed April 25, 2008.

EIA (Energy Information Administration). 2008. Annual Energy Review 2007. DOE/EIA-0384(2007). Washington, D.C.: U.S. Department of Energy, Energy Information Administration.

English, B.C., D.G. de la Torre Ugarte, K. Jensen, C. Hellwinckel, J. Menard, B. Wilson, R. Roberts, and M. Walsh. 2006. 25% Renewable Energy for the United States by 2025: Agricultural and Economic Impacts. University of Tennessee, Knoxville. Available at http://www.25x25.org/storage/25x25/documents/RANDandUT/UTEXECsummary25X25FINALFF.pdf. Accessed April 25, 2008.

EPA (U.S. Environmental Protection Agency). 2007. Municipal Solid Waste Generation, Recycling, and Disposal in the United States: Facts and Figures for 2006. Available at http://www.epa.gov/epaoswer/nonhw/muncpl/pubs/msw06.pdf. Accessed April 25, 2008.

Ernsting, A., and A. Boswell. 2007. Agrofuels: Towards a Reality Check in Nine Key Areas. Available at http://www.tni.org/reports/ctw/agrofuels.pdf. Accessed April 25, 2009.

Fargione, J., J. Hill, D. Tilman, S. Polasky, and P. Hawthorne. 2008. Land clearing and the biofuel carbon debt. Science 319:1235-1238.

Farrell, A., R. Plevin, B. Turner, A. Jones, M. O'Hare, and D. Kammen. 2006. Ethanol can contribute to energy and environmental goals. Science 311:506-509.

Ferguson, R.B., G.W. Hergert, J.S. Schepers, C.A. Gotway, J.E. Cahoon, and T.A. Peterson. 2002. Site-specific management of irrigated maize: Yield and soil residual nitrate effects. Soil Science Society of America Journal 66:544-553.

Field, C.B., J.E. Campbell, and D.B. Lobell. 2008. Biomass energy: The scale of the potential resource. Trends in Ecology and Evolution 23:65-72.

Fight, R.D., and R.J. Barbour. 2005. Financial Analysis of Fuel Treatments. Portland, Oreg.: U.S. Department of Agriculture, Forest Service.

Fike, J., D. Parrish, D. Wolf, J. Balasko, J. Green, Jr., M. Rasnake, and J. Reynolds. 2006. Long-term yield potential of switchgrass-for-biofuel systems. Biomass and Bioenergy 30(3):198-206.

Firbank, Les G. 2008. Assessing ecological impacts of bioenergy projects. Bioenergy Research 1(1):12-19.

Fornara, D.A., and D. Tilman. 2008. Plant functional composition influences rates of soil carbon and nitrogen accumulation. Journal of Ecology 96(2):314-322.

French, B. 1960. Some considerations in estimating assembly cost functions for agricultural processing operations. Journal of Farm Economics 62:767-778.

Giles, D.K., and D.C. Slaughter. 1997. Precision band sprayer with machine-vision guidance and adjustable yaw nozzles. Transactions of the American Society of Agricultural and Biological Engineers 40:29-36.

Graham, R.L., and M.E. Walsh. 1999. A National Assessment of Promising Areas for Switchgrass, Hybrid Poplar, or Willow Energy Crop Production. Oak Ridge, Tenn.: Oak Ridge National Laboratory.

Groom, M.J., E.M. Gray, and P.A. Townsend. 2007. Biofuels and biodiversity: Principles for creating better policies for biofuel production. Conservation Biology 22:602-609.

Heaton, E., T. Voight, and S.P. Long. 2004a. A quantitative review comparing the yields of two candidate C4 perennial biomass crops in relation to nitrogen, temperature and water. Biomass and Bioenergy 27:21-30.

Heaton, E.A., J. Clifton-Brown, T.B. Voight, M.B. Jones, and S.P. Long. 2004b. *Miscanthus* for renewable energy generation: European Union experience and projections for Illinois. Mitigation and Adaptation Strategies for Global Change 9:433-451.

Heaton, E.A., F.G. Dohleman, and S.P. Long. 2008. Meeting US biofuel goals with less land: The potential of *Miscanthus*. Global Change Biology 14(9):2000-2014.

Heggenstaller, A.H., R.P. Anex, M. Liebman, D.N. Sundberg, and L.R. Gibson. 2008. Productivity and nutrient dynamics in bioenergy double-cropping systems. Agronomy Journal 100:1740-1748.

Hess, J.R., C.T. Wright, and K.L. Kenney. 2007. Cellulosic biomass feedstocks and logistics for ethanol production. Biomass, Bioproduction and Biorefining 1:181-190.

Hill, J., E. Nelson, D. Tilman, S. Polasky, and D. Tiffany. 2006. Environmental, economic, and energetic costs and benefits of biodiesel and ethanol biofuels. Proceedings of the National Academy of Sciences USA 103(30):11206-11210.

Hill, J., S. Polasky, E. Nelson, D. Tilman, H. Huo, L. Ludwig, J. Neumann, H. Zheng, and D. Bonta. 2009. Climate change and health costs of air emissions from biofuels and gasoline. Proceedings of the National Academy of Sciences USA 106:2077-2082.

Hoskinson, R.L., D.L. Karlen, S.J. Birrell, C.W. Radtke, and W.W. Wilhelm. 2007. Engineering, nutrient removal, and feedstock conversion evaluations of four corn stover harvest scenarios. Biomass and Bioenergy 31:126-136.

IPCC (Intergovernmental Panel on Climate Change). 2006. Guidelines for National Greenhouse Gas Inventories, Volume 4, Agriculture, Forestry and Other Land Use. Available at http://www.ipccnggip.iges.or.jp/public/2006gl/vol4.html. Accessed April 24, 2009.

Johnson, J.M.F., R.R. Allmaras, and D.C. Reicosky. 2006a. Estimating source carbon from crop residues, roots, and rhizodeposits using the national grain-yield database. Agronomy Journal 98:622-636.

Johnson, J.M.F., D. Reicosky, R. Allmaras, D. Archer, and W. Wilhelm. 2006b. A matter of balance: Conservation and renewable energy. Journal of Soil and Water Conservation 63:121-125.

Johnson, J.M.F., M.D. Coleman, R.W. Gesch, A.A. Jaradat, R. Mitchell, D.C. Reicosky, and W.W. Wilhelm. 2007. Biomass-bioenergy crops in the United States: A changing paradigm. American Journal of Plant Science and Biotechnology 1(1):1-28.

Kaliyan, N., and R.V. Morey. 2009. Factors affecting strength and durability of densified biomass products. Biomass and Bioenergy 33:337-359.

Karlen, D.L., and S.J. Birrell. Unpublished. Crop Residue—What's It Worth? U.S. Department of Agriculture and Iowa State University. Available at http://www1.eere. energy.gov/biomass/pdfs/Biomass_2009_Sustainabiliy_III_Karlen.pdf. Accessed April 25, 2009.

Kaylen, M., D.L. Van Dyne, Y.S. Choi, and M. Blase. 2000. Economic feasibility of producing ethanol from lignocellulosic feedstocks. Bioresource Technology 72:19-32.

Khanna, M., and B. Dhungana. 2007. Economics of Alternative Feedstocks in Corn-Based Ethanol in Illinois and the US: A Report from Department of Agricultural and Consumer Economics. Urbana-Champaign: University of Illinois. Available at http://www.farmdoc.uiuc.edu/policy/research_reports/ethanol_report/Ethanol%20Report.pdf. Accessed April 25, 2009.

Khanna, M., B. Dhungana, and J. Clifton-Brown. 2008. Costs of producing Miscanthus and switchgrass for bioenergy in Illinois. Biomass and Bioenergy 32(6):482-493.

Khosla, R., K. Fleming, J.A. Delgado, T.M. Shaver, and D.G. Westfall. 2002. Use of site-specific management zones to improve nitrogen management for precision agriculture. Journal of Soil and Water Conservation 57:513-518.

Kumar, A., J. Cameron, and P. Flynn. 2003. Biomass power cost and optimum plant size in western Canada. Biomass and Bioenergy 24(6):445-464.

Kumar, A., J. Cameron, and P. Flynn. 2005. Pipeline transport and simultaneous saccharification of corn stover. Bioresource Technology 96(7):819-829.

Kumar, A., and S. Sokhansanj. 2007. Switchgrass (Panicum vigratum, L.) delivery to a biorefinery using Integrated Biomass Supply Analysis and Logistics (IBSAL) model. Bioresource Technology 98:1033-1044.

Lal, R. 2007. Biofuels from crop residues. Soil and Tillage Research 93:237-238.

Landis, D.A., M.M. Gardiner, W. van der Werf, and S.M. Swinton. 2008. Increasing corn for biofuel production reduces biocontrol services in agricultural landscapes. Proceedings of the National Academy of Sciences USA 105:20552-20557.

Lewandowski, I., J. Scurlock, E. Lindvall, and M. Christou. 2003. The development and current status of perennial rhizomatous grasses as energy cops in the US and Europe. Biomass and Bioenergy 25(4):335-361.

Liang, H.Y., C.J. Frost, X.P. Wei, N.R. Brown, J.E. Carlson, and M. Tien. 2008. Improved sugar release from lignocellulosic material by introducing a tyrosinerich cell wall peptide gene in poplar. CLEAN—Soil, Air, Water 36:662-668.

Liebig, M.A., S.L. Kronberg, and J.R. Gross. 2008. Effects of normal and altered cattle urine on short-term greenhouse gas flux from mixed-grass prairie in the northern Great Plains. Agriculture Ecosystems and Environment 125:57-64.

Low, S.A., and A.M. Isserman. 2009. Ethanol and the local economy: Industry trends, location factors, economic impacts, and risks. Economic Development Quarterly 23:71-88.

Mapemba, L.D., F.M. Epplin, R.L. Huhnke, and C.M. Taliaferro. 2008. Herbaceous plant biomass harvest and delivery cost with harvest segmented by month and number of harvest machines endogenously determined. Biomass and Bioenergy 32(11)1016-1027.

Mapemba, L., F. Epplin, C. Taliaferro, and R. Huhnke. 2007. Biorefinery feedstock production on conservation reserve program land. Review of Agricultural Economics 29(2):227-246.

McAloon, A., F. Taylor, W. Yee, K. Ibsen, and R. Wooley. 2000. Determining the Cost of Producing Ethanol from Corn Starch and Lignocellulosic Feedstocks. Golden, Colo.: National Renewable Energy Laboratory.

McCarl, B., D. Adams, R. Alig, and J. Chmelik. 2000. Competitiveness of biomass fueled electrical power plants. Annals of Operations Research 94:37-55.

McLaughlin, S.B., D.G. de la Torre Ugarte, C.T. Garten, Jr., L.R. Lynd, M.A. Sanderson, V.R. Tolbert, and D.D. Wolf. 2002. High-value renewable energy from prairie grasses. Environmental Science and Technology 36:2122-2129.

McLaughlin, S.B., and L.A. Kszos. 2005. Development of switchgrass (Panicum virgatum) as a bioenergy feedstock in the United States. Biomass and Bioenergy 28:515-535.

Meksem, K., E. Ruben, D. Hyten, M. Schmidt, and D.A. Lightfoot. 2001. High-throughput detection of polymorphism physically linked soybean cyst nematode resistance gene Rhg4 using Taqman probes. Molecular Breeding 7:63-71.

Milbrandt, A. 2005. A geographic perspective on the current biomass resource availability in the United States. Available at http://www.osti.gov/bridge. Accessed November 9, 2008.

Monti, A., S. Fazio, and G. Venturi. 2009. The discrepancy between plot and field yields: Harvest and storage losses of switchgrass. Biomass & Bioenergy 33(5):841-847.

Morey, R.V., D. Tiffany, and D. Hatfield. 2005. Biomass for Electricity and Process Heat at Ethanol Plants. St. Joseph, Mich: American Society for Agricultural and Biological Engineers.

Murray, B.C., B.A. McCarl, and H.C. Lee. 2004. Estimating leakage from forest carbon sequestration programs. Land Economics 80:109-124.

Myers, N., and J. Kent. 2003. New consumers: The influence of affluence on the environment. Proceedings of the National Academy of Sciences USA 100:4963-4968.

NBB (National Biodiesel Board). 2008. U.S. Biodiesel Production Capacity. Jefferson City, Mo.: NBB.

NCGA (National Corn Growers Association). 2008. World of Corn. Chesterfield, Mo.: NGCA.

Nelson, R.G. 2002. Resource assessment and removal analysis for corn stover and wheat straw in the eastern and midwestern United States: Rainfall and wind-induced soil erosion methodology. Biomass and Bioenergy 22:349-363.

NRC (National Research Council). 1989. Alternative Agriculture. Washington, D.C.: National Academy Press.

NRC. 2008a. Achievements of the National Plant Genome Initiative and New Horizons in Plant Biology. Washington, D.C.: The National Academies Press.

NRC. 2008b. Water Implications of Biofuels Production in the United States. Washington, D.C.: The National Academies Press.

Perlack, R., and A. Turhollow. 2002. Assessment of Options for the Collection, Handling, and Transport of Corn Stover. Oak Ridge, Tenn.: Oak Ridge National Laboratory.

Perlack, R., and A. Turhollow. 2003. Feedstock cost analysis of corn stover residues for further processing. Energy 28:1395-1403.

Perlack, R.D., L.L. Wright, A.F. Turhollow, R.L. Graham, B.J. Stokes, and D.C. Erbach. 2005. Biomass as feedstock for a bioenergy and bioproducts industry: The technical feasibility of a billion-ton annual supply. Washington, D.C.: U.S. Department of Agriculture and U.S. Department of Energy.

Perrin, R., K. Vogel, M. Schmer, and R. Mitchell. 2008. Farm-scale production cost of switchgrass for biomass. BioEnergy Research 1(1):91-97.

Petrolia, D.R. 2008. The economics of harvesting and transporting corn stover to fuel ethanol: A case study for Minnesota. Biomass and Bioenergy 32(7):603-612.

Pineiro, G., E.G. Jobbagy, J. Baker, B.C. Murray, and R.B. Jackson. 2009. Set-asides can be better climate investment than corn ethanol. Ecological Applications 19:277-282.

Raghu, S., R.C. Anderson, C.C. Daehler, A.S. Davis, R.N. Wiedenmann, D. Simberloff, and R.N. Mack. 2006. Adding biofuels to the invasive species fire? Science 313:1742.

Reijnders, L. 2004. Conditions for the sustainability of biomass based fuel use. Energy Policy 34:863-876.

Rice Chromosomes 11 and 12 Sequencing Consortia. 2005. The sequence of rice chromosomes 11 and 12, rich in disease resistance genes and recent gene duplications. BMC Biology 3:20.

Robert, P.C. 2002. Precision agriculture: A challenge for crop nutrition management. Plant and Soil 247:143-149.

Robertson, G.P, V.H. Dale, O.C. Doering, S.P. Hamburg, J.M. Mclillo, M.M. Wander, W.J. Parton, P.R. Adler, J.N. Barney, R.M. Cruse, C.S. Duke, P.M. Fearnside, R.F. Follett, H.K. Gibbs, J. Goldemberg, D.J. Mladenoff, D. Ojima, M.W. Palmer, A. Sharpley, L. Wallace, K.C. Weathers, J.A. Wiens, and W.W. Wilhelm. 2008. Sustainable biofuels redux. Science 322:49.

Schilling, K.E., M.K. Jha, Y-K. Zhang, P.W. Gassman, and C.F. Wolter. 2008. Impact of land use and land cover change on the water balance of a large agricultural watershed: Historical effects and future directions. Water Resources Research 44: W00A09.

Schlesinger, W.H. 1997. Biogeochemistry: An Analysis of Global Change. 2nd ed. San Diego, Calif.: Academic Press.

Schmer, M.R., K.P. Vogel, R.B. Mitchell, L.E. Moser, K.M. Eskridge, and R.K. Perrin. 2006. Establishment stand thresholds for switchgrass grown as a bioenergy crop. Crop Science 46:157-161.

Schmer, M.R., K.P. Vogel, R.B. Mitchell, and R.K. Perrin. 2008. Net energy of cellulosic ethanol from switchgrass. Proceedings of the National Academy of Sciences USA 105:464-469.

Searchinger, T., R. Heimlich, R.A. Houghton, F. Dong, A. Elobeid, J. Fabiosa, S. Tokgoz, D. Hayes, and T-H. Yu. 2008. Use of U.S. croplands for biofuels increases greenhouse gases through emissions from land use change. Science 319:1238-1240.

Searcy, E., P. Flynn, E. Ghafoori, and A. Kumar. 2007. The relative cost of biomass energy transport. Applied Biochemistry and Biotechnology 137-140(1-12):639-652.

Secchi, S., and B.A. Babcock. 2007. Impact of high crop prices on environmental quality: A case of Iowa and the Conservation Reserve Program. Ames: Iowa State University. Available at http://www.card.iastate.edu/publications/DBS/PDFFiles/07wp447.pdf. Accessed April 25, 2009.

Shinners, K.J., G.C. Boettcher, R.E. Muck, P.J. Weimer, M.D. Casler. 2006. Drying, harvesting, and storage characteristics of perennial grasses as biomass feedstocks. ASABE Paper No. 061012. St. Joseph, Mich.: American Society of Agricultural and Biological Engineers.

Skinner, J.S., J. von Zitzewitz, L. Marquez-Cedillo, T. Filichkin, P. Szücs, K. Amundsen, E.J. Stockinger, M.F. Thomashow, T.H.H. Chen, and P.M. Hayes. 2005. Barley contains a large cbf gene family associated with quantitative cold tolerance traits. In Advances in Plant Cold Hardiness: Molecular Genetics and Transgenics, T.H.H. Chen, M. Uemura, and S. Fujikawa, eds. Oxon, United Kingdom: CAB International.

Sokhansanj, S., and A. Turhollow. 2002. Baseline cost for corn stover collection. Applied Engineering and Agriculture 18:525-530.

Stevenson, F.J. 1994. Humic Chemistry: Genesis, Composition, Reactions. 2nd ed. New York: Wiley.

Summit Ridge Investments, LLC. 2007. Eastern Hardwood Forest Region Woody Biomass Energy Opportunity. Granville, Vt: Summit Ridge Investments.

Suzuki, Y. 2006. Estimating the Cost of Transporting Corn Stalks in the Midwest. Ames: Iowa State University College of Business, Business and Partnership Development.

Tian, L., J.F. Reid, and J.W. Hummel. 1999. Development of a precision sprayer for site-specific weed management. Transactions of the ASABE 42:893-900.

Tilman, D., J. Hill, and C. Lehman. 2006. Carbon-negative biofuels from low-input high diversity grassland biomass. Science 314:1598-1600.

U.S. Census Bureau. 2008. Projections of the Population and Components of Change for the United States: 2010 to 2050. Washington, D.C.: U.S. Census Bureau.

USDA-FSA (U.S. Department of Agriculture, Farm Service Agency). 2008a. Conservation Reserve Program: Summary and Enrollment Statistics—Fiscal Year 2007. Washington, D.C.: USDA-FSA.

USDA-FSA. 2008b. Conservation Reserve Program: Monthly Summary—June 2008. Washington, D.C.: U.S. Department of Agriculture.

USDA-NASS (U.S. Department of Agriculture, National Agricultural Statistics Service). 2007a. Agricultural Prices 2006 Summary. Washington, D.C.: U.S. Department of Agriculture, National Agricultural Statistics Service.

USDA-NASS. 2007b. Agricultural Prices December 2007. Washington, D.C.: U.S. Department of Agriculture, National Agricultural Statistics Service.

USDA-NASS. 2008a. Quick Stats. Available at http://www.nass.usda.gov/QuickStats/ PullData_US.jsp. Accessed October 6, 2008.

USDA-NASS. 2008b. Crops and Plants. Available at http://www.nass.usda.gov. Accessed August 22, 2008.

USDA-NRCS (U.S. Department of Agriculture, Natural Resources Conservation Service). 2003. Costs Associated with Development and Implementation of Comprehensive Nutrient Management Plans. Part I: Nutrient Management, Land Treatment, Manure and Wastewater Handling and Storage, and Recordkeeping. Available at http://www. nrcs.usda.gov/technical/NRI/pubs/cnmp1full.pdf. Accessed October 13, 2008.

Vadas, P.A., K.H. Barnett, and D.J. Undersander. 2008. Economics and energy of ethanol production from alfalfa, corn, and switchgrass in the upper Midwest, USA. BioEnergy Research 1:44-55.

Vogel, K. 2007. Switchgrass for Biomass Energy: Status and Progress. Available at http:// ageconsearch.umn.edu/bitstream/8030/1/fo07vo01.pdf. Accessed March 2, 2009.

Vogel, K., J. Brejda, D. Walters, and D. Buxton. 2002. Switchgrass biomass production in the Midwest USA: Harvest and nitrogen management. Agronomy Journal 94:413-420.

Wang, M., W. May, and H. Huo. 2007. Life-cycle energy and greenhouse gas emission impacts of different corn ethanol plant types. Environmental Research Letters 2:1-9.

Wilhelm, W.W., J.M-F. Johnson, D.L. Karlen, and D.T. Lightle. 2007. Corn stover to sustain soil organic carbon further constrains biomass supply. Agronomy Journal 99:1665-1667.

Wu, M., Y. Wu, and M. Wang. 2006. Energy and emissions benefits of alternative transportation liquid fuels derived from switchgrass: A fuel life cycle assessment. Biotechnology Progress 22(4):1012-1024.

York News-Times. 2009. Seed companies look to increase corn yields. Available at http://www.yorknewstimes.com/articles/2009/02/04/news/doc49891c8e093e3392930640.txt. Accessed March 4, 2009.

Zhang, N., M. Wang, and N. Wang. 2002. Precision agriculture—A worldwide overview. Computers and Electronics in Agriculture 36:113-132.

3

Biochemical Conversion of Biomass

This chapter focuses on the biochemical conversion of biomass to liquid transportation fuels. It addresses the questions raised in the statement of task related to the application of biochemical conversion to the production of alternative liquid transportation fuels from biomass by discussing the following:

- The technology alternatives for converting biomass to liquid transportation fuels.
- The status of development of biochemical conversion of lignocellulosic biomass to ethanol.
- The projected costs, performance, environmental impact, and barriers to deployment of biochemical conversion of lignocellulosic biomass to ethanol.
- Challenges and needs in research and development (R&D), including basic-research needs for the long term.
- Other technologies for converting biomass to liquid fuels that are not likely to be ready for commercial deployment before 2020.

TECHNOLOGY ALTERNATIVES

Liquid fuels can be derived from biomass through biochemical processing, chemical catalysis, or thermochemical conversion. Biochemical conversion and chemical conversion typically transform the biomass into sugars as intermediates. In contrast, thermochemical conversion uses heat to convert the biomass into building

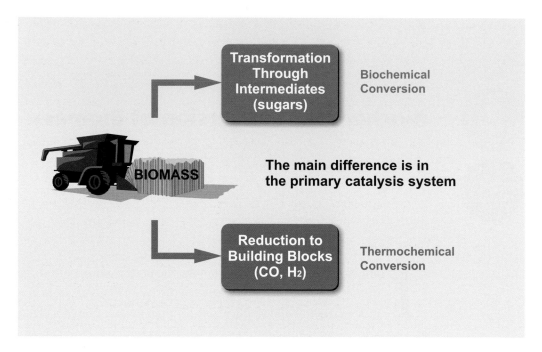

FIGURE 3.1 *Comparison of biochemical and thermochemical routes for converting biomass to fuels.*
Source: Dayton, 2007.

blocks, such as carbon monoxide (CO) and hydrogen (H_2), which can be used for the synthesis of fuels (Figure 3.1). Other thermochemical conversion processes include pyrolysis and liquefaction.

Biochemical Conversion to Fuels

Biochemical conversion uses enzymes to break down structural carbohydrates (for example, the cellulose[1] and hemicellulose[2] found in plant cell walls) into sugars, which are transformed into alcohols, organic acids, or hydrocarbons by microorganisms in fermentation. The conversions typically take place at atmospheric pressure and temperatures ranging from ambient to 70°C.

Early ethanol production technology based on biochemical conversion of sugar and starch has been deployed commercially. In that technology, ethanol is

---

[1]A complex carbohydrate $(C_6H_{10}O_5)_n$ that forms cell walls of most plants.
[2]A matrix of polysaccharides present in almost all plant cell walls with cellulose.

produced when wild-type yeast ferments six-carbon sugars. Sugar can be obtained directly from sugarcane (Brazil) and sugar beets (Europe) or indirectly from the hydrolysis of starch-based grains, such as corn (United States) and wheat (Canada and Europe). In the latter case, the starch feedstock needs to be ground to a meal that is hydrolyzed to glucose by enzymes. The resulting mash is fermented by natural yeast and bacteria. Finally, the fermented mash is separated into ethanol and residues (for feed production) via distillation and dehydration (Figure 3.2).

Corn grain is the major source of ethanol in the United States, and its potential for growth is defined by production efficiencies, food-versus-fuel debates, and the question of sustainability and carbon footprint. Developments aimed at future processes are targeting cellulose conversions that could address those issues by providing a growth potential, a low carbon footprint, and sustainability. The infrastructure that was established by the corn grain ethanol industry will benefit the future cellulosic-ethanol industry because the use of ethanol as a transportation fuel has been proved to be feasible, a distribution system exists, and automobiles with internal-combustion engines that use ethanol efficiently are on the road.

Recent analyses of the full life cycle of corn grain ethanol have indicated

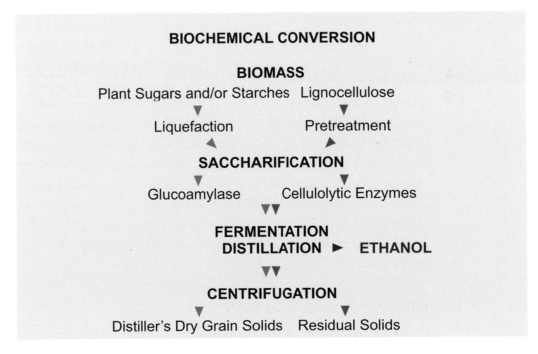

FIGURE 3.2 *Schematic representation of bioprocessing elements.*

that it provides society with small net energy gains over the fossil energy needed to produce it (Farrell et al., 2006; Hill et al., 2006) and might lead to only small net greenhouse gas advantages (Farrell et al., 2006; Hill et al., 2006; Fargione et al., 2008; Searchinger et al., 2008) or might release more greenhouse gas than do production and combustion of an energetically equivalent amount of gasoline once direct and indirect land-use changes are taken into account (Fargione et al., 2008; Searchinger et al., 2008). Issues with corn grain ethanol have led to increased interest in second-generation biofuel feedstocks—including switchgrass, *Miscanthus*, hybrid poplar, and the other lignocellulosic feedstocks—and in conversion methods that potentially can make biofuels that, relative to corn ethanol, offer larger energy gains and greenhouse gas benefits and reduced competition with food crops.

The development of biofuels needs to move toward conversion of lignocellulosic materials (so-called second-generation biofuels) that are unused agricultural or forestry residues, agricultural cover crops, dedicated perennial crops grown on marginal lands that are not suitable for commodity-crop production even with high commodity prices, or municipal solid wastes. The need to move away from corn grain ethanol is highlighted by the renewable fuel standard (RFS) as amended in the 2007 Energy Independence and Security Act. The RFS mandates that production of ethanol from corn grain level off from 2008 to 2015 and that production of cellulosic and other advanced biofuels increase from 2008 to 2020. The key differences in production between grain ethanol and cellulosic ethanol are the pretreatment of the biomass and the use of by-products (Figure 3.2). This chapter focuses on the conceptual design, conversion technologies, and economics of the biochemical conversion of cellulosic biomass to ethanol. It will also discuss other technologies to produce advanced biofuels that use nonfood renewable feedstocks. The other technologies could produce fuels more desirable than ethanol—for example, lipids, higher alcohols, hydrocarbons, and other products that can be separated by low-energy distillation. New routes of biochemical conversion of biomass to liquid fuels will probably encounter complications as they are being developed and scaled up; these issues will have to be addressed in a continuous R&D program.

Chemical Conversion to Fuels

In contrast with biochemical conversion, chemical conversion uses inorganic catalysts in a series of aqueous-phase reactions to convert sugars to hydrocarbons that can be used as fuels. It is a developing technology that will not be ready for commercial deployment by 2020, but it is discussed later in this chapter.

Thermochemical Conversion to Fuels

In what is currently the most developed thermochemical route, biomass is initially converted into CO and H_2 via gasification. The gas stream can be cleaned of impurities and shifted to the needed H_2:CO ratio, and CO_2 can be removed to produce a gas stream that can be catalytically converted to liquid fuels by several routes, including Fischer-Tropsch (FT) and methanol synthesis followed by methanol-to-gasoline (MTG) conversion. Thermochemical conversion is discussed in Chapter 4. Other thermochemical conversion routes involve production of bio-oil by pyrolysis or liquefaction and refinement of the bio-oil (Huber et al., 2006); this technology is not as well developed as FT or MTG.

BIOCHEMICAL CONVERSION OF CELLULOSIC BIOMASS

This section discusses the biochemical processes for converting cellulosic biomass to ethanol in a biorefinery. The processes discussed here occur at the end of the supply chain, when the biomass has been delivered to the biorefinery (Figure 3.3). The process economies are those within the biorefinery.

Process Overview

The biochemical conversion of cellulosic biomass involves six major steps: feedstock preparation, pretreatment to release cellulose from the lignin shield, saccharification (breaking down of the cellulose and hemicellulose by hydrolysis to sugars, such as glucose and xylose), fermentation of sugar to ethanol, and distillation to separate the ethanol from the dilute aqueous solution (Figure 3.4). In the sixth step, the residues, primarily lignin, can be combusted to provide energy (Figure 3.4). The integration of those steps with each other and with the living microorganisms and enzymes that carry out the catalytic conversions in a biorefinery is essential to the development of cost-effective processes.

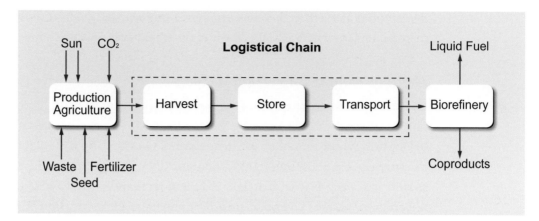

FIGURE 3.3 *Logistics of bioprocessing to convert cellulosic biomass to ethanol.*

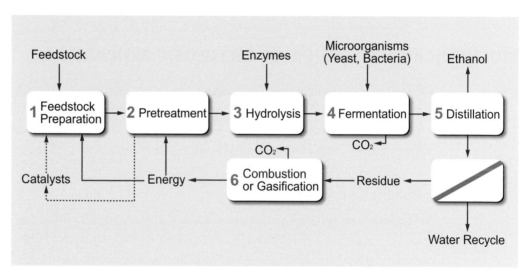

FIGURE 3.4 *Unit operations of a biorefinery. A biomass-based biorefinery should be energy self-sufficient or could even sell excess power to the grid. CO_2 is recycled into plant matter through biomass production.*

Feedstock Preparation

Some feedstocks have to be washed to remove inorganic and other undesirable materials before pretreatment. Whether washing is needed depends on the source and the manner of storage before the feedstock is delivered to the conversion facility. The biomass is then chopped or ground to the desirable size range to feed into the pretreatment stage. The extent of grinding and size reduction will depend on the type of biomass and the pretreatment technology being used. Cellulosic feedstock can be chopped or ground with existing forestry or agricultural techniques.

Pretreatment

Producing fuel ethanol from lignocellulosic feedstocks has been challenging because of the recalcitrant nature of the cellulose that is embedded in the plant cell-wall structure. Therefore, pretreatment is a key step in production of cellulosic ethanol. Pretreatment greatly increases the rates and extents of enzyme action in breaking down cellulose to fermentable sugars (Ladisch et al., 1978) by improving the accessibility of the structural carbohydrates in the cell wall (Figure 3.5). Yields of fermentable sugars from untreated native lignocellulosic materials are low because of the highly packed cellulose structure and the presence of hemicellulose and lignin, which shield cellulose from acid or enzymatic hydrolysis.

Maximizing the use of all lignocellulosic material that is capable of yielding simple (six- and five-carbon) sugars is essential for improving ethanol yield and

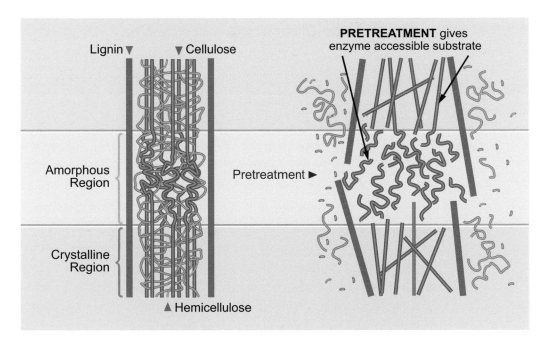

FIGURE 3.5 *Schematic of pretreatment to disrupt the physical structure of biomass. Reprinted from Mosier et al., 2005. Copyright 2005, with permission from Elsevier.*

lowering the cost of ethanol production. Hence, pretreatment of lignocellulosic material is required to improve the hydrolytic efficiency of cellulose by removing and hydrolyzing hemicellulose, by separating the cellulose from the lignin, and by loosening the structure of cellulose and thereby increasing its porosity. The pretreatment of lignocellulosics is particularly important for enzymatic hydrolysis to reduce the amount of enzyme and the time required to convert cellulose to glucose.

Among the various pretreatment methods, hydrothermolysis with steam or water has been shown to be effective in removing and solubilizing hemicellulose and thus in improving hydrolytic efficiency (Mosier et al., 2005; Wyman et al., 2005a,b). Hot-water pretreatment of lignocellulosic biomass at a controlled pH effectively dissolves hemicellulose and some of the lignin and minimizes the formation of monosaccharides and other coproducts that could interfere with biological processes downstream (Yang and Wyman, 2008). For example, monosaccharides inhibit cellulase in the hydrolysis of cellulose downstream. The sugars could degrade further to form such toxic substances as furfural during the pretreatment step (Ladisch et al., 1998; Kim and Ladisch, 2008; Hendriks and Zeeman, 2009). Other pretreatments are similarly effective, and they use acid, bases, ammonia, or other materials (Mosier et al., 2005; Jorgensen et al., 2007; Murnen et al., 2007; Sendich et al., 2008; Yang and Wyman, 2008; Hendriks and Zeeman, 2009). Several of the promising pretreatment methods have been demonstrated on a pilot scale, but the lowest-cost approach is yet to be determined.

Saccharification

In the saccharification step, the cellulose polymers (long chains of sugar) are broken down by hydrolysis into five-carbon and six-carbon sugars (xylose and glucose) for fermentation into alcohol (Figure 3.6). The enzymes used for hydrolysis are referred to as cellulolytic enzymes, and they are classified into three main groups: cellobiohydrolases, endoglucanases, and beta-glucosidases. The cellobiohydrolases and endoglucanases are modular proteins with two distinct independent domains; the first domain is responsible for the hydrolysis of the cellulose chain, and the second is a cellulose-binding domain (CBD) that has the dual activity of increasing adsorption of cellulolytic enzymes onto insoluble cellulose and affecting cellulose structure. A schematic of the action is shown in Figure 3.7. By intercalating between fibrils and surface irregularities of the cellulose surface, CBDs help to reduce particle size and increase specific surface area. Microscopy of cellulose

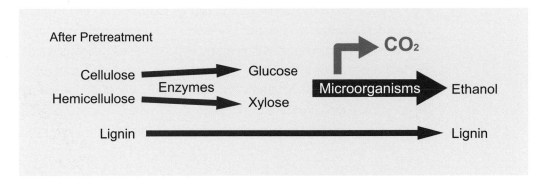

FIGURE 3.6 *Schematic diagram of bioprocessing of sugars to ethanol through enzymatic hydrolysis (catalytic step that frees the sugars) and microbial conversion of sugars to ethanol and CO_2, which are formed in approximately equal parts. Lignin remains unconverted.*

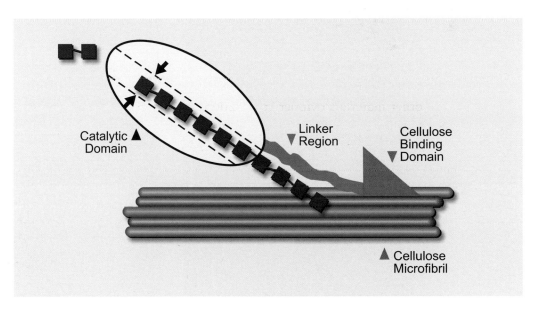

FIGURE 3.7 *Schematic representation of mechanisms of enzyme action. Source: Reprinted from Mosier et al., 1999. Copyright 1999, with permission from Springer.*

treated with isolated CBDs generated from recombinant organisms has shown the release of small particles from insoluble cellulose with no detectable hydrolytic activity and an increase in the roughness of highly crystalline fibers.

The cellobiohydrolases are the most important cellulolytic enzyme group in that cellobiohydrolase I makes up 60 percent of the protein mass of the cellulolytic system of *Trichoderma reesei*, and its removal by gene deletion reduces overall cellulase system activity on crystalline cellulose by 70 percent. The concerted effects of pretreatment and enzymatic hydrolysis affect the plant at the cellular level as illustrated in Figure 3.8. According to the prevailing understanding, cellobiohydrolases attack the chain ends of cellulose polymers to release cellobiose, the repeat unit of cellulose. Endoglucanases decrease the degree of polymerization of cellulose by attacking amorphous regions of cellulose through random scission of

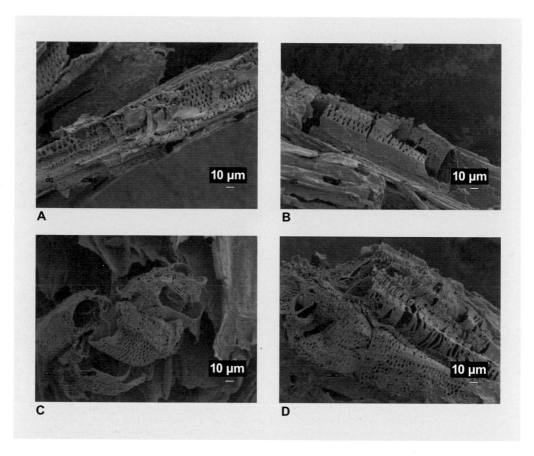

FIGURE 3.8 *Scanning electron microscopic images of enzymatically hydrolyzed 425–710 mm corn stover pretreated with hot water (at 500X magnification). (A) 3-h enzymatic hydrolysis, 43.3 percent glucose conversion. (B) 24-h enzymatic hydrolysis; 56.8 percent glucose conversion. (C) 72-hour enzymatic hydrolysis; 64.2 percent glucose conversion. (D) 168-h enzymatic hydrolysis; 63.1 percent glucose conversion. The images from a laboratory experiment illustrate how enzymatic hydrolysis of corn stover pretreated with hot water is connected to pore formation (during pretreatment) and enlargement (during hydrolysis).*
Reprinted from Zeng et al., 2007. Copyright 2007, with permission from Wiley-Blackwell.

the cellulose chains. Beta-glucosidase completes the process by hydrolyzing cellobiose to glucose. Cellulolytic systems, such as those in filamentaous *T. reesei,* have enzymes in all three groups: two cellobiohydrolases, four endoglucanases, and one beta-glucosidase (Mosier et al., 1999). The mechanism by which cellulolytic enzymes act in hydrolyzing cellulose is complex and requires a system of different enzymes to achieve deploymerization of the oligosaccharides to monosaccharides, such as glucose and xylose. Most studies have been done with cellulases, which are produced industrially from *T. reesei.*

The development of the cellulases has resulted in effective systems that are capable of hydrolyzing cellulose to glucose almost completely. Similar studies are being done on hemicellulases, enzymes that are responsible for breaking down hemicellulose to xylose. Hemicellulases are not as well developed as cellulases.

Despite the complexities, much progress has occurred in the development of enzymes for the hydrolysis of pretreated cellulose. Costs are being reduced, with the ultimate goal of combining cellulases with glucose- and xylose-fermenting microorganisms in a concept referred to as consolidated bioprocessing (Lynd et al., 2005). Hydrolysis and fermentation (combined bioprocessing) are being demonstrated on a pilot scale with the goal of reducing costs.

Fermentation

Pretreatment and enzymatic hydrolysis of plant matter—such as wood, corn stover, or grasses—result in a mixture of five-carbon and six-carbon sugars. Many microorganisms, particularly yeasts, will ferment glucose to ethanol. Typically, however, 25–30 percent of the sugar derived from likely candidates as cellulosic feedstocks for bioprocessing (for example, hardwoods, agricultural residues, and some types of grasses) are pentoses—sugars that have five carbon atoms rather than six carbon atoms. Other potential sources of biomass, such as softwoods, have a lower proportion of hemicelluloses and hence fewer pentoses. Pentoses are not readily fermented to ethanol, so yeasts or bacteria that have been genetically modified to ferment both hexoses (six-carbon sugars) and pentoses are needed to maximize the yield of ethanol from cellulosic materials. Some researchers have been successful in engineering microorganisms that are able to use pentose efficiently but cannot do so naturally to produce ethanol. An alternative would be to supply ethanol-producing microorganisms with pentose-using pathways (Nevoigt, 2008). Development of such microorganisms presents a number of challenges. They have to be capable of fermenting the sugars to ethanol, and they have to be sufficiently robust to withstand antimicrobial agents released by the pretreatment and hydrolysis of the lignocellulose and to withstand relatively high alcohol concentrations.

The glucose and xylose that result from the saccharification step are fermented into ethanol by microorganisms. Traditionally, fermentation is a separate step from saccharification. As noted above, the ideal development would be organisms that could do both simultaneously. Although *Saccharomyces cerevisiae* (wild-type yeast) has been used for fermentation in ethanol production from corn grain, it cannot ferment xylose sugars obtained from lignocellulose unless it is genetically modified (Sonderegger et al., 2004). Organisms that ferment xylose and glucose have been developed through metabolic engineering (Aristidou and Penttila, 2000).

The composition of the biomass feedstock determines the amount of ethanol that can be produced per ton of biomass (Table 3.1). The ultimate yield is determined by the maximum yield of sugars that can be obtained from a given biomass type, and the yield of sugars is determined by the combined starch, cellulose, and hemicellulose content of the biomass. The ethanol yield can range from 105 to 135 gal/ton (on a dry-weight basis) if all bioprocessing steps occur at 100 percent efficiency, that is, all the structural carbohydrates—starch, cellulose, and hemicellulose—are used to produce ethanol (Table 3.1). Because the efficiency is typically less than 100 percent—ranging from about 95 percent overall for a corn-grain-to-ethanol plant to 50 percent with some current cellulose-conversion technologies—the actual yields are substantially lower. Ethanol yields can be improved with a combination of advanced pretreatments and enzymes to improve cellulose-conversion efficiency and with the application of fermenting microorganisms that are able to convert glucose, xylose, and other pentoses to ethanol (Wyman et al., 2005a).

Cellulolytic enzymes are subject to product inhibition in which the rate of an enzymatic reaction is inhibited by the end products of the reaction (Gong et al., 1977, 1979; Ladisch et al., 1981). Some work has been done on process strategies and modified enzymes to reduce that inhibition, but more is needed. Simultaneous saccharification, in which the enzyme and the microorganism are used in the same tank, could enable reduced enzyme use because the reaction is not inhibited by the product of enzymatic action, glucose or cellobiose (Gauss et al., 1976; Takagi et al., 1977; Wright et al., 1977; Saddler et al., 1982a,b; Wyman et al., 1986; Spindler et al., 1987; Lynd et al., 2002). The ethanol product that is formed also inhibits enzymatic activity, but to a smaller extent than glucose and cellobiose do.

TABLE 3.1 Compositions of Corn Grain, Corn Cob, Corn Stover, and Poplar

Type of Material	Grain[a]	Cob[b]	Stover[c]	Poplar[c]
Starch	71.7%	n/m	n/m	n/m
Cellulose	2.4%	42.0%	36.0%	40.3%
Hemicellulose	5.5%	33.0%	26.0%	22.0%
Protein	10.3%	n/m	5.0%	n/m
Oil	43%	n/m	n/m	n/m
Lignin	0.2%	18.0%	19.0%	23.7%
Ash	1.4%	1.5%	12.0%	0.6%
Other	4.2%	5.5%	2.0%	13.4%
Total	100.0%	100.0%	100.0%	100.0%
Maximum yield of monosaccharides (lb/ton with 100% efficiency)	1,778	1,684	1,392	1,396
Calculated best-case ethanol yield (gal/ton with 100% efficiency)	135	128	105	106
Dry weight[d]	52%	10%	48%	52%

Note: n/m, not measured.
[a]Gulati et al., 1996.
[b]Corn-cob composition measured at Laboratory of Renewable Resources Engineering, Purdue University.
[c]DOE, 2007a.
[d]Pordesimo et al., 2005. Absolute weight of corn grain is based on corn grain data provided by the U.S. Department of Agriculture, National Agricultural Statistics Service (USDA-NASS, 2005), which were used for calculation of ethanol yields. Absolute weights of corn cob and corn stover are derived from the given weight percentages based on the absolute weight of corn grain.
Source: Adapted from Schwietzke et al., 2008b.

The reduction in inhibition allows the reaction to proceed to the final desired product—ethanol in this case—more rapidly.

Distillation

The ethanol solution from the fermentation step is distilled to produce 95 percent ethanol. Ethanol is further dried with molecular sieves to produce the required purity. The distillation requires large quantities of energy (Katzen et al., 1981; Shapouri et al., 2002). Distillation of ethanol is a well-established technology and is used in commercial corn grain ethanol plants. However, recovery of solids will require improved engineering design.

Combustion of Residual Solids

The bottoms from distillation are centrifuged to concentrate them. Solids from other portions of the process might also be added. The residual solids are rich in lignin and can be burned in a boiler to generate steam and electricity for the biorefinery. Pathways of using lignin as a fuel with low carbon emission are well defined.

Current Status

A substantial amount of ethanol has been produced in the United States since the 1970s, and the feedstock of choice has been corn grain. The process of producing ethanol from corn grain has been steadily improved in efficiency, and costs have been greatly reduced. The process can be considered fully commercial, well understood, and optimized. There were more than 130 corn grain ethanol plants in the United States in 2008. In contrast, there are no large-scale commercial cellulosic-ethanol plants, although many of the components are in pilot demonstration. In February 2007, the Department of Energy announced that it would invest up to $385 million for six biorefinery projects over 4 years to bring cellulosic ethanol to market (DOE, 2007b), and a number of private companies are actively pursuing pilot demonstration of integrated cellulosic plants, which could lead to commercial-scale demonstrations and eventual deployment of cellulosic-ethanol plants. If the current effort in development and demonstration of cellulosic ethanol is sustained or accelerated, cellulosic-ethanol plants could be ready for commercial demonstration in the next 3–5 years, and the first plant could be built by 2015.

Technical Challenges

As discussed above, the key issues associated with cellulosic ethanol are related to the ability to develop enzymes and microorganisms that can break cellulose bundles and depolymerize cellulose and hemicellulose to produce soluble sugars, and also to the ability to produce the enzymes and microorganisms at a reasonable cost. Microorganisms capable of fermenting five-carbon and six-carbon sugars in separate steps are available, and their performance is rapidly being proved in the laboratory. The challenge will be to demonstrate their robustness under industrial conditions. A longer-term challenge involves development of microorganisms and enzymes that can break down cellulose structures, depolymerize them to sugars, and ferment the sugars in the same vessel. If hydrolysis and fermentation are to be carried out separately, the development of microorganisms that can withstand and ferment sugars at higher temperatures would provide opportunities to increase the rate of fermentation and reduce the need for cooling the fermentation itself. The amount of energy that would be required to heat the broth to distillation temperature would also be reduced.

Research, Development, and Demonstration

One research gap is an incomplete understanding of the molecular biology of the plants that provide the feedstock and of the microorganisms that provide the biocatalysts and enzymes for transforming the feedstocks into ethanol and other biofuels. Fundamental studies on the structure and function of the enzymes that catalyze the breakdown of cellulosic components to fermentable sugars are key to improving the rates at which the enzymes operate and to reducing product inhibition. The nature and chemical structure of inhibitors and how they interact with the active catalytic sites of the enzymes need to be defined so that strategies to mitigate the inhibition can be developed at the catalytic level or through engineering approaches to remove the inhibitors by bioseparation.

Another research gap is in the fundamental understanding of how the plant cell-wall structure resists enzyme attack. Genomic studies could provide insights into how lignin (the key agent in resistance of the cell wall to enzymatic conversion) might be modified at the molecular level to enable facile transformation but retain the properties that provide resistance to pathogens. The molecular biology and metabolic pathways involved in directing the flow of carbon into ethanol or other advanced biofuels and the manner in which modified bacteria and other microorganisms are able to break down cellulose are also important topics. Research will help to reduce enzyme costs, a major component of cellulose-conversion costs. Research needs to be pursued on pretreatments that exploit knowledge gained and biochemical engineering that will define design principles of low-cost manufacturing through enzyme-catalyzed reactions, thermal processing, fermentation, and advanced bioseparation. Fundamental research would also address how temperature, pressure, and pretreatment media interact at the microscopic and macroscopic levels so that plant cells are readily attacked by biological or chemical catalysts.

COST AND PERFORMANCE

Prior Work

The corn grain ethanol plants have created an industry, and they provide a basis for estimating some of the capital and operating costs of cellulosic-ethanol plants. Several key studies also provide information on and analysis of the design and performance of cellulosic-ethanol plants (Aden et al., 2002; Johnson, 2006; Kwiatkowski et al., 2006; Kim et al., 2008).

Modeling

On the basis of experience gained from corn-ethanol plants, published studies, and its own expertise, the panel used SuperPro Designer® to estimate the capital and operating costs of ethanol-production plants (Intelligen, 2009). SuperPro Designer is a chemical-process simulation software that contains a set of unit operations that can be customized to the specific modeling needs of the corn grain-to-ethanol and cellulosic biomass-to-ethanol processes. It has a well-developed economic-evaluation package, which includes an extensive database of chemical components and mixtures, extensive equipment and resource databases, equipment sizing and costing information, and economic-evaluation parameters—such as financing, depreciation, running royalty expenses, inflation rate, and taxes—for cost estimates.

A grain-ethanol plant was used as the baseline to validate the model and input parameter values. Once validated, the model was applied to estimate capital and operating costs for cellulosic-ethanol production. Most of the unit operations for cellulosic-ethanol production are similar to those for grain-ethanol production.

The panel decided to develop its own model for cost estimates. The process-flow diagram used for the panel's modeling is shown in Figure 3.9. The panel conducted three sets of analyses with the cellulosic-ethanol plant model. First, it assessed the sensitivity of the capital and manufacturing cost estimates with respect to such process parameters as enzyme cost, types of feedstock, plant size, solids loading, and yields in the pretreatment, hydrolysis, and fermentation steps. The results are presented in the form of process cost estimates for three representative scenarios: the most pessimistic, with little advancement in technologies or process efficiencies from where they are today (2008) to reduce costs (low); reasonable advancement (medium); and the most optimistic (high) in terms of technology and process improvement. Second, the panel assessed the sensitivity

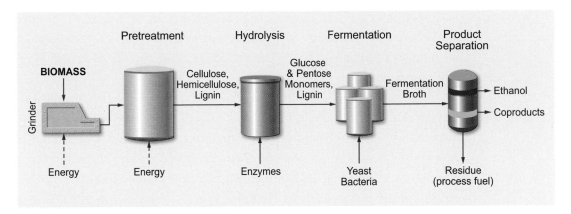

FIGURE 3.9 *Process-flow diagram for ethanol production from lignocellulose. Source: Schwietzke et al., 2008a. Reprinted with permission from IEA Bioenergy.*

of operating costs to different types of feedstock. As mentioned earlier, process efficiency depends on feedstock composition. Assuming reasonable advancement in process technologies (that is, using the medium case), the panel compared the costs of ethanol produced from different biomass alternatives. Third, the panel assessed whether and how the capital and operating costs of a cellulosic-ethanol plant vary with size; in this analysis, the panel used the medium case to estimate costs and assumed that plants of different size used poplar chips as a feedstock.

Design Criteria

The panel used a biorefinery with a capacity of 40 million gallons of ethanol per year as the basis of its calculations and assessed the effect of biorefinery size later. The capacity was used because it is comparable with that of grain-ethanol plants, matches well with previous studies and current cellulosic-ethanol projects in the planning stage, and could serve well those geographic areas that have a lower concentration of biomass. The distribution of biomass could pose a limit on the volumes that could be economically supplied to a biorefinery in some areas, and transportation of large volumes of biomass over a long distance could be expensive and consume more fossil fuel (see Chapter 2).

Producing 40 million gallons of ethanol per year with a fermentor effluent concentration of 4–5 percent ethanol by weight requires nine fermentors, each of which has a capacity of 800,000 gal if a fermentation time of 72 h is assumed. The 800,000-gal capacity of the fermentors was selected on the basis of typical industry practice. The run time was assumed to be 80 h/batch. The current analysis is based on an output of 40 million gallons per year. Further economic benefits as a result of scale will be realized as the technologies develop and larger plants are constructed.

In the model, each fermentor has a train of pretreatment and saccharification tanks that work in batch mode. Each train supplies its own fermentor, thereby avoiding the need for holding tanks for intermediate storage and minimizing the potential for fouling the saccharified mash. The pretreatment unit, saccharification tank, and fermentor, therefore, scale linearly with the size of the plant. Other units—such as distillation, centrifugation, shredding, drum drying, and boiler and steam generator—operate continuously. The continuously operating units were sized so that they can supply and process the contents of a single fermentor over a period equal to t/N, where t is the duration of each batch and N is the number of processing units. Because the fermentations are staggered for efficient loading and emptying, overall batch time is set at 80 h, which is the time required for cycling the full set of 8–10 fermentors. Although fermentations operate in batch mode, using multiple vessels enables a staged operation so that the upstream processing, including pretreatment, can be carried out continuously.

The major process determinants that affect the size and operation of a plant are

- Yield in saccharification,
- Yield in pretreatment,
- Fermentation time and yield, and
- Solids loading.

With the exception of fermentation time, which was set at 72 h for all simulations, the parameters listed above were varied, with the ranges indicated in Table 3.2 representing low-, medium-, and high-performance scenarios. Once those parameters were defined, the design of the various units, the sizing, and the cost calculations were automatically handled by SuperPro Designer. Details of the modeling and analyses are presented in Appendix I.

Core Assumptions

The following assumptions were made to simplify the modeling:

- Total dry weight is the sum of the weight of soluble and insoluble components.
- Ash, fat, protein, and lignin do not react in the process.
- There is no contamination with lactic acid or acetic acid bacteria.
- Added enzymes add to protein mass balance.
- All batch fermentations take 72 h.
- Distillation and dehydration recover 100 percent of the ethanol.
- The final concentration of ethanol is 99.5 percent by weight.

Some of the assumptions represent ideal operating conditions that might not be achievable all the time. For example, some ash, fat, protein, and lignin could react, or there could be occasional contamination with lactic acid or acetic acid bacteria. However, most of the assumptions do not affect cost substantially.

A solids loading of 30 percent was used in initial modeling. However, for the majority of the modeling, a range of 18–25 percent was used to provide a more realistic assessment. At the time the report was written, 25 percent solids by weight was considered optimistic. The yield range for pretreatment was assumed to be 80–95 percent; the best case assumes that after pretreatment 95 percent of glucans and five-carbon sugar polymers, such as xylans, are accessible to cellulases for saccharification. For saccharification, 90–95 percent of the accessible sugar

TABLE 3.2 Assumptions for Low-, Medium-, and High-Performance Cases

Variable	Low	Medium	High
Size of reactors	800,000 gal	800,000 gal	800,000 gal
Solids loading for fermentation and hydrolysis	18%	21%	25%
Pretreatment yield	80%	85%	95%
Saccharification extent	90%	95%	95%
Fermentation glucose	85%	90%	90%
Fermentation xylose	75%	81%	81%
Cellulose cost	$0.40/gal	$0.25/gal	$0.10/gal
Cost of biomass (wet weight)	$44/ton	$35/ton	$25/ton
Cost of biomass[a] (dry weight)	$88/ton	$70/ton	$50/ton

[a]Biomass costs in this table are used as illustrations and do not represent the range of costs estimated by the panel.

polymers[3] are assumed to be hydrolyzed into single sugars (monomers[4]) given the concentration of enzymes and the solids loading. For fermentation, in which the final conversion is a multiple of sugar conversion into products and the percentage of maximum theoretical yield of ethanol produced by yeast, the value for sugar conversion into ethanol was assumed to range from 85 to 90 percent for glucose and from 75 to 81 percent for xylose.

Cellulase cost was calculated as cents per gallon of ethanol produced by using the required amount of cellulase per unit biomass dry weight, the ethanol yield per ton of biomass, and the price per kilogram of cellulose. That gave a range of $0.10–0.40/gal of ethanol produced.[5] That range was validated by discussions with an industry representative (B. Foody, Iogen Corporation, personal communication, March 2008).

Costs

Three scenarios were developed in the form of process cost estimates: current technology (low), reasonable evolutionary advances in technology (medium), and the most optimistic technology advances (high) for the biochemical conversion of cellulosic feedstocks (Table 3.3). The panel judges that the reasonable-advances case best represents where the technology would be for 2020 deployment. The optimistic case shows that considerable potential will remain. The results of the calculations were validated via interviews with representatives of three cellulosic-ethanol companies. For proprietary reasons, the level of detail provided in the validation ranged from the expected cost of individual units and materials to overall plant cost. The work was also validated by comparing plant configurations and cost estimates generated herein with those previously reported in the literature.

Sensitivity of Capital and Operating Costs

The initial analysis is a comparison of three assumed scenarios, from most pessimistic to most optimistic as outlined in Table 3.2. Initially, poplar woodchips were used as the feedstock for all cases to illustrate the differences between scenarios in

---

[3]Large molecules composed of repeating structural units.

[4]Small molecules that may become chemically bonded to other monomers to form a polymer.

[5]Enzyme companies project enzyme costs to be about $0.40–0.50/gal of ethanol produced by 2010 and about $0.20–0.30 by 2020 (Jensen, 2008). The cost of cellulase per gallon of ethanol produced in the Nth plant is modeled to be about $0.10–0.20 (DOE/EERE, 2007).

TABLE 3.3 Comparison of Costs Under Three Scenarios That Represent Little (Low), Reasonable (Medium), and Major (High) Improvements in Technology and Process Efficiency in a Biorefinery

	Poplar			HGBM		
	Low	Medium	High	Low	Medium	High
Total capital ($ millions)	223	194	174	166	140	128
Total capital ($/annual gallon)	5.65	4.85	4.34	4.15	3.49	3.20
Total capital ($/bbl per day)	87,000	75,000	67,000	64,000	54,000	49,000
Biomass used (dry tons)	593,000	514,000	461,000	485,000	421,000	378,000
Yield (gal/dry ton of biomass)	67	78	87	83	95	106
Ethanol operating cost ($/gal)	1.95	1.40	0.90	1.70	1.20	0.80
Ethanol production cost ($/gal)	2.70	2.00	1.50	2.30	1.70	1.20
Facility-dependent (fraction of cost)	34%	39%	48%	30%	34%	43%
Raw-materials-dependent (fraction of cost)	57%	51%	40%	61%	55%	43%

Note: Poplar woodchips and high-sugar/glucan biomass (HGBM) are used as feedstocks for these scenarios. The capacity of the biorefinery is 40 million gallons of ethanol per year (2,600 bbl/d).

process economics. Those analyses, however, generated some unexpected results (for example, high electricity generation from the lignin-rich residue of poplar). Poplar woodchips, rich in lignin, can be considered an outlier as a feedstock with respect to the other biomass types. Therefore, cases were run with a high-sugar/glucan biomass (HGBM) that has a composition closer to those of most of the other cellulosic biomass types. The cost estimates are summarized in Table 3.3. The detailed breakdown of costs of major equipment, fixed capital, materials, and other items are detailed in Appendix I.

Capital Costs

The total capital cost required to build a 40-million-gallon biorefinery and put it into operation ranges from $174 million to $223 million ($4.34 to $5.65 per annual gallon) if poplar woodchips are used as feedstock and from $128 to $166 ($3.20 to $4.15 per annual gallon) if HGBM is used as feedstock (Table 3.3). The capital-cost requirements for a biorefinery that uses poplar woodchips is about 35 percent higher than the requirements for one that uses HGBM primarily because of the increased cost of the boiler and steam electrical generator associated with the increased lignin content of the feedstock. However, the higher capital cost is offset by the sale of excess electricity generated by burning the lignin residue. (See discussion on energy cost later in this section.)

Operating Costs

For the materials costs, two items have important effects on the cost of production for the plant: the cellulases for the saccharification step and the lignocellulosic feedstock. If the cellulases are produced in-house in an enzyme-propagation unit for the plant, the cost of cellulases would be equivalent to the operating cost of producing the enzymes. That cost includes the cost of sugar water and of cofactor additives, the cost of utilities consumed by the propagation unit, and the cost of cleaning and sterilizing the seed fermentors after every batch. The panel estimated enzyme cost to be about $0.10–0.40/gal and used the midpoint of $0.25/gal as the medium case.

The purpose of this analysis is to illustrate the sensitivity of the process economics of biochemical conversion itself. The assumed feedstock costs listed in Table 3.2 are used here. As discussed in Chapter 2, they represent the low cost of feedstock. However, the largest input cost for ethanol production is the cost of cellulosic feedstock, which represents about 75 percent of material costs. Feedstock cost represents about 30–45 percent of overall plant operating costs. Discussions with industry representatives confirmed that even though in some rare cases cellulosic biomass can be obtained at low cost, the price is likely to increase as demand increases.

The external energy requirements of a cellulosic biorefinery are expected to be zero because the lignin and other unconverted residue from processing can be used to generate energy. For a lignin content of 20–30 percent of the feedstock (on a dry-weight basis), the steam generated from the lignin-residue combustion is enough to supply the energy requirements for the entire biorefinery and to export electricity to the grid. The lignin is assumed to have an energy content of 11,500 Btu/lb (26.7 kJ/g). The conversion of steam to electricity by the steam turbine generator was assumed to be 80 percent efficient, considering the steam turbine and generator only. If the process efficiencies of all steps from lignin to electricity are taken into account, the conversion of lignin to electricity is estimated to be about 25 percent efficient. That resulted in an estimated 43 MW of gross electrical power that could be sold to the grid, which is higher than the Aden et al. (2002) estimate of 29 MW of gross electrical power.

The higher estimate of energy generation is a result of the use of poplar woodchips, which have a high lignin content. The difference between the panel's estimate and the Aden et al. (2002) estimate was smaller when HGBM was used as a feedstock. The credit gained from the export of excess electricity generated by the biorefinery could be $8–16 million per year and could reduce the manufacturing cost by $0.20–0.40/gal of ethanol.

The estimated labor cost was $3–4 million per year on the basis of discussions with industry representatives. Within the range of production capacity of 20–100 million gallons of ethanol per year, the labor cost is expected to be essentially independent of the size of the biorefinery. Other material costs or consumables presumably could affect production costs if a different pretreatment process were used instead of hot-water pretreatment (used in this model). However, the choice of pretreatment process has a large effect on the overall cost because of its relationship to reducing the costs of cellulase. Nonetheless, biomass feedstock costs are likely to dominate overall product cost unless the conversion process is expensive. Water use could also affect costs, but plants will probably be designed to economize on water use and be located with that variable in mind.

On the basis of those considerations, the model projected the production cost of a biorefinery that uses poplar woodchips to be $1.50–2.70/gal of ethanol (including capital charge) with a middle value of $2.00/gal (Table 3.3). The cost estimates assume that excess electricity generated from the lignin residue will be sold at $0.05/kWh.

Solids loading emerges from the model study as a key parameter for the following reasons:

- The theoretical ideal solids loading of 30 percent has the physical consistency of a solid paste, and even though it might be achieved with substantial technological breakthroughs, many operational issues would have to be overcome—for example, achieving good mixing between the pretreated biomass and the cellulases and achieving greater than 80 or 85 percent saccharification.
- Solids loading affects the capital cost of the plant because for equal production, the equipment volumes required for the pretreatment, saccharification, and fermentation processes vary approximately inversely with solids loading.

- Higher solids loading translates into increased ethanol concentration in the fermentations, and to accommodate this increase, ethanol-tolerant yeasts will have to be developed; this is one of the R&D objectives for more efficient future ethanol processes.

Effects of Feedstock Choice on Capital and Operating Costs

As illustrated above, the capital and operating costs vary with the type of feedstock used in the biorefinery. Therefore, the costs were estimated for poplar woodchips, wheat straw, switchgrass, corn stover, and *Miscanthus* (Table 3.4). A constant feedstock cost is used in this analysis to demonstrate the effect of feedstock type on capital costs and operating costs of the conversion process. The total capital costs range from $123 to $194 million for a capacity of 40 million gallons of ethanol per year. Because cellulosic-ethanol plants will receive biomass from specific and relatively limited geographic areas, plant designs will be specific to the expected biomass type. If cellulosic feedstock other than poplar is used, the capital costs could decrease by up to 37 percent.

TABLE 3.4 Comparison of Capital and Operating Costs of Biorefineries Using Different Feedstocks

	Poplar	*Miscanthus*	Switchgrass	Corn Stover	Wheat Straw
Total capital ($ millions)	194	176	156	150	123
Total capital ($/annual gallon)	4.85	4.40	3.90	3.80	3.10
Total capital ($/bbl per day)	75,000	68,000	60,000	58,000	47,000
Biomass used (dry tons/year)	514,000	507,000	504,000	471,000	375,000
Yield (gal/dry ton of biomass)	78	79	80	76	88
Ethanol operating cost ($/gal)	1.40	1.35	1.35	1.25	1.20
Ethanol production cost ($/gal)	2.00	1.90	1.80	1.80	1.60
Revenues from electricity sales ($/gal)	0.40	0.28	0.02	0.08	0.00
Facility-dependent (fraction of cost)	39%	37%	34%	34%	31%
Raw-materials-dependent (fraction of cost)	51%	53%	56%	55%	58%

Note: Cost of biomass is held constant to highlight the variations in other cost parameters. The cost estimates for different biomass types are discussed in Chapter 2. Reasonable (medium) improvements in technology and process efficiencies were assumed in all cases.

Effects of Biorefinery Size on Capital and Operating Costs

Some aspects of a biorefinery can benefit from economies of scale. Table 3.5 summarizes how sensitive capital and operating costs are to the size (capacity) of a biorefinery. Poplar is used as a feedstock, and reasonable improvement in technologies and process efficiencies are assumed in this illustration. In this analysis, a facility with a capacity of 20 million gallons per year would consume feedstock at about 700 tons/day, and a facility with a capacity of 100 million gallons per year would consume about 3500 tons/day. The yield of ethanol per dry ton of feedstock and the expected revenue from electricity do not vary with the size of the biorefinery. In contrast, capital cost is sensitive to the size of the biorefinery and benefits from economies of scale because larger equipment and construction of a larger facility cost less per unit size than do the equipment for and construction of a smaller one. The capital cost per annual gallon of ethanol produced was estimated to be $5.84 for a 20-million-gallon biorefinery compared with $3.49 for a 100-million-gallon biofinery. The operating costs also vary with size and range from about $1.52/gal for a 20-million-gallon facility to $1.30/gal for a 100-million-gallon facility. Increasing the size of a 20-million-gallon biorefinery by a factor of 5 decreases the annual capital costs per annual gallon by about 40 percent and the operating costs by about 15 percent.

TABLE 3.5 Comparison of Costs Among Biorefineries of Different Sizes

	Biorefinery Capacity (millions of gallons of ethanol per year)				
	20	40	60	80	100
Total capital ($ millions)	117	194	264	329	349
Total capital ($/annual gallon)	5.84	4.85	4.40	4.12	3.49
Plant size (bbl/d)	1,300	2,600	3,900	5,200	6,500
Total capital ($/bbl per day)	90,000	75,000	68,000	63,000	61,000
Biomass used (dry tons/year)	258,000	514,000	772,000	1,029,000	1,286,000
Yield (gal/dry ton of biomass)	78	78	78	78	78
Ethanol operating cost ($/gal)	1.52	1.40	1.34	1.31	1.30
Ethanol production cost ($/gal)	2.30	2.00	1.92	1.86	1.82
Facility-dependent (fraction of cost)	41%	39%	38%	36%	36%
Raw-material-dependent (fraction of cost)	45%	51%	54%	55%	57%

Note: Poplar woodchips are used as feedstock for these scenarios. Reasonable (medium) improvements in technology and process efficiencies were assumed in all cases.

Greenhouse Gas Performance

Estimates of carbon dioxide (CO_2) emission in different scenarios or with different feedstocks show that it varies widely from 8 kg to 16 kg of CO_2 per gallon of ethanol produced (Tables 3.6 and 3.7). Those values represent CO_2 emission during the biochemical conversion process and do not take into account CO_2 uptake by photosynthesis during the growth of the feedstock. Net CO_2 emissions from cellulosic ethanol from field to wheel are presented in Chapter 6. The range of values presented in Tables 3.6 and 3.7 is tied to the burning of lignin and other residual materials for generating steam and power. The most effective use of the residual lignin is to generate steam and electricity to power the plant. Wheat straw as a feedstock has the lowest lignin content (Table 3.7) and thus produces the lowest CO_2 emission, but it does not provide enough energy for the biorefinery.

Biochemical conversion processes, as configured today, produce a stream of pure CO_2 from the fermentor that can be dried, compressed, and stored in geologic formations. The concept of using carbon capture and storage (CCS) in

TABLE 3.6 Comparison of CO_2 Emission from a Biorefinery in Three Scenarios That Represent Little (Low), Reasonable (Medium), and Major (High) Improvements in Technology and Process Efficiency

	Poplar			HGBM		
	Low	Medium	High	Low	Medium	High
Tonnes per year	650,210	543,682	460,913	472,128	390,644	323,812
Tonnes per day	1,781	1,489	1,262	1,293	1,070	887
Kg per gallon	16.3	13.6	11.5	11.8	9.8	8.1

TABLE 3.7 Comparison of CO_2 Emission from a Biorefinery with Different Feedstocks and with Reasonable Improvements in Technologies and Process Efficiencies Assumed in All Cases

	Poplar	*Miscanthus*	Switchgrass	Corn Stover	Wheat
Tonnes per year	543,682	478,757	393,298	389,040	350,385
Tonnes per day	1,490	1,312	1,078	1,066	960
Kg per gallon	14	12	10	10	9

cellulosic-to-liquid fuels biochemical conversion processes, although not discussed before, follows directly from the modeling work done for CCS technology applied to thermochemical conversion in Chapter 4. The panel did not analyze that possibility from a technology standpoint but believes that the required CCS technologies could be integrated into the biochemical-conversion-plant design. The potential issues are scale, capital-cost efficiency, and logistics of the biochemical conversion plant relative to the storage site.

In addition to greenhouse gases emitted by the biorefinery during processing, exhaust emissions are associated with ethanol. They are less toxic than those associated with gasoline and have lower atmospheric reactivity (Worldwatch Institute, 2007). Brazil presents an informative case study. The use of ethanol in Brazil reduced CO emissions from 50 g/km to less than 5.8 g/km by 1995. However, aldehyde emission was found to increase with the use of hydrous-ethanol engines. Total aldehyde emission was higher from engines using ethanol (both neat and blended) than from engines using gasoline. Nonetheless, present ambient concentrations of aldehyde in São Paulo are below the recommended reference for human health (Goldemberg et al., 2008).

Environmental Impacts

Water Use and Discharge

The production of cellulosic ethanol requires process water for mixing with fermentation substrates and for cooling, heat, electricity, and reagents that are associated with hydrolysis and fermentation. In the case of thermal processing of cellulose, process water is required primarily for cooling. The amount of water required for processing biomass into ethanol or other biofuels is estimated to be 2–6 gal per gallon of ethanol produced (Aden et al., 2002; Pate et al., 2007; Cornell, 2008). The lower value would be approached if the plant were designed for recycling process water. The processing of cellulosics to ethanol will result in a residual water stream that will need to undergo treatment. However, by definition an efficient process will ferment most of the sugars to ethanol and leave only small amounts of organic residue.

Air Emissions

Air emissions will result from either bioprocessing or thermal processing. Fermentation processes release CO_2 as a consequence of microbial metabolism; 1 mole of CO_2 will be formed for each mole of ethanol produced. Thermal processes release CO_2 as a result of partial combustion of the biomass but also form other volatiles, CO, and H_2. CO and H_2 are the desired intermediates in some processes because they can be passed over a catalyst or fermented to biofuels. Other emissions include water vapor, particularly if the lignin coproduct is dried before being shipped from the plant for use as boiler fuel at an off-site power-generation facility. Sulfur and nitrogen content of the fermentation residues would be expected to be low unless chemicals are used in the pretreatment of biomass. Chemicals used in pretreatment would need to be recovered or otherwise to be in the wastewater or in the solid residues. Because the residues are expected to be burned and used as boiler fuel, air emissions (sulfur in the case of acid pretreatment, nitrogen if nitrogen-containing reagents are used for pretreatment) would increase if the solid residues contain pretreatment chemicals. Emissions from the cellulosic feedstock itself are low, and if processes that transform it to ethanol add chemicals only minimally, emissions will be minimized. Minerals in the cellulosic biomass will end up in the wastewater.

Product Characteristics

Ethanol has 66 percent as much energy as gasoline does. Ethanol is hygroscopic and cannot be transported in existing fuel-infrastructure pipelines because of its affinity for water. (See Chapter 5 for details on distribution of ethanol.)

Barriers and Challenges

There are three key challenges that need to be overcome before widespread commercialization: (1) improving the effectiveness of pretreatment—removing and hydrolyzing the hemicellulose, separating the cellulose from the lignin, and loosening the cellulose structure or using other pretreatment methods (Murnen et al., 2007; Sendich et al., 2008); (2) reducing the production cost of enzymes for converting cellulose to sugars; and (3) reducing capital costs by developing more efficient microorganisms for converting the sugar products of biomass deconstruction to biofuels. The size of some biorefineries could be limited by the supply of biomass. The limit in size could result in loss of the economies of scale that can be achieved with large plants. The costs of products of the first-generation commercial plants could be higher than estimated. One company that has been operating a fully integrated demonstration reported that the capital cost of the plant was substantially higher than predicted by models because of unanticipated problems, such as the complexity of handling mineral matter in feedstocks (B. Foody, Iogen Corporation, personal communication, March 17, 2008). Costs are expected to decrease as experience is gained.

In addition to the mandated 16 billion gallons of cellulosic biofuel, the RFS as amended in the 2007 Energy Independence and Security Act states that advanced biofuels must reduce net greenhouse gas emissions by at least 50 percent relative to those from gasoline. To meet the goals established by the 2007 act, economically viable lignocellulosic-ethanol production is essential, and technological progress needs to be made. If broad deployment (implementation) is to occur by 2020, the facilities would have to obtain permits by about 2015. Although a series of evolutionary changes are likely to occur between 2008 and 2015, successful deployment by 2020 would require a large, sustained effort to improve technologies and process efficiencies. Demonstrations on a commercial scale have to occur at an aggressive pace to ensure sufficient learning from the activity. Much engineering, technical, and operational knowledge can be gained only from designing and building integrated facilities and then operating them for a reasonable period.

TECHNOLOGY FORECAST

The growing biofuel industry is based on well-established technology for producing ethanol via fermentation and separating it by distillation. Fuel ethanol cannot be added to gasoline before pipeline transportation. The cost of fuel-ethanol transportation is estimated to range from $0.13 to $0.18 per gallon, which is as much as 6 times the cost of transporting traditional petroleum-based fuels (GAO, 2007). Cellulosic ethanol derived from nonfood renewable feedstocks could be a transition to or one of many contributors to a diverse portfolio of alternative fuels as other biofuels are proved, demonstrated, and commercialized. This section discusses some of the technologies for producing other biofuels. As is the case with ethanol derived from renewable cellulosic feedstocks, the technologies that would use sugars as part of the conversion process would be attractive only if the feedstocks and the sugars obtained from them are inexpensive compared with ethanol production.

Hydrocarbon Fuels from Biomass

Approaches to produce hydrocarbon fuels directly from biomass that are analogous to production of fuels from petroleum are being explored (Huber et al., 2006).

Gasoline Blend Stock

One approach produces straight-chain hydrocarbons, mostly hexane, via aqueous-phase hydrogenation of biomass-derived sugars followed by dehydration. All the hydrogen consumed in the process can be obtained from biomass processing. The process is exothermic as a result of oxidation of a portion of the biomass-derived carbohydrates. Because the reactants are dissolved in water, the hydrocarbons produced form a separate phase, and distillation is not required. This process has the potential of higher energy efficiency and shorter residence times than the fermentation and distillation steps used in ethanol production, but considerable development is required to confirm that the potential can be realized in a commercially viable process (Huber et al., 2005).

The product consisting of linear hydrocarbons can be isomerized in a conventional refining process to form branched hydrocarbons with higher octane more suitable for gasoline blending. Conventional refinery alkylation technology can be used to process the low-boiling-point straight-chain hydrocarbons to increase octane and boiling point to the extent needed for gasoline blending. If this kind of production of hydrocarbons from biomass were widely commercialized, refining capability for isomerization and alkylation would probably need to be increased.

Diesel Fuel Components

Another approach to biohydrocarbon fuels produces high-cetane diesel (Huber et al., 2005). Sugars are first dehydrated and then hydrogenated to form cyclic oxygenated molecules that can undergo aldol condensation (self-addition) to form larger oxygenated molecules that remain soluble in water. The condensation products are hydrogenated and then dehydrated to form mostly straight-chain hydrocarbons ranging from 7 to 15 carbons. The final hydrogenation and dehydration reactions are carried out in a four-phase reactor. The feed streams to the reactor in the four phases are water with dissolved oxygenated hydrocarbon reactants, gaseous hydrogen, solid catalyst, and hydrocarbon (required to reduce coke formation on the catalyst). The process can be modified to produce oxygenated compounds in the diesel boiling range that are soluble in the fuel.

Status

Although those processes have been shown to be feasible in the laboratory with pure feedstocks, much development beyond what has been reported remains. The concepts need to be tested by using biomass-derived feedstocks with recycling and with reactors that can be scaled for commercial operation. The keys to success in the processes appear to be achieving sufficient yield of the hydrocarbon product, developing high-activity catalysts with long-term stability, and minimizing coking reactions. There remains a large amount of R&D to be done on these concepts before commercial applications can be undertaken.

Biobutanol

Butanol is another potential entrant into the light-duty-vehicle biofuel market. Butanol is a four-carbon alcohol (ethanol has two carbons). When butanol is made from biomass, it is referred to as biobutanol. The longer hydrocarbon chain makes it fairly nonpolar and thus more similar to gasoline. Biobutanol has a number of attractive features as a fuel: its energy content is close to that of gasoline, it has a low vapor pressure, it is not sensitive to water, it is less hazardous to handle and less flammable than gasoline, and it has a slightly higher octane content than gasoline. Thus, it can go directly into the existing distribution system. It has been shown to work in gasoline engines without modification (DuPont, 2008).

Several technologies to produce biobutanol are in the R&D phase. The one receiving the most attention is the acetone-butanol-ethanol process, which uses the bacterium *Clostridium acetobutylicum*. This process was used initially to produce acetone for making cordite in 1916. The process produced about twice as much butanol as acetone and also produced acetic, lactic, and propionic acids and ethanol and isopropanol. As currently being commercialized, the process involves the biochemical conversion of sugars or starches from sugar beets, sugar cane, corn, wheat, or cassava to butanol. When the mutated strain *Clostridium beijernickii* BA101 is used, greater selectivity for butanol is achieved. There are also efforts to develop microorganisms that have an increased rate and selectivity in the conversion of sugars to butanol. These include microorganisms that can efficiently convert the different sugars that are obtained from cellulose and hemicelluloses. Because butanol is toxic to the producing organism, its concentration is limited to about 15–18 g/L even in the native organism that produces it (for example, clostridia). Isobutanol is less toxic and is also a good fuel component, so a more promising approach to improve the process is to seek or engineer organisms that produce isobutanol.

The obvious extension of that technology is the conversion of cellulose to biobutanol. It depends on the development of biotechnologies for the effective, efficient depolymerization of cellulose and hemicelluloses into the basic sugars, which can then be converted to butanol. The most important development would be metabolic engineering of microorganisms that could depolymerize the biomass components into sugars and then convert the sugars to butanol in the same reactor to reduce capital cost. The cellulose approach to biobutanol is being studied, but the technology is in the research stage and is far from commercial.

In another variation, fermentation in a fixed-bed bioreactor using *Clostridium tyrobutyricum* produces primarily butyric acid and hydrogen. When the butyric acid is fed into another bioreactor with *C. acetobutylicum*, the butyric acid is converted to butanol with high selectivity.

Biobutanol's main challenge now is cost. To address the cost and initiate market entry, some companies are working on retrofitting an existing sugar-based bioethanol plant to produce biobutanol (Chase, 2006; DuPont, 2008). An improved next-generation bioengineered organism is projected to be available by 2010 (Chase, 2006).

Algal Approaches to Fuel Production

Large-scale production of photosynthetic microorganisms (algae) to be used as biomass feedstock for liquid transportation fuels has been contemplated for many years, but uncertainties surrounding production costs have resulted in smaller investments in R&D compared with that in cellulosic biofuels. (See Appendix J for details of systems, strains, and resource requirements for production of microbial biomass.) A major program in this field was funded and managed by the U.S. Department of Energy National Renewable Energy Laboratory (Sheehan et al., 1998). Research in the development of algae that have high lipid productivity is being conducted (Briggs, 2004; Pacheco, 2006). Advances in the metabolic engineering and genomics of algae are leading to new strategies for increasing the utility of algae for fuel production.

Several types of fuel potentially can be produced by photosynthetic microorganisms. To date, the emphasis has been on producing biodiesel via transesterification of algal glycerolipids to produce fatty acid methyl (or ethyl) esters, which can be used in diesel engines. Cellular lipids can also be converted via catalytic hydrocracking to a mixture of alkanes suitable for use as a jet fuel or gasoline ingredient. Some algae, such as *Botryococcus*, produce long-chain hydrocarbons that are potentially usable as fuel after hydrocracking to reduce the chain length of the molecules. Production of ethanol from recombinant photosynthetic microorganisms is also being considered. It involves introducing foreign genes that encode ethanol biosynthetic enzymes into cyanobacteria or microalgae. Tens of thousands of species of cyanobacteria and microalgae occur naturally in a wide variety of habitats, but only a small fraction of them are available in public culture collections. Strain collection and characterization programs are needed to enable large-scale production of biofuels from photosynthetic microorganisms.

The basic resources required for large-scale cultivation of photosynthetic microorganisms for fuel production are land with suitable topography, climate, and sunlight; water of acceptable quality and abundance; and concentrated sources of CO_2 (Maxwell et al., 1985). The co-location of the three resources is important for minimizing production costs. One study estimated that 18–34 liters of water is needed to produce 1 MJ of energy from algae (Dismuskes et al., 2008). That water requirement is similar to that of corn grain ethanol (33 L/MJ). Large quantities of saline groundwater are present in the southwestern United States, much of which could be used to support large-scale cultivation of photosynthetic microorganisms for biofuel production. It is imperative, however, to gain a better understanding of how much water can be removed from saline aquifers without adversely affecting the flow and quality of contiguous freshwater aquifers (if present) and without creating wastewater-disposal issues. Closed photoreactors for culturing algae would use less water than would open ponds.

Many of the challenges related to algal biofuel production are engineering matters associated with how and where to grow the algae to achieve needed productivity. In most production schemes, the algal oil is extracted from harvested algae. Because of the metabolic burden associated with the biosynthesis of high-energy lipids, production strains that accumulate large amounts of oil tend to grow and reproduce more slowly than strains that do not accumulate oil. Open cultures are therefore prone to contamination with undesirable species unless the production strain is able to grow in specialized conditions that restrict the growth of other species (for example, high pH). Alternatively, production strains selected for high growth rates and high biomass yields without regard for oil content can often compete satisfactorily with contaminating strains, but the chemical composition of the algae would be better suited for anaerobic digestion than to liquid-fuel production. The use of closed photobioreactors can lower the risk of culture contamination substantially, but capital costs of such systems are high. Maxwell et al. (1985) made a rough estimate of the sites in the southwestern United States that have suitable terrain, climate, and water availability. About 1 percent of the area considered (that is, 2 million of the 200 million hectares considered) could be suitable sites of algal-production facilities.

Large quantities of biofuels potentially could be produced by photosynthetic microorganisms. With an average productivity of 0.046 lb/yd² per day (25 g/m² per day) and lipid content of 20 percent (both values have been demonstrated by numerous groups), 44 lb (20 kg) of lipid could be produced per acre per day. If the growth facility were in operation 300 days/year, total biomass productivity would be 1,730 gal/acre per year. Thus, to produce enough fuel to replace 10 percent of the current U.S. gasoline consumption, about 8.1 million acres (12,700 square miles) of algal-growth facilities would be required. That is slightly less than 5 percent of the area of Texas. (In comparison, about 300 million acres of soybean or 32 million acres of corn would be required to produce the equivalent volume of vegetable oil or ethanol, respectively.) Coal-fired and gas-fired power plants theoretically could supply all the CO_2 necessary to produce the microorganisms. According to Feinberg and Karpuk (1990), 80 million tonnes of CO_2 are required per quad (10^{15} Btu) of algal lipid produced, so the production of 14 billion gallons of lipid (equivalent to 1.8 quads) would require 144 million tonnes of CO_2. That represents about half the CO_2 emitted by electricity-generating power plants in Texas (see http://carma.org). Technologies for developing algal strains with desirable traits for biofuel production that encompass classical strain improvement, metabolic engineering, and synthetic biology will probably enhance biofuel productivity in the future.

Few detailed economic analyses of costs of producing biofuels from algae have been completed. Results of the few analyses varied widely because of differences in such input variables as production-system design, cell- and product-recovery procedures, fuel type, and site characteristics. In one case, the reported cost of algal biofuel was well over $4.00/gal, indicating that much progress in R&D is needed to reduce production cost if this technology is to have utility in the foreseeable future (Pacheco, 2006). In another case, however, fuel-production costs with existing technologies were estimated to be $2.00/gal (Huntley and Redalje, 2006). Large-scale testing will be necessary to validate the assumptions used in those and similar economic analyses.

Bacteria-Based and Yeast-Based Direct Routes to Biofuels

With the rapid growth of synthetic biology and the enhanced ability to engineer metabolic pathways into organisms to produce specific chemical or fuel products, synthetic biology and metabolic engineering for renewable fuel production have great potential and are receiving renewed interest (Savage, 2007). The approaches being taken include using well-established recombinant-DNA techniques to insert genes into microorganisms to make specific fuel precursors or even direct synthesis of hydrocarbon fuel components. Another approach involves redesigning genes with computer assistance to perform specific reactions and then synthesizing the desired genes for insertion into microorganisms. Yeasts can also be engineered to produce larger amounts of lipids, which with additional metabolic engineering can be converted to useful products, potentially fuels. That work has not progressed as far as the work on bacteria.

Those techniques might make it possible to modify bacteria to produce and excrete specific hydrocarbon molecules that have desired fuel or other chemical properties. Microorganisms that produce and excrete specific hydrocarbons minimize the costs of energy-consuming separation, although developing organisms that excrete the fuel products is a major challenge in that most synthesis products, including hydrocarbons, accumulate in the cell. No specific processes can be considered to be approaching commercial production at this point, but the magnitude of activity and the current rate of progress could change that in the not-too-distant future. Several companies are using synthetic biology to produce bacteria that make increased amounts of fatty acids or other lipids that are then converted to hydrocarbons and excreted. The bacteria make and excrete hydrocarbons of any desired length and structure. The phase-separation of hydrocarbons from the growth medium markedly reduces separation costs. The feedstock for the bacteria is renewable sugars, which can be obtained from sugar cane or grain or from cellulosic biomass (LS9, 2008). It is difficult to project future developments. Some companies are producing fuels, but projected costs of fuels have not been reported (Service, 2008).

Technologies to Improve Biochemical Conversion

Important advances are being made in genomics, molecular breeding, synthetic biology, and metabolic and bioprocess engineering that will probably enable discontinuous innovation and advancement in alternative transportation fuels. Those advances and related technologies have the potential to accelerate the creation of dedicated or dual-purpose energy crops and microorganisms that can be used for both biofuel production and feedstock conversion.

Genomics

The sequencing of full genomes continues to become faster and less expensive, and this is enabling the sequencing of energy crops, such as trees, perennial grasses, and such nonedible oilseeds as castor and jatropha. Their sequence data are extremely important for improving overall yields, for enabling improved nutrient and water use, and for understanding and manipulating biochemical pathways to enhance the production of desired materials. Sequence data can also be used to target specific genes for downregulation by classical methods, such as antisense and RNA interference, and via complete inactivation with new and evolving procedures for homologous recombination-based gene disruption. Rapid sequencing of breeding populations of energy crops will enable marker-assisted selection to accelerate breeding programs in ways previously not possible. Furthermore, rapid and inexpensive sequencing of fermentative and photosynthetic microorganisms is redefining and shortening the timelines associated with strain-development programs for converting sugars, lignocellulosic materials, and CO_2 to alternative liquid fuels. Strains generated through classical mutagenesis that have improved biocatalytic properties can now be analyzed at the molecular level to determine the specific genetic changes that result in the improved phenotype, and this allows the changes to be implemented in additional strains. In addition, "metagenome" sequence data obtained by randomly sequencing DNA isolated from environmental samples is providing huge numbers of new gene sequences that can be used in genetic engineering to improve crops and microorganisms.

Synthetic Biology and Synthetic Genomics

Improved technologies for synthesizing megabase DNA molecules are being developed to allow the introduction of entire biochemical pathways into energy crops and biofuel-producing microorganisms. The technologies could have a great effect on scientists' ability to generate plants and microorganisms with specific desirable traits. For example, it is becoming conceivable to replace large portions of, or even complete, chromosomes of microorganisms (including photosynthetic microorganisms) in ways that will focus the vast majority of their cells' biochemical machinery toward production of next-generation biofuel molecules and thus provide cost and product advantages. Maintaining the purity of such cultures, and finding ways to put at a disadvantage mutants that gain competitive ability by producing less of the desired secondary chemicals, could be serious hurdles.

Metabolic and Bioprocess Engineering

In addition to genetic manipulation, new bioengineering technologies that will lower the cost of biofuel formation and recovery are coming on line. Synthetic biology can now provide synthetic DNA for transferring heterologous genes into suitable host cells, but metabolic engineering is the enabling technology for constructing functional and optimal pathways for microbial fuel synthesis. This field has matured in only a few years and has an impressive record of accomplishments, many already being applied in industry (for example, in the production of biopolymers, alcohols, 1,3-propanediol, oils, and hydrocarbons). Microbial strains that secrete hydrophobic fuels that are similar to constituents of diesel fuel and gasoline into the culture medium have been developed. The fuels can be separated from the aqueous phase in a manner that simplifies distillation and thereby reduces energy inputs and facilitates continuous production. By taking a systems view of metabolism, metabolic engineering developed tools for overall biosystems optimization that are now facilitating the optimal construction of biosynthetic pathways and elicitation of novel multigenic cellular properties of critical importance for biofuels production, such as tolerance of fuel toxicity. In bioprocessing, the successful development of membrane-based alcohol separation would greatly reduce energy costs relative to the typically used distillation process (Vane, 2008). Gas-stripping, liquid-liquid extractions of secreted fuel molecules or new adsorbent materials that will allow continuous production modes for fermentation-based products are also being developed (Vane, 2008). For photosynthetic production of biofuels, the development of low-cost photobioreactors and associated recovery systems for algal biofuel production is of great interest and could have substantial beneficial effects on overall process economics.

FINDINGS AND RECOMMENDATIONS

Grain-based ethanol is a bridge to advanced biofuels that has important potential for greenhouse gas displacement. Advanced biofuels do not directly compete with food and feed supply, and they minimize indirect land-use change if appropriate feedstocks are selected and sustainable practices are used in their production.

Grain ethanol has initiated public awareness of the use of ethanol in the current and future transportation fleet and of the pitfalls of feedstock supply for a new industry. Grain ethanol has helped to establish an industrial infrastructure for advanced biofuels and for distribution and use of fuel ethanol.

Lignocellulosic feedstocks for production of advanced biofuels could be agricultural or forestry residues, agricultural cover crops, dedicated perennial crops grown on marginal lands that are not suitable for commodity-crop production even with high commodity prices, or municipal solid wastes. Biochemical conversion of cellulose to liquid fuels emulates commercial corn grain-to-ethanol technology but might require additional processing steps and could result in other types of alcohols and hydrocarbon-rich fuels.

The technologies for biochemical conversion of cellulosic biomass to ethanol are in the early stages of demonstration and commercial development. Several demonstration plants are expected to be operational by 2012. The panel judges that cellulosic bioethanol will be commercially deployable before 2020, and other advanced biofuels are likely to emerge after 2020.

Finding 3.1

Engineering and operational knowledge can be gained only from designing and building commercial-scale, integrated cellulosic-ethanol facilities and then operating them for a reasonable period. The first few commercial plants will be more expensive than commercial facilities that follow because of the learning that occurs with a first-of-its-kind facility. The initial learning that occurs with first-of-a-kind plants will lead to further cost-reducing improvements in commercial facilities deployed thereafter. The pace of learning is expected to be similar to that in the chemical industry, in which costs have historically decreased by 30–40 percent over several cycles of deployment and concurrent process improvement.

Recommendation 3.1

The federal government and industry should aggressively pursue technology demonstration or small-scale commercial plants, which will lead to full-scale commercial production of cellulosic ethanol to define its potential and to provide data on engineering and cost performance to help in preparation for full commercial deployment.

In the immediate term, pretreatment and enzymatic hydrolysis, fermentation, or combined enzymatic hydrolysis and fermentation need to be substantially improved to allow efficient deconstruction of carbohydrate polymers to simple sugars and fermentation of the sugars to ethanol. Research in and improvement of pretreatment, with engineering of appropriate microorganisms for optimal use of the resulting simple sugars in an adverse fermentation environment, will have a direct impact on reducing the cost of transforming cellulosic feedstocks to ethanol. The cost of producing sugars directly affects the cost of ethanol. In addition, the sugars have to be converted to ethanol efficiently to minimize feedstock and operational costs.

Feedstock, pretreatment, and enzymes are key components of a cellulose-to-ethanol process, and they are all related to the goal of preparing lignocellulosic feedstocks (through agronomics, plant molecular genetics, and pretreatment) so that they are readily transformed to sugars and ethanol at low cost. Other targets for improvement include increasing solids loading and developing engineered microorganisms and enzymes that have increased tolerance of toxic compounds in biomass hydrolysates and of the biofuel products themselves. Incremental improvements in biochemical conversion technologies and the learning and experience gained from R&D and demonstration can be expected to reduce non-feedstock processing costs by 25 percent by 2020 and 40 percent by 2035 (see Table 3.3).

Finding 3.2

Process improvements in cellulosic-ethanol technology are expected to be able to reduce the plant-related costs associated with ethanol production by up to 40 percent over the next 25 years. Over the next decade, process improvements and cost reductions are expected to come from evolutionary developments in technology, from learning gained through commercial experience and increases in scale of operation, and from research and engineering in advanced chemical and biochemical catalysts that will enable their deployment on a large scale.

Recommendation 3.2

The federal government should continue to support research and development to advance cellulosic-ethanol technologies. R&D programs should be pursued to resolve the major technical challenges facing ethanol production from cellulosic biomass: pretreatment, enzymes, tolerance to toxic compounds and products, solids loading, engineering microorganisms, and novel separations for ethanol and other biofuels. A long-term perspective on the design of the programs and allocation of limited resources is needed; high priority should be placed on programs that address current problems at a fundamental level but with visible industrial goals.

Recommendation 3.3

The pilot and commercial-scale demonstrations of cellulosic-ethanol plants should be complemented by a closely coupled research and development program. R&D is necessary to resolve issues that are identified during demonstration and to reduce costs of sustainable feedstock acquisition. Industrial experience shows that such reductions typically occur as processes go through multiple phases of implementation and expansion.

Finding 3.3

Future improvements in cellulosic technology that entail invention of biocatalysts and biological processes could produce fuels that supplement ethanol production in the next 15 years. In addition to ethanol, advanced biofuels (such as lipids, higher alcohols, hydrocarbons, and other products that are easier to separate than ethanol) should be investigated because they could have higher energy content and would be less hygroscopic than ethanol and therefore could fit more smoothly into the current petroleum infrastructure than ethanol could.

Recommendation 3.4

The federal government should ensure that there is adequate research support to focus advances in bioengineering and the expanding biotechnologies on developing advanced biofuels. The research should focus on advanced biosciences—genomics, molecular biology, and genetics—and biotechnologies that could convert biomass directly to produce lipids, higher alcohols, and hydrocarbons fuels that can be directly integrated into the existing transportation infrastructure. The translation of those technologies into large-scale commercial practice poses many challenges that need to be resolved by R&D and demonstration if major effects on production of alternative liquid fuels from renewable resources are to be realized.

Finding 3.4

Biochemical conversion processes, as configured in cellulosic-ethanol plants, produce a stream of relatively pure CO_2 from the fermentor that can be dried, compressed, and made ready for geologic storage or used in enhanced oil recovery with little additional cost. Geologic storage of the CO_2 from biochemical conversion of plant matter (such as cellulosic biomass) further reduces greenhouse gas life-cycle emissions from advanced biofuels, so their greenhouse gas life-cycle emissions would become highly negative.

Recommendations 3.5

Because geologic storage of CO_2 from biochemical conversion of biomass to fuels could be important in reducing greenhouse gas emissions in the transportation sector, it should be evaluated and demonstrated in parallel with the program of geologic storage of CO_2 from coal-based fuels.

REFERENCES

Aden, A., M. Ruth, K. Ibsen, J. Jechura, K. Neeves, J. Sheehan, B. Wallace, L. Montague, A. Slayton, and J. Lukas. 2002. Lignocellulosic Biomass to Ethanol Process. Design and Economics Utilizing Co-Current Dilute Acid Prehydrolysis and Enzymatic Hydrolysis for Corn Stover. Golden, Colo.: National Renewable Energy Laboratory.

Aristidou, A., and M. Penttila. 2000. Metabolic engineering applications to renewable resource utilization. Current Opinion in Biotechnology 11:187-198.

Briggs, Michael. 2004. Widescale biodiesel production from algae. Available at www.unh.edu/p2/biodiesel/article_alagae.html. Accessed October 1, 2008.

Chase, R. 2006. DuPont, BP join to make butanol: They say it outperforms ethanol as a fuel additive. USA Today, June 26. Available at http://www.usatoday.com/money/industries/energy/2006-06-20-butanol_x.htm. Accessed October 11, 2008.

Cornell, C.B. 2008. GM announces new cellulosic ethanol partnership with Mascoma Corp. Available at http://gas2.org/2008/05/01/gm-announcesnew-cellulosic-ethanol-partnership-with-mascoma-corp/. Accessed March 10, 2009.

Dayton, D. 2007. R&D needs for integrated biorefineries: The 30 × 30 Vision. In 4th Annual California Biomass Collaborative Forum, Sacramento, Calif.

Dismuskes, G.C., D. Carrieri, N. Bennette, G.M. Ananyev, and M.C. Posewitz. 2008. Aquatic phototrophs: Efficient alternatives to land-based crops for biofuels. Current Opinion in Biotechnology 19:235-240.

DOE (U.S. Department of Energy). 2007a. Biomass Feedstock Composition and Property Database. Available at http://www1.eere.energy.gov/biomass/feedstock_databases.html. Accessed December 13, 2007.

DOE. 2007b. DOE selects six cellulosic ethanol plants for up to $385 million in federal funding. Available at http://www.energy.gov/print/4827.htm. Accessed October 16, 2008.

DOE/EERE (U.S. Department of Energy, Office of Energy Efficiency and Renewable Energy). 2007. Biomass Program 2007 Accomplishments Report. Available at http://www1.eere.energy.gov/biomass/pdfs/program_accomplishments_report.pdf. Accessed February 10, 2009.

DuPont. 2008. DuPont Fact Sheet on Biobutanol. Available at http://www.dupont.com/ag/news/releases/BP_DuPont_Fact_Sheet_Biobutanol.pdf. Accessed October 16, 2008.

Fargione, J., J. Hill, D. Tilman, S. Polasky, and P. Hawthorne. 2008. Land clearing and the biofuel carbon debt. Science 319:1235-1238.

Farrell, A., R. Plevin, B. Turner, A. Jones, M. O'Hare, and D. Kammen. 2006. Ethanol can contribute to energy and environmental goals. Science 311:506-509.

Feinberg, D., and M. Karpuk. 1990. CO_2 sources for microalgae-based liquid fuel production. Golden, Colo.: Solar Energy Research Institute and National Renewable Energy Laboratory.

GAO (U.S. Government Accountability Office). 2007. Biofuels: DOE Lacks a Strategic Approach to Coordinate Increasing Production with Infrastructure Development and Vehicle Needs. Washington, D.C.: GAO.

Gauss, W.F., S. Suzuki, and M. Takagi, inventors. 1976. Manufacture of Alcohol from Cellulosic Materials Using Plural Ferments. U.S. Patent 3990944.

Goldemberg J., S.T. Coelho, and P.M. Guardabassi. 2008. The sustainability of ethanol production from sugarcane. Energy Policy 36:2086-2097.

Gong, C.S., M.R. Ladisch, and G.T. Tsao. 1977. Cellobiase from *Trichoderma viride*: Purification properties, kinetics and mechanism. Journal of Biotechnology and Bioengineering 19:959-981.

Gong, C.S., M.R. Ladisch, and G.T. Tsao. 1979. Hydrolysis of cellulose: Mechanisms of enzymatic and chemical catalysis. In Advances in Chemistry Series 181, R. Brown and L. Jurasek, eds. Washington, D.C.: American Chemical Society.

Gulati, M., K. Kohlmann, M.R. Ladisch, R. Hespell, and R.J. Bothast. 1996. Assessment of ethanol production options for corn products. Bioresource Technology 58:253-264.

Hendriks, A.T.W.M., and G. Zeeman. 2009. Pretreatments to enhance the digestibility of lignocellulosic biomass. Bioresource Technology 100:10-18.

Hill, J., E. Nelson, D. Tilman, S. Polasky, and D. Tiffany. 2006. Environmental, economic, and energetic costs and benefits of biodiesel and ethanol biofuels. Proceedings of the National Academy of Sciences USA 103:11206-11210.

Huber, G.W., J.N. Chheda, C.J. Barrett, and J.A. Dumesic. 2005. Production of liquid alkanes by aqueous-phase processing of biomass-derived carbohydrates. Science 308:1446-1450.

Huber, G.W., S. Ibora, and A. Corma. 2006. Synthesis of transportation fuels from biomass: Chemistry, catalysts, and engineering. Chemical Review 106:4044-4098.

Huntley, M.E., and D.G. Redalje. 2006. CO_2 Mitigation and Renewable Oil from Photosynthetic Microbes: A New Appraisal. Mitigation and Adaptation Strategies for Global Change 12:573-608.

Intelligen, Inc. 2009. SuperPro Designer: Cost analysis and project economic evaluation. Available at http://www.intelligen.com/costanalysis.shtml. Accessed January 28, 2009.

Jensen, T.H. 2008. Race for key biofuel breakthrough intensifies. Reuters, June 2. Available at http://www.reuters.com/article/reutersEdge/idUSL0248503420080602. Accessed February 4, 2009.

Johnson, J.C. 2006. Technology Assessment of Biomass Ethanol: A Multi-objective, Life Cycle Approach Under Uncertainty. Ph.D. Thesis. Department of Chemical Engineering. Cambridge: Massachusetts Institute of Technology.

Jorgensen, H., J.B. Kristensen, and C. Felby. 2007. Enzymatic conversion of lignocellulose into fermentable sugars: Challenges and opportunities. Biofuels, Bioproducts and Biorefining 1:119-134.

Katzen, R., W.R. Ackley, Jr., G.D. Moon, J.R. Messick, B.F. Brush, and K.F. Kaupisch. 1981. Low-energy distillation systems. In Fuels from Biomass and Wastes, D.L. Klass and G.H. Emert, eds. Ann Arbor, Mich.: Ann Arbor Science Publishers.

Kim, Y., and M.R. Ladisch. 2008. Liquid hot water pretreatment of cellulosic biomass. In Methods in Molecular Biology, J. Meilanz, ed. Totowa, N.J.: Humana Press.

Kim, Y., N. Mosier, and M.R. Ladisch. 2008. Process simulation of modified dry grind ethanol plant with recycle of pretreated and enzymatically hydrolyzed distillers' grains. Bioresource Technology 99:5177-5192.

Kwiatkowski, J., A.J. McAllon, F. Taylor, and D.B. Johnston. 2006. Modeling the process and costs of fuel ethanol production by the corn dry-grind process. Industrial Crops and Products 23:288-296.

Ladisch, M.R., J. Hong, M. Voloch, and G. Tsao. 1981. Cellulase kinetics: Trends in the biology of fermentations for fuel and chemicals. Basic Life Sciences 18:55-83.

Ladisch, M.R., K. Kohlmann, P. Westgate, J. Weil, and Y. Yang, inventors. 1998. Purdue Research Foundation Office of Technology Transfer, assignee. Processes for Treating Cellulosic Material. U.S. Patent 5,846,787.

Ladisch, M.R., M.C. Flickinger, and G.T. Tsao. 1978. Fuels and chemicals from biomass. Energy 4:263-275.

LS9. 2008. Renewable Petroleum Technology. Available at http://www.ls9.com/technology/. Accessed May 1, 2008.

Lynd, L.R., P.J. Weimer, W.H. van Zyl, and I.S. Pretorius. 2002. Microbial cellulose utilization: Fundamentals and biotechnology. Microbiology and Molecular Biology Review 66:506-577.

Lynd, L.R., W.H. van Zyl, J.E. McBride, and M. Laser. 2005. Consolidated bioprocessing of cellulosic biomass: An update. Current Opinion in Biotechnology 16:577-583.

Maxwell, E.L., A.G. Folger, and S.E. Hogg. 1985. Resource evaluation and site selection for microalgae production systems. Golden, Colo.: Solar Energy Research Institute.

Mosier, N., C. Wyman, B. Dale, R. Elander, Y.Y. Lee, M. Holtzapple, and M.R. Ladisch. 2005. Features of promising technologies for pretreatment of lignocellulosic biomass. Bioresource Technology 96:673-686.

Mosier, N.S., P. Hall, C.M. Ladisch, and M.R. Ladisch. 1999. Reaction kinetics, molecular action, and mechanisms of cellulolytic proteins. Advances in Biochemical Engineering and Biotechnology 65:24-40.

Murnen, H.K., V. Balan, S.P.S. Chundawat, B. Bals, L.D. Sousa, and B.E. Dale. 2007. Optimization of ammonia fiber expansion (AFEX) pretreatment and enzymatic hydrolysis of *Miscanthus x Giganteus* to fermentable sugars. Biotechnology Progress 23:846-850.

Nevoigt, E. 2008. Progress in metabolic engineering of *Saccharomyces cerevisiae*. Microbiology and Molecular Biology Reviews 72:379-412.

Pacheco, Michael. 2006. Potential of Biofuels to Meet Commercial and Military Needs. Presentation to the National Research Council Committee on Assessment of Resource Needs for Fuel Cell and Hydrogen Technologies.

Pate, R., M. Hightower, C. Cameron, and W. Einfield. 2007. Overview of Energy-Water Interdependencies and the Emerging Energy Demands on Water Resources. Los Alamos, N.Mex.: Sandia National Laboratories.

Pordesimo, L.O., B.R. Hames, S. Sokhansanj, and W.C. Edens. 2005. Variation in corn stover composition and energy content with crop maturity. Biomass and Bioenergy 28:366-374.

Saddler, J.N., H.H. Brownell, L.P. Clermont, and N. Levitin. 1982a. Enzymatic hydrolysis of cellulose and various pretreated wood fractions. Biotechnology and Bioengineering 24:1389-1402.

Saddler, J.N., C. Hogan, M.K.H. Chan, and G. Louisseize. 1982b. Ethanol fermentation of enzymatically hydrolyzed pretreated wood fractions using Trichoderma Cellulases, *Zymomonas-Mobilis*, and *Saccharomyces-Cerevisiae*. Canadian Journal of Microbiology 28:1311-1319.

Savage, N. 2007. Better biofuels. Technology Review 110:1.

Schwietzke, S., M.R. Ladisch, L. Russo, K. Kwant, T. Makinen, B. Kavalov, K. Maniatis, R. Zwart, G. Shahanan, K. Sipila, P. Grabowski, B. Telenius, M. White, and A. Brown. 2008a. Gaps in the research of 2nd generation transportation biofuels. IEA Bioenergy T41:2008:2001.

Schwietzke, S., Y. Kim, E. Ximenez, N. Mosier, and M. Ladisch. 2008b. Ethanol production from maize. In Molecular, Genetic Approaches to Maize Improvement, Biotechnology in Agriculture and Forestry, A.L. Kriz and B.A. Larkins, eds. Heidelberg, Germany: Springer-Verlag Berlin.

Searchinger, T., R. Heimlich, R.A. Houghton, F. Dong, A. Elobeid, J. Fabiosa, S. Tokgoz, D. Hayes, and T-H. Yu. 2008. Use of U.S. croplands for biofuels increases greenhouse gases through emissions from land use change. Science 319:1238-1240.

Sendich, E., M. Laser, S. Kim, H. Alizadeh, L. Laureano-Perez, B. Dale, and L. Lynd. 2008. Recent process improvements for the ammonia fiber expansion (AFEX) process and resulting reductions in minimum ethanol selling price. Bioresource Technology 99:8429-8435.

Service, R.F. 2008. Eyeing oil, synthetic biologists mine microbes for black gold. Science 322:522-523.

Shapouri, H., J.A. Duffield, and M. Wang. 2002. The Energy Balance of Corn Ethanol: An Update. Washington, D.C.: U.S. Department of Agriculture.

Sheehan, J., T. Dunahay, J. Benemann, and P. Roessler. 1998. A Look Back at the US Department of Energy's Aquatic Species Program: Biodiesel from Algae. Golden, Colo.: U.S. Department of Energy.

Sonderegger, M., M. Jeppsson, C. Larsson, M.F. Gorwa-Grauslund, E. Boles, L. Olsson, I. Spencer-Martins, B. Hahn-Hagerdal, and U. Sauer. 2004. Fermentation performance of engineered and evolved xylose-fermenting Saccharomyces cerevisiae strains. Biotechnology and Bioengineering 87:90-98.

Spindler, D.D., C.E. Wyman, K. Grohmann, and A. Mohagheghi. 1987. Thermotolerant yeast for simultaneous saccharification and fermentation of cellulose to ethanol. Applied Biochemistry Symposium 17:279-293.

Takagi, M., S. Abe, S. Suzuki, G.H. Emert, and N. Yata. 1977. A method for production of alcohol directly from cellulose using cellulase and yeast. In Proceedings of the Bioconversion Symposium. New Delhi: Indian Institute of Technology.

USDA-NASS (U.S. Department of Agriculture, National Agricultural Statistics Service). 2005. Quick Stats, Iowa County Data—Crops. Available at http://www.nass.usda.gov. Accessed December 18, 2007.

Vane, L.M. 2008. Separation technologies for the recovery and dehydration of alcohols from fermentation broths. Biofuels, Bioproducts and Biorefining 2:553-588.

Worldwatch Institute. 2007. Biofuels for Transport: Global Potential and Implications for Sustainable Energy and Agriculture. London: Earthscan.

Wright, J.D., C.E. Wyman, K. Grohmann, and N. Yata. 1977. Simultaneous saccharification and fermentation of lignocellulose. Applied Biochemistry and Biotechnology 18:75-90.

Wyman, C.E., B.E. Dale, R.T. Elander, M. Holtzapple, M.R. Ladisch, and Y.Y. Lee. 2005a. Comparative sugar recovery data from laboratory scale application of lead pretreatment technologies to corn stover. Bioresource Technology 96:2026-2032.

Wyman, C.E., B.E. Dale, R.T. Elander, M. Holtzapple, M.R. Ladisch, and Y.Y. Lee. 2005b. Coordinated development of leading biomass pretreatment technologies. Bioresource Technology 96:1959-1966.

Wyman, C.E., D.D. Spindler, K. Grohmann, and S.M. Lastick. 1986. Simultaneous saccharification and fermentation with the yeast *Brettanomyces clausenii*. Biotechnology Bioengineering Symposium 17:221-238.

Yang, B., and C.E. Wyman. 2008. Pretreatment: The key to unlocking low-cost cellulosic ethanol. Biofuels, Bioproducts and Biorefining 2:26-40.

Zeng, M., N.S. Mosier, C-P. Huang, D.M. Sherman, and M.R. Ladisch. 2007. Microscopic examination of changes of plant cell structure in corn stover due to cellulase activity and hot water pretreatment. Biotechnology and Bioengineering 97:265-278.

4

Thermochemical Conversion of Coal and Biomass

This chapter reviews the thermochemical conversion of coal, biomass, and combined coal and biomass to liquid transportation fuels. It addresses the questions raised in the statement of task related to the application of thermochemical conversion to the production of alternative liquid transportation fuels from those feedstocks by discussing the following:

- The development status of each major technology with estimated times of commercial deployment.
- Projected costs, performance, environmental impact, and barriers to deployment by 2020.
- Potential supply capability, plant carbon dioxide (CO_2) emissions, and life-cycle greenhouse gas emissions.
- Challenges and needs in research and development (R&D), including basic-research needs for the long term.

The available technologies are described first, and their status and technical and commercial readiness are assessed. Detailed cost and performance analysis, R&D and demonstration needs, environmental impacts, and analysis of greenhouse gas life-cycle emissions of the key technologies are discussed.

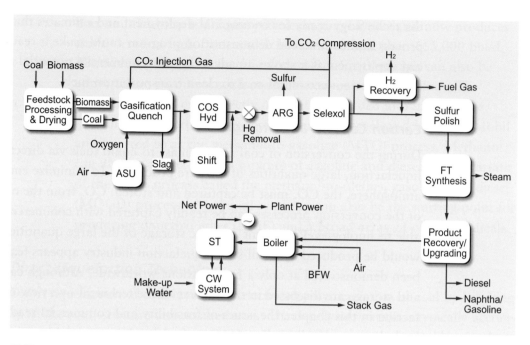

FIGURE 4.2 *Schematic of generic plant for indirect conversion of coal and/or biomass. Source: Tomlinson and Gray, 2007.*

synthesis gas. They are generally smaller and operate at lower pressures and temperatures than do coal gasifiers. Although there are many fixed-bed biomass gasifiers, fluid-bed and recirculating-bed systems have been developed.

A 3000 tons/day coal gasifier would produce enough synthesis gas to yield transportation fuel at about 6000 bbl/d by indirect liquefaction. After being ground into very small particles, the coal can be slurried with water or fed dry into the gasifier with a controlled amount of air or oxygen and steam. Temperatures in a gasifier range from 1400°F to 2800°F. At such high temperatures in the gasifier, steam reacts with the carbonaceous material of the feedstock to form syngas.

Coal Gasification

A number of technologies have been developed for coal gasification; they include moving-bed, fluid-bed, circulating-bed (transport), and entrained-flow gasifiers (MIT, 2007). The operating temperature and the size of coal feed vary with the type of gasifier. The moving-bed gasifier was developed by Lurgi and improved by Sasol. It operates at 425–600°C and accepts coal feed sizes of 6–50 mm. The

Sasol–Lurgi gasifier has been used extensively at the Sasol commercial plant in South Africa. Entrained-flow gasifiers operate at 1250–1600°C and accept coal-feed particles smaller than 100 μm. Those oxygen-blown, high-pressure gasifiers have been developed by General Electric (it was formerly referred to as the Texaco gasifier), Shell, Conoco Phillips (E-Gas), and Siemens (formerly referred to as the Future Energy gasifier). Fluid-bed gasifiers are less developed than the other two types. They operate at 900–1050°C and can use coal feed of 6–10 mm. In most types of gasifiers, avoiding soft ash particles is essential because the particles stick together, stick to process equipment, and typically lead to shutdown (MIT, 2007).

Coal gasification is commercially deployable today by using any one of several gasification systems that are being commercially used. Producing coal-to-liquid (CTL) fuels and other applications of gasification will lead to further improvements in the technology so that it would become more robust and efficient by 2020. Those improvements are part of the usual evolution of any new technology.

Coal and Biomass Gasification

Adding sustainably grown and harvested biomass to the coal feedstock would allow an increase in domestic fuel supply while reducing total greenhouse gas emissions in two ways. First, the emission of carbon in the burning of the fuels made from biomass is countered by the removal of carbon from the atmosphere by the biomass through photosynthesis during its growth. Second, the biomass and coal carbon that is converted to CO_2 during the conversion to transportation fuels could be captured and stored.

The notion of gasifying mixtures of coal and biomass to produce liquid fuels is relatively new, and there has been little commercial experience. Many gasifiers can gasify biomass, but most of them are small in scale, use air instead of oxygen, and operate at lower temperatures and at low or atmospheric pressure. Under those less severe conditions, pyrolysis dominates, and the main products, in addition to syngas, are light hydrocarbons, bio-oils, tars, and char. Those products make such gasifiers less suitable for producing FT liquid fuels.

The NUON Shell 253-megawatt electric (253-MW$_e$) integrated gasification combined-cycle (IGCC) facility in the Netherlands has proved that gasification of combined wood (30 percent by weight) and coal can be achieved for the generation of electric power. It has also gasified other biomass feedstocks, including chicken litter.

The operation of a combined coal-and-biomass-to-liquids (CBTL) plant would be similar to that of a CTL plant, except that biomass is gasified in addition to coal (Figure 4.2). Separate gasifiers could be used for the biomass and the coal, but it might be more efficient and cost-effective if the same gasifier could convert both feeds simultaneously. That would be similar to the situation at the NUON discussed above in which the Shell gasifier was able to gasify both wood and other biomass with the same lock-hopper high-pressure feeding system.

Combined coal and biomass gasification is deployable today, although the amount of biomass relative to coal feed is small, as discussed above. Further commercial development of the technology will make it more robust and efficient and enhance its ability to use higher fractions of biomass by 2020.

Biomass Gasification

Published data on high-pressure biomass gasifiers are sparse. Because of the fibrous nature of most biomass sources, the material is difficult to pretreat and feed into a high-pressure gasifier. Typical problems include clumping and bridging.

Biomass gasification exhibits many similarities to coal gasification, including the variety of gasifier types and different available approaches to gasification technology. However, the reaction conditions are generally milder than those for coal gasification because of the higher reactivity of biomass.

Gasification with direct firing with oxygen at higher pressures and temperatures produces a relatively pure syngas stream with small quantities of CO_2 and other gases. For temperatures greater than 1000°C, little or no methane, higher hydrocarbons, or tar is present.

A major difference between biomass gasification and coal gasification is that the former generally involves smaller units than the latter because of the limits on the availability of biomass in a reasonable harvesting area. Biomass gasification therefore will not have the benefit of economies of scale that larger-scale coal gasification has. The lack of economies of scale will increase the cost per unit product of biomass gasification unless major process simplification and capital-cost reduction can be achieved. Like coal gasifiers, biomass gasifiers can be lumped into specific types, each of which has many variations.

Several U.S. and European organizations are developing advanced biomass gasification technologies, and about 10 biomass gasifiers have a capacity greater than 100 tons/day operating in the United States, Europe, and Japan (IEA, 2007; Cobb, 2007). Those units have a broad variety of feedstocks, feed capabilities, characteristics, product-gas cleanup approaches, and primary products. The Biomass Technology Group lists more than 90 installations (most are small) and more than 60 suppliers of equipment that is used in gasification (Knoef, 2005). Although several of the available technologies have been commercially demonstrated, they have yet to be fully demonstrated commercially for integrated biomass gasification and transportation-fuel production. The panel considers biomass gasification to be technically ready for aggressive commercial demonstration but not yet well enough understood to ensure efficient, effective commercial deployment today. Many variations require understanding and improvement. With an aggressive commercial development program, biomass gasification technology could be ready for full-scale commercial deployment by 2015. The major issues to be resolved are related to engineering, particularly the extent of biomass pretreatment necessary and effective feeding of biomass to high-pressure gasification reactors. An example of the conversion of biomass into liquid transportation fuels is the partnership of Choren Industries and Shell. Choren provides the Carbo V gasification process, and Shell provides the FT synthesis technology.

Most of the gasification technologies present technical or operational challenges, most of which can probably be resolved or managed with commercial experience. Gasifier choice depends on the type of biomass feed and on the specific application of the gasification or pyrolysis products. The gasifier units will generally be smaller than large-scale coal gasifiers because of the economics and logistics of the feed supply. The most persistent problem appears to be related to biomass feeding, processing, and handling, particularly if a gasifier has to contend with different biomass feeds.

Syngas Cleanup and Conditioning

The raw syngas produced in the gasification of coal and biomass contains many impurities, such as CO_2, hydrogen sulfide, carbonyl sulfide, ammonia, chlorine, mercury, and other toxic chemicals. Biomass has much lower sulfur content than coal does, and sulfur impurities in the syngas are correspondingly lower. However, biomass ash can contain high concentrations of sodium, potassium, and silicon that might pose additional requirements for the cleanup system. The impurities have to be removed before the syngas is allowed to contact the synthesis catalysts; otherwise, catalyst poisoning and deactivation will result. For example, in the conceptual configuration shown in Figure 4.2, carbonyl sulfide is hydrolyzed to hydrogen sulfide. Ammonia is scrubbed out and mercury is removed with activated carbon, and CO_2 and hydrogen sulfide are removed with Selexol or another acid-gas removal system. The processes for removing the contaminants are all commercially available.

In addition to cleaning, the H_2:CO ratio is adjusted to be compatible with the synthesis process by using the water–gas shift process. In this process, CO is converted by reaction with steam to H_2 and CO_2. The CO_2 can then be removed in the acid–gas removal system to produce a concentrated stream of CO_2 that is suitable for storage. The same is true for biochemical conversion of biomass to ethanol. The fermentation step produces a stream of pure CO_2 that can be compressed and geologically stored. The transport and storage costs will be somewhat higher because the amount of CO_2 will typically be smaller for the biochemical conversion route than for a thermochemical conversion route with an equal biomass feed rate. Because synthesis catalysts are readily poisoned by minute quantities of sulfur, a polishing reactor that removes sulfur down to parts per billion is included before the synthesis reactor. Ultimately, the hydrogen and carbonyl sulfides are converted (99.99 percent) to elemental sulfur, and the mercury is removed.

Syngas cleanup and conditioning technology is ready for full-scale commercial deployment today. It will undergo substantial improvement as a result of normal process evolution and become more robust and efficient by 2020.

Synthesis

Once the syngas produced by gasification of the carbonaceous feed has been cleaned of impurities and shifted to the desired H_2:CO ratio, it can be used to synthesize liquid transportation fuels. Two major commercial synthesis processes can be used to produce transportation fuels, such as gasoline, diesel, and jet fuel. These are FT and methanol synthesis followed by MTG. DME can also be produced by dehydration of methanol, but it is not a liquid fuel under ambient conditions. DME is discussed in Chapter 9.

Fischer-Tropsch Synthesis

The clean synthesis gas is sent to FT reactors, where most of the clean gas is converted into zero-sulfur liquid hydrocarbon fuels. If the major required product is distillate or diesel boiling-range fractions, slurry-phase reactors are used. One of the limitations of FT synthesis is that it produces a wide array of hydrocarbon products in addition to some oxygenates. The array of products depends on the probability of chain growth relative to chain termination. The probability function can theoretically be modeled with the Schultz–Flory–Anderson relationship, in which the parameter alpha determines the shape of the probability curve; the higher the alpha, the longer the hydrocarbon chains. To maximize liquid products in the naphtha and diesel boiling range, it is best to produce waxes first and then to crack the wax selectively to lower-boiling-point materials.

The low-temperature FT process produces about 10 percent hydrocarbon gases, 25 percent liquid naphtha, 22 percent distillate, and 46 percent wax and heavy oil. The wax can then be selectively hydrocracked into distillate. With this approach, the overall product distribution can be skewed in favor of diesel. The clean fuels are recovered, and the wax is hydrocracked into more diesel fuel and naphtha. The naphtha can be upgraded into gasoline, but substantial refining is necessary to produce high-octane material because of the paraffinic nature of naphtha. The CO_2 in the FT tail gas is removed for storage, and the remaining synthesis gas is returned to the FT reactors for additional liquid production.

The FT process has been used for decades by Sasol and involves reacting synthesis gas over metal-based catalysts to yield a variety of hydrocarbons that can be converted to high-quality transportation fuels (gasoline, diesel, and jet fuel). The first such plant, known as Sasol I, used a combination of fixed-bed and circulating-fluid-bed FT reactors to produce the fuels. Recently, the Sasol I plant changed from coal to natural gas as feedstock, and it is now a gas-to-liquid (GTL) plant. In the early 1980s, Sasol built two large FT-based indirect coal-liquefaction facilities that together produce transportation fuels at over 160,000 bbl/d. The plants were designated Sasol II and III. Twenty years later, the plants are profitable, but they received government subsidies for several years after start-up. They would not have been economically viable in a market economy with relatively cheap oil and without government assistance.

FT synthesis is continuously being improved; since the building of the large Sasol plants, there have been substantial advances both in coal-gasification technologies that produce synthesis gas and in FT technology that produces clean fuels. The Sasol II and III plants originally used circulating-fluid-bed synthol reactors, which were later replaced by fixed-fluid-bed Sasol advanced synthol reactors. These are less expensive, are easier to operate, and have a much greater fuel-production capacity than synthol reactors. Research and development (R&D) at Sasol started experimenting with slurry-phase FT reactors in the early 1980s and built a 2,500-bbl/d prototype reactor at Sasol I to demonstrate and develop the technology. These reactors, which have operated on both iron and cobalt FT catalysts, formed the basis for the huge slurry reactors that have been installed at the Oryx GTL plant in Qatar. The slurry reactors, with a diameter of about 36 ft, are each capable of producing fuels at 17,000 bbl/d.

Other companies are also developing FT reactor technology. Shell has developed the fixed-bed FT process known as the Shell middle-distillate synthesis process. Its GTL plant in Bintulu has been operating since the late 1980s, and recent improvements in the reactors and catalysts have increased the fuel-production rate substantially. ExxonMobil has developed a slurry-bed FT process with a patented cobalt-catalyst system that was the basis of its Qatar GTL plant design. The company withdrew that project from consideration in 2007 in favor of a liquefied-natural-gas plant. Conoco Phillips also has developed a FT system that was demonstrated on a pilot scale in Oklahoma. Syntroleum, another U.S. company, has also developed a somewhat different FT process for its GTL system. It has produced sufficient quantities of FT jet fuel for testing by the U.S. Air Force. The U.S. company Rentech has been developing an FT technology based on a slurry-bed reactor for a number of years and has recently built a pilot facility in Colorado. Other experimental FT systems are under development, including a microchannel reactor being tested by Velocys.

No commercial plant that combines advanced[2] coal gasification with advanced FT technologies has been built. The only operating commercial-scale indirect CTL plants in the world are the Sasol plants. China—a country with increasing consumption of liquid fuels, a scarcity of domestic petroleum, and large coal resources—is moving rapidly toward commercialization of CTL technologies. The Shenhua direct-liquefaction process in Inner Mongolia launched its first trial operation of fuel production in December 2008.

FT synthesis technology can be considered commercially deployable today. Like several other ready-to-deploy technologies, it will undergo substantial process improvement by 2020, which will lead to more robust and efficient technology for producing liquid transportation fuels.

---

[2]Advanced technologies are technologies that are developed or have been improved since the Sasol plants were deployed. Examples of advanced technologies include the use of cobalt catalysts and improved reactor designs.

Methanol Synthesis and Conversion to Gasoline

The other major indirect liquefaction route involves the synthesis of methanol and its conversion to liquid transportation fuels. Methanol synthesis is a large-scale, commercial technology that can be supplied by several license holders and is used commercially to produce methanol from coal. It is well developed, is highly selective, and is used primarily to convert synthesis gas made from natural gas. The largest methanol plants can each produce about 5000 tons/day. Methanol is a feedstock for the manufacture of many chemicals and can be used as a fuel itself. Because of the ubiquity of methanol manufacture (Kung, 1980), that process is not discussed in detail here.

The MTG technology developed by Mobil Oil Corporation was demonstrated in a commercial plant in New Zealand (S. Tabak, ExxonMobil Research and Engineering Company, presentation to the panel on February 19, 2008). MTG technology produces mainly high-octane gasoline. A variant of MTG involves the conversion of methanol to olefins and their conversion to gasoline and diesel fuel and is referred to as MOGD. It has not been demonstrated commercially.

The key to the MTG process was the development of shape-selective zeolite catalysts that produce hydrocarbon molecules in the gasoline size range. The principal product is high-octane gasoline, and the secondary product is LPG. A plant with a capacity of 14,500 bbl/d was started in 1985 in New Zealand. It used natural gas as the feedstock and operated successfully for about 10 years. The drop in crude-oil and gasoline prices at the time resulted in curtailment of gasoline production and conversion of the plant to production of chemical-grade methanol. However, the improvements learned from the commercial operation in New Zealand are being incorporated into a second-generation plant under construction in Shanxi, China, by Jincheng Anthracite Coal Mining Company. The plant will feed coal-derived methanol and was scheduled to start in late 2008. The process uses gas-phase conventional fixed-bed reactors. A coal-to-fuels project in the United States is also planning to use MTG: a small-scale plant is under development by Consol Energy and Synthesis Energy Systems to convert West Virginia coal into gasoline at about 6000 bbl/d with the U-GAS® process followed by MTG. The development of that plant, however, was on hold in 2008 because of unfavorable economic conditions.

Figure 4.3 shows the schematic flow diagram of the New Zealand natural-gas-to-gasoline complex (Tomlinson et al., 1989), which converts methanol to 38.7 percent gasoline, 0.7 percent fuel gas, 4.6 percent LPG, and 56 percent water

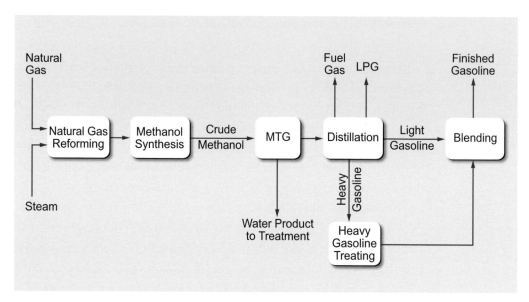

FIGURE 4.3 *Schematic of New Zealand MTG complex.*
Source: Adapted from Tomlinson et al., 1989.

by weight. The water is recycled as process water. Gasoline produced by the process is completely compatible with the conventional gasoline infrastructure, and it contains zero sulfur and is low in benzene (Tomlinson et al., 1989).

The panel considers standard MTG technology to be commercially deployable today, and, as indicated above, several projects are moving toward commercial deployment. Several variations of the technology are ready for commercial demonstration and could provide improvements in the standard MTG technology. They will evolve with commercial application and become more robust and efficient by 2020.

Challenges and Barriers to Deployment

Because the nation has more than 250 billion tons of recoverable coal reserves and because there is a considerable potential to provide large quantities of biomass, there is an opportunity to use the technologies described above to enhance U.S. energy security by producing clean, fungible transportation fuels to supplement the conventional petroleum supply. In spite of the large quantity of coal and continued high oil prices, there were no coal-liquefaction plants in the United

States in 2008, but several potential plants are in the development phase. This section discusses the environmental, economic, commercial, and social barriers to deployment.

The key components of CTL fuel technology have been commercially demonstrated and are ready for commercial deployment. However, from a technical and engineering standpoint, the integration of advanced entrained coal-gasification technologies, an advanced syngas cleanup process, and advanced slurry-phase FT synthesis technologies has never been demonstrated on the scale of a large synthetic liquid-fuel plant. The lack of experience poses a degree of technical risk that would be considered unacceptable by potential process developers and project funders. The panel believes that the technical barriers will be substantially reduced as soon as several first-mover plants become operational. The financial barriers will still be of concern because of the potential high variability of the energy markets. The technology is expected to evolve and improve with commercial experience and to become more robust and efficient.

Because of concerns about global climate change and greenhouse gas emissions, another major technical barrier is the demonstration that captured CO_2 can be stored in geologic formations for extended periods in a safe, effective, and efficient manner. Resolving issues of potential long-term leakage and safety will require an aggressive program to demonstrate geologic storage and to develop data and procedures related to evaluation, permitting, injection, monitoring, and closure. That will also be needed to gain the political and popular support required to make geologic storage ready for multiple commercial deployments. The current status of the technology and the desired work remaining suggest that it will not be commercially deployable on a broad scale before 2015. Ideally, funds and programs for the design, construction, and operation of three commercial demonstrations of geologic storage in different geologic formations focused on gaining the CO_2 storage information outlined above will be available soon. Such programs could be linked to indirect-liquefaction plants that use advanced technologies or coal-based power plants as total commercial demonstration of technologies with integrated carbon capture and storage (CCS). Two of the integrated facilities would be fed by coal of different rank and one by coal and biomass. One or two of the facilities could be operated to demonstrate geologic CO_2 storage independently if integration of generation and storage causes a substantial delay in the demonstration of geologic storage.

Carbon Capture and Storage

The central issue in using coal in a carbon-constrained world is its inherently low ratio of hydrogen to carbon, which results in large CO_2 emissions. Unless the resulting CO_2 is captured during conversion and stored permanently (underground or by incorporating it in some other product), the life-cycle greenhouse gas emissions in converting coal to liquid fuels are about twice as great as those in producing and using fuels based on petroleum (Jaramillo et al., 2008; Bartis et al., 2008). Therefore, use of coal to produce liquid fuels in quantities needed to substitute for transportation fuels will require developing and demonstrating CCS on a large scale, which involves efficient and economic capture of CO_2 and safe and efficient geologic storage. Demonstrating "the technical, economic, and environmental performance of the technologies that make up all of the major components of a large-scale integrated CCS project" (MIT, 2007, p. xi) will take billion-dollar investments by industry and government and could take a decade. Therefore, it is critical to start those demonstrations, with research involving multiple fully integrated monitoring and data-gathering activities, immediately (MIT, 2007). To date, few demonstrations of geologic storage of CO_2 have been carried out on the needed scale. Governments and private companies have been hesitant to make the necessary investment that would ensure that the United States has a robust set of technologies that could be used for its energy future in the absence of any clear CO_2-management policy. Without policy, there are no strong drivers, and the economics are negative.

Carbon dioxide capture is commercially deployable technology. Such processes as Selexol, Rectisol, and amine scrubbing are common in the petroleum and chemical industries. In indirect liquefaction, carbon dioxide is removed from the synthesis gas before production of liquid fuels, so the production of a concentrated stream of CO_2 is an integral part of the processes. The concentrated CO_2 stream can then be dehydrated and compressed for storage. Thus, it is possible to demonstrate any of several coal or coal–biomass processes commercially and to produce a concentrated CO_2 stream for geologic storage with little impact on the cost of the liquid fuels produced.

Several projects are injecting megatonne quantities of CO_2 each year into geologic formations, and no problems have been observed; but none of the projects is in the United States. Demonstrations in the United States that address the issues peculiar to the country and that are well planned and monitored are needed. In Norway, at the Sleipner Field in the North Sea, more than 1 million tonnes of CO_2 per year has been injected into a deep saline aquifer for more than a decade with no identifiable problems. The FutureGen program of the U.S. Department of Energy (DOE) intended to demonstrate CCS, but it was abandoned in 2007 when the estimated price doubled unexpectedly. However, DOE Regional Partnership projects to demonstrate larger-scale CO_2 storage in various geologic formations are under way. Those projects need to be increased to true commercial size and moved forward much more rapidly if geologic storage is to be adequately demonstrated so that it can be used to store CO_2 captured from coal plants in the needed timeframe. The government of the United Kingdom just announced a plan to carry out a demonstration on its soil, according to news accounts (Lovell, 2008), but has not taken steps beyond the announcement.

In the technical scenarios that are compared earlier in this chapter, the panel assumes that CO_2 capture uses state-of-the-art technology, such as would be used in conventional refining and IGCC power plants. Such processes as Selexol, Rectisol, and amine scrubbing would be used. The processes considered produce a concentrated stream of CO_2 as an integral part, so CO_2 storage can be readily and more cheaply achievable. CO_2 transport by pipeline is a well-demonstrated technology; such CO_2 is used in enhanced oil-recovery (EOR) operations at about 35 million tons/year. Pipelining of CO_2 poses no technical issues, but permitting issues are associated with obtaining rights-of-way, as is the case with most infrastructure projects. However, the technical and legal issues associated with the storage of captured CO_2 still need to be clarified and resolved.

The estimates of potential costs of CCS in an earlier section are "bottom-up" and are based largely on engineering estimates of expense for transport, land purchase, permitting, drilling, all required capital equipment, storing, capping wells, and monitoring for an additional 50 years. However, experience suggests that the full cost of storage might not be captured by such an approach in light of barriers to implementation that increase cost. Uncertainties in the regulatory environment arising from concerns of the general public and policy makers are likely to evolve under the influence of future events (Palmgren et al., 2004). It is difficult to estimate such costs without some commercial-scale geologic storage experience, as outlined above. A reliable estimate of future cost of storage would contain, at least qualitatively, the uncertainty arising from such factors. Accordingly, quantified costs based on engineering analysis would probably represent a lower bound on future costs. (See Appendix K for a more detailed discussion.)

If liquid fuels produced by thermochemical conversion of coal or coal and biomass with CCS are to meet a sizable portion of U.S. demand for transportation fuels, more than a gigatonne of CO_2 captured from such processes would have to be stored each year. CO_2 capture and transport entail a potential *health risk* associated with acute leaks and with exposure of workers or populations to hazardous concentrations of CO_2 near facilities. Geologic storage has the potential of an *ecological risk* to soils and groundwater as a result of chronic leakage and a *warming risk* associated with either sudden or chronic leaks that might partially or entirely vitiate the climatic value of a storage site (Anderson and Newell, 2004; Socolow, 2005). The public and policy makers are likely to anticipate those risks and require that they be taken into account in the design, monitoring, and carbon-accounting procedures and associated regulatory frameworks that would be part and parcel to storage (Wilson et al., 2007). As a result, timing estimates need to recognize the potential for delay in initiating demonstration projects because of lags in conception and development of the overall regulatory regime for storage and in licensing of each specific project, both in the demonstration phase and beyond. Some issues, such as liability insurance for near-term operation and long-term site maintenance, require political resolution that could introduce additional delays (IRGC, 2008). Uncertainty over the likelihood of long-term leaks could translate into regulations that require the sources that plan to store carbon to purchase allowances equivalent to fractions of the carbon stored. Such a requirement would increase the net cost of CCS. All those issues need to be evaluated as part of the several geologic-storage demonstration projects mentioned above to provide the best information for the evaluation of future commercial activities.

Once CCS attains full commercial-scale operation, delays could arise because of accidents that cause or threaten releases. Because the technologies, monitoring, and regulation of storage are likely to be closely related, if not identical among sites, interruption of operations at one site could affect operations at other sites; broadly reduce or temporarily eliminate storage; undermine the credibility of the technology among investors, regulators, policy makers, and the general population; and add a substantial risk premium to investment in CCS. Continuous storage operations might be subject to multiple regulatory regimes (and varied siting, licensing, and monitoring requirements) at various government levels. Those issues and potential causes of delay apply to other major commercial operations, including production, pipelining, and refining of crude oil.

CO_2 is being used for EOR at the Weyburn oil field in Canada; CO_2 from the Great Plains lignite gasification plant is used at almost 1 million tons/year. Statoil has been successfully injecting CO_2 from the Sleipner gas field into the Utsira Formation, a deep saline aquifer, at more than 1 million tonnes/year for over a decade.

CO_2 is also being reinjected at the Salah liquefied-natural-gas (LNG) project in Algeria at about 1 million tonnes/year. There has been no indication of problems arising from any of those projects, and the CO_2 storage shows no sign of leakage.

EOR can present an opportunity for early CCS and can reduce the cost of CCS by providing a net return. Use of CO_2 in EOR has been safe and has not raised any questions about the ability to store CO_2 in proper geological formations safely over the long term. EOR in the United States uses CO_2 at 35–40 million tonnes/year. There are opportunities for additional EOR, but those storages are small compared with the large amounts of CO_2 that would be captured if CTL becomes widely deployed, potentially in the gigatonnes-per-year range. CO_2 could be stored in deep coal seams, where it can displace methane for use in the natural-gas pipeline. CO_2 binds more strongly to coal than does methane and thus replaces it; and just as the methane is permanently locked in the coal seam for extremely long times, the CO_2 will be permanently stored there. Again, however, the use of CO_2 in coal-bed displacement is small in relation to the total amounts that need to be stored.

With adequate demonstration and long-term monitoring, CCS could offer a way to use the nation's wealth of fossil fuel while limiting adverse effects on climate. What is now needed is aggressive demonstration on a commercial scale in several U.S. geologic formations to develop the needed data and to understand and resolve issues.

Supply of Feedstock

Deployment of such facilities will require the use of large quantities of coal and thus an expansion of the coal-mining industry. For example, a 50,000-bbl/d plant will use about 7 million tons of coal per year, and 100 such plants producing liquid transportation fuels at 5 million bbl/yr would require about 700 million tons of coal per year—a 70 percent increase in coal consumption. That would require major increases in coal mining and transportation infrastructure to move coal to the plants and fuel from the plants to the market. Those issues could pose major challenges, but they could be overcome.

The next question is whether sufficient coal is available in the United States to support such increased use. The National Research Council evaluated domestic coal resources (NRC, 2007) and concluded:

> Federal policy makers require accurate and complete estimates of national coal reserves to formulate coherent national energy policies. Despite significant uncertainties in existing

reserve estimates, it is clear that there is sufficient coal at current rates of production to meet anticipated needs through 2030. Further into the future, there is probably sufficient coal to meet the nation's needs for more than 100 years at current rates of consumption. . . . A combination of increased rates of production with more detailed reserve analyses that take into account location, quality, recoverability, and transportation issues may substantially reduce the number of years of supply. Future policy will continue to be developed in the absence of accurate estimates until more detailed reserve analyses—which take into account the full suite of geographical, geological, economic, legal, and environmental characteristics—are completed.

The Energy Information Administration (EIA) recently estimated the proven U.S. coal reserves to be about 260 billion tons (EIA, 2009). A key conclusion of the NRC and EIA studies is that there are sufficient coal reserves in the United States to meet the nation's needs for more than 100 years at current rates of consumption. Even with increased rates of consumption, ramped up over time, the reserves could support our needs for 100 years. The primary issue probably is not the reserves but the increase in mining of coal and the opening of many new mines. Increased mining has numerous environmental effects that will need to be dealt with in an environmentally acceptable way. Public opposition to increased coal mining is to be expected because of the need to open new mines and the environmental implications of mining more coal. Increasing coal use will undoubtedly increase the cost of coal, but coal costs are relatively low, and substantial amounts of coal can probably be produced at current or slightly higher prices.

A particular barrier to the establishment of biomass-to-liquids (BTL) plants is the availability of sufficient quantities of feedstock in a reasonable area. Because only small quantities of biomass (3000 tons/day) can be gathered, such plants will be limited in size by feedstock constraints. That leads to small-scale plants and hence diseconomies of scale and high capital cost. Another challenge is the successful feeding of raw biomass to high-pressure gasification systems. Biomass, unlike coal, is soft and fibrous and difficult to reduce to the small sizes necessary for gasification. A third challenge is to reduce the high costs of biomass feed, including the costs of growing, harvesting, and transportation to the conversion plant. Biomass has very low energy density when raw, so transportation costs are high compared with the cost of coal, which is high in energy density.

Efforts to increase the energy density of raw biomass by pyrolysis are under way. Lurgi and Air Liquide have an interesting concept for conversion of low-energy-density biomass to liquid fuels. Biomass, such as switchgrass or woody biomass, is pyrolized in a double-screw retort with hot sand as the heat-transfer medium. The biomass degrades to form pyrolysis oil and char. The pyrolysis oil and char are mixed together to form a "bio-syncrude," which has an energy density 13 times that of the unprocessed biomass and contains 80 percent of the energy in the biomass. The bio-syncrude can be readily transported and fed to the Lurgi multipurpose gasification (MPG) process or other gasification processes to produce syngas, which can then be cleaned and used to synthesize liquid transportation fuels by FT. The concept appears to overcome the problems of transporting low-energy-density raw biomass and feeding raw or pretreated biomass to high-pressure gasification. The initial pyrolysis could conceivably be done on a field scale, and the high-energy-density bio-syncrude could be shipped to a central gasification facility for production of transportation fuels.

Economics and Investment

The uncertainty of future oil prices is an important barrier to deployment of CTL, as is the high capital expenditure needed for commercial CTL plants. A 50,000-bbl/d plant could cost $4–5 billion, so the plants could be expected to approach $100,000 per daily barrel, which is about 6 times as high as deepwater Gulf of Mexico crude-oil capital investment costs. The investment risk for such a large expenditure is considerable. In that context, it should be noted again that biorefineries for converting cellulosic biomass to ethanol have an estimated capital cost of about $120,000 per daily barrel of gasoline equivalent and that about 30 biorefineries (with production capacity of 40 million gallons of ethanol per year) are required for a 50,000-bbl/d output.

Infrastructure and Labor

If many plants are built worldwide at the same time, there will be competition for critical process equipment and engineering and labor skills. On the basis of parallels with the indirect-liquefaction industry, the timeline for commercial deployment in the United States would be long. Permitting and the usual public reluctance to accept the need for new facilities, especially coal-based plants, are issues. The proposed FT plant for conversion of anthracite residue to clean diesel fuel, to be built in Gilberton, Pennsylvania, has been in gestation for 12 years, and construction apparently has yet to begin. The Dakota gasification plant (which produces substitute natural gas from lignite) in Beulah, North Dakota, was originally proposed in the late 1960s and came on line in the early 1980s—a time span of some 12–15 years. Even if permitting and other legal issues do not impose a delay, it would still take at least 6 years to construct an indirect-liquefaction plant.

For example, the Sasol II and III complexes in South Africa required 6 years to construct from the time the South African government approved the plans.

Technology Forecast

Technologies Deployable in 2008–2020

The discussion above is related to technologies that are deployable now or potentially deployable in the near future. CTL plants that use gasification followed by FT or MTG synthesis can be built today. Although integration of advanced entrained coal gasification with FT has not been commercially demonstrated, the technical risk associated with such a venture is low because of the separate experience with commercial gasification for other applications and commercial use of FT in CTL and GTL processes. Because of the challenges listed above and the long lead time required for planning, detailed design, permitting, and construction, it is unlikely that any CTL plants will be in commercial operation in the United States before the 2015–2020 timeframe. CTL plants with CCS will probably take longer to be commercialized because of the need for commercial demonstration of carbon dioxide storage and monitoring before it can be applied broadly in commercial operation.

With some additional R&D focused on biomass pretreatment and feeding to gasification reactors, CBTL plants that coprocess small amounts of biomass (up to 30 weight percent) could be deployed today. Their rate of deployment would be subject to the same restrictions as the rate of deployment of CTL plants, and there is the additional issue of biomass availability and suitable plant site location. With the benefit of successful biomass pretreatment, small-scale thermochemical BTL plants using current biomass gasification and FT or MTG technology could also be deployed today.

With respect to deployment of future technologies, the panel's review of the thermochemical-conversion technologies has separated them into two groups: those likely to be deployable in 2020–2035 and those requiring longer-term R&D.

Technologies Likely to Be Deployable in 2020–2035

Continued advances in both coal and biomass gasification technologies after 2020 are likely. For example, Pratt and Whitney Rocketdyne is developing a compact gasifier based on rocket-engine technology that, if proved successful, could reduce costs and improve efficiencies. The production of synthesis gas is the most capital-intensive section of a thermochemical conversion plant, so cost reductions in that component would greatly improve overall economics.

As long as industrial interest in alternative fuels continues, the synthesis process—whether FT, MTG, or MOGD—is likely to undergo continuing improvement. For example, Velocys is developing a microchannel FT process that could improve synthesis gas conversion and reduce costs. With continued emphasis on climate change, successful demonstration and practice of CCS is likely to be attained, greatly accelerating the ability to deploy thermochemical fuel plants with safe CCS.

Another option for the conversion of syngas to liquids, other than FT, is catalysis (Chu et al., 1995; Herman, 2000). Syngas can be converted catalytically through the chain-growing process to such higher alcohols as isobutanol in a slurry-phase reactor. Better catalysts and reactor design are needed to improve the yield and selectivity of the catalytic conversion of syngas to higher alcohols (Herman, 2000; Li et al., 2005). The development and deployment of improved syngas cleanup, including reduction of hydrogen sulfide to parts-per-billion concentrations, are required to minimize catalyst poisoning. Then, the technology needs to be demonstrated on a semi-work scale for commercial deployment.

A novel approach that has potential for commercialization is chemical-looping gasification. In that process, a metal oxide is used as an oxygen carrier and is itself reduced to metal. The metal can then be reacted with steam to produce hydrogen and/or carbon monoxide, which can then be used to produce liquid fuels, chemicals, and electricity (Fan and Iyer, 2006; Fan and Li, 2007; Gupta et al., 2007). An example is the syngas chemical-looping process that has the potential to convert coal to hydrogen at 7–10 percent higher efficiency than conventional coal-to-hydrogen processes (Gupta et al., 2007). Furthermore, the syngas chemical-looping scheme can be integrated into the conventional CTL process (Gupta et al., 2007; Tomlinson and Gray, 2007), allowing the by-products of liquid-fuel synthesis to be converted to hydrogen. Such integration can lead to a 10 percent increase in liquid-fuel yield and a 19 percent decrease in carbon emission (Tomlinson and Gray, 2007). The full operability of the new process needs to be tested on a pilot scale. The feasibility of the technology will then have to be shown in a demonstration plant for later commercial deployment.

Combining technologies in a plant could result in improvements in the product slate, reductions in greenhouse gas emissions, or other benefits compared with a plant that uses a single technology. For example, it is well known that indirect liquefaction with FT produces an excellent high-cetane diesel fuel, but FT naphtha is not well suited for gasoline. In contrast, the MTG process produces a high-octane gasoline with very high selectivity. Therefore, one might envision a plant in which syngas is split between FT and MTG to obtain the best of both: high-quality gasoline *and* diesel. Another example is potential reduction in greenhouse gas emissions through use of nuclear process heat as the source of process heat for thermochemical conversion of coal, biomass, or combined coal and biomass. Coupling a nuclear power plant with a synthetic-liquid-fuel facility could have, as one benefit, the elimination of greenhouse gas emissions from furnaces and other heaters throughout the synthetic-fuel production side of the plant.

Technologies Likely to Be Deployable After 2035

Technologies presented in this section are ones for which substantial R&D effort is still needed, but they could potentially provide drastic improvement to thermochemical conversion. Those technologies will probably not be realized until after 2035. Despite many apparent differences among process strategies, virtually all processes for thermochemical conversion of biomass and coal have several characteristics in common. They rely on the thermal breakdown of the feedstock (typically at 350°C or above) to produce a population of free-radical intermediates that undergo a complex sequence of reactions, they tend to produce a mixture of products rather than showing high selectivity to a single desired product, and they yield 2 ± 0.5 bbl of liquid product per ton of feedstock.

Technological developments that are beyond incremental improvements will probably have to be based on different ways of breaking apart the macromolecules in the feedstock rather than relying on thermally driven bond-breaking. There are several potential developments. One is changing the reaction intermediates from radicals to positively charged carbon atoms (carbocations); this could be done with Lewis acid catalysts, for example. Consolidation Coal Company has investigated the direct liquefaction of coal in molten zinc chloride and reported high selectivities to gasoline. A second is enzymatic bond cleavage with fungi, bacteria, or other organisms engineered to have enzymes with high activity and selectivity for cleavage of particular kinds of bonds. A third is the application of energy to cleave bonds in much more targeted fashion with, for example, microwave heating or ultrafast (femtosecond) lasers tuned to specific bonds.

A major step forward will need to be based on a thorough understanding of the molecular structures of the feedstocks and of the specific kinds of bonds to be broken. The molecular features of coal, in particular, are not well understood and are thought to vary from one kind of coal to another.

Pennsylvania State University has developed two approaches for introducing coal or coal extracts into oil refineries (Clifford and Schobert, 2007). One involves extraction of coal with a petroleum solvent, such as light-cycle oil, followed by two-stage hydrotreating of the extract mixture. Fractionation after hydrotreating provides mainly clean jet fuel and diesel as products and smaller amounts of gasoline and heating oil. The second approach blends coal with the feed to delayed cokers. The coker liquid is mainly in the fuel-oil range with smaller amounts of lighter distillates. The university has licensed the technology to CoalStar Industries, Johnstown, Pennsylvania. CoalStar Industries is planning to build a 10,000-bbl/d demonstration and is in the final stage of selecting a site for the plant, which will probably be in southwestern Pennsylvania (D. Fyock, CoalStar Industries, personal communication, November 6, 2008).

Research, Development, and Demonstration

If the goal is to increase production of domestic liquid transportation fuels in the next several decades to enhance energy security, it is important to rapidly advance technologies that are commercially deployable today if their economics justify the deployment. Those first movers would need to have an associated applied R&D program to ensure success and to develop learning. An R&D program that addresses step-out technology improvements and developments and that develops new technologies also needs to be supported. Engaging in a new research program on the assumption that it will provide energy solutions in the near to middle term is unwise.

For thermochemical technologies that are deployable now, the financing hurdle remains serious primarily because of the volatility of the energy markets; but deployment is also affected by uncertainties in climate-change policy and by lack of full-scale commercial demonstration. The energy market's uncertainty is illustrated by the price of crude oil over the last 3 years and its decrease from a high of $147/bbl to a low of $32/bbl in 5 months. The projects in question have a multiyear timeline from planning to operation, and they require capital of $1–5 billion or more. They face what has often been referred to as a valley of death in getting from development and demonstration to commercial deployment. Reaching commercial deployment will probably require a number of commercial first-mover projects combined with geologic storage of CO_2 to gain commercial experience and to move the technology to robustness and to substantial cost reductions for the Nth plant, where N is a small number. The commercial first-mover projects would include a major R&D component to focus on solving problems and to develop technology for specific improvements. That would improve the technologies, quantify their relative costs, and reduce the risk associated with their commercial deployment if they show economic competitiveness. The panel considers this phase critically important for facilitating commercial deployment of thermochemical technologies.

An R&D program should be associated with commercial-scale demonstrations of geologic CO_2 storage. The demonstrations need to involve detailed geologic research and a broad array of monitoring tools and techniques before initiation, as they proceed, and after they are closed to provide the understanding and data on which future commercial projects will depend. Because of the scale of geologic storage, research and monitoring need to be continued at a steady rate, after the demonstration projects are declared completed. Increased research efforts on the coal-mining end of the value chain are also warranted to improve understanding of the immediate and longer-term environmental effects of increased coal mining and use.

On the gasification and gas-treatment side, the current research program focuses on broadly applicable improvements. Continuation of that program would provide improved coal pumps, ion-transport membranes for oxygen separation, membranes for other separations, and various other technology improvements.

New catalysts and catalytic routes to liquid transportation fuels need continued study because those step-out technologies offer much potential. Likewise, new reactor concepts or separation concepts offer much potential. As new ideas come along, they need to be evaluated and their economic potential analyzed. The section "Technologies Likely to Be Deployable After 2035" above contains a number of new process concepts that require focused R&D. The ones that meet needs can be advanced to the process-demonstration stage to obtain data for evaluating commercial potential.

Costs and Performance

Between now and 2020, technologies for the thermochemical conversion of coal, biomass, and coal–biomass mixtures by gasification followed by FT synthesis or methanol synthesis followed by an MTG process will probably be commercially deployed in the United States and in other countries that have large coal resources, such as China, Russia, India, and Australia. To reduce the CO_2 footprint of CTL plants, CCS technologies will have to be used. Capture of CO_2 from CTL plants uses the same state-of-the-art technology used in conventional refining, natural-gas processing, and IGCC facilities—for example, Selexol, Rectisol, and amine scrubbing. CTL plant configurations produce a concentrated stream of CO_2 as an integral part of the process, so CO_2 capture can be readily and more cheaply achievable than, for example, in IGCC or pulverized-coal plants. The higher cost of CO_2 avoided[3] with IGCC is a result of the fact that an IGCC plant without CCS would use a different configuration from one with CCS. An IGCC plant without CCS does not have water–gas shift and does not separate CO_2 in the gasification–purification train. In contrast, water–gas shift and CO_2-separation equipment has to be included in an IGCC plant that practices CCS, and this increases the cost of the plant. The higher cost and added energy use of an IGCC plant with CCS results in a much higher cost of CO_2 avoided. In contrast, the only difference between CTL plants that vent CO_2 and CTL plants that use CCS is the need to dehydrate and compress the concentrated CO_2 stream that would otherwise be vented.

Because there are no thermochemical-conversion plants in the United States, this section provides a detailed technical and economic analysis of conceptual plants simulated with Aspen Plus software. Both indirect- and direct-liquefaction models have been developed.

To evaluate the commercial potential of coal conversion to liquid transportation fuels, the panel carried out a series of evaluations of various conversion processes and options. They all used a consistent capital-cost basis and the same set of economic and operational parameters.[4] Thus, the relative costs of fuels produced with different processes and among different options for a given process are quite accurate, although substantial uncertainty may be associated with the absolute cost. Details of this approach and the capital-cost basis and the economic and operational parameters used are given elsewhere (Kreutz et al., 2008).

Indirect-liquefaction models include CTL, BTL, CBTL, and combined electric-power and fuel generation (polygeneration). To keep the extent of work

---

[3]Cost of CO_2 equivalent avoided is estimated as {[levelized FT liquid product cost (in $/GJ) for CCS design] – [levelized FT liquid product cost for vent design]}/{[greenhouse gas emissions (in tonnes CO_2 eq per GJ of liquid product) for vent design] – [greenhouse gas emissions per GJ of product for CCS design]}.

[4]Key economic and operating parameters used in all of the analyses include the middle of 2007 as the base year for capital-cost estimates (Gulf Coast), a 14.4 percent capital charge rate based on the total plant cost per year, 7.16 percent of total plant cost as interest charged during construction (3 years), 4 percent of total plant cost as the operation and maintenance cost per year, and a 90 percent capacity factor.

and the number of cases evaluated within reason, a number of parameters were fixed, such as gasifier type, coal type, and location. For example, the analyses were all based on a Texaco–GE entrained-flow gasifier and Illinois no. 6 coal.[5] Equipment capital costs were from recent detailed design studies and were updated to 2007 dollars on the basis of the *Chemical Engineering Plant Construction Cost Index*. Plant design involved material-balance and energy-balance calculations with Aspen Plus. Most of the studies were based on a synthesis-process configuration that involved recycling of unconverted synthesis gas leaving the reactor back to the reactor to achieve maximum synthesis of hydrocarbons. The configuration will be referred to as recycling. The recycling cases included designs both with and without CCS. The designs involved generation of power from fuel-gas streams for use in the plant, and excess power was sold to the grid. Some of the designs involved passage of synthesis gas through the synthesis reactor without recycling of the unconverted fraction and with generation of power from the unconverted gases and are referred to as once-through cases. Those cases typically produced large quantities of power. They also included designs with and without CCS. The costs and performance estimates cited here correspond to those in a workbook that is available at http://cmi.princeton.edu/NRC_AEF_workbook.

Coal to Liquid Fuels

Table 4.1 summarizes the results of the analysis of conceptual CTL plants operating in the recycle mode with and without CCS (Kreutz et al., 2008; Larson et al., 2008). Each column shows the performance, cost, and greenhouse gas life-cycle emissions for the indicated process configuration. Figure 4.4 shows the plant configuration with the main process units indicated for diesel and gasoline production using FT synthesis. The plant has FT reactor tail-gas recycling and venting of the CO_2 recovered from the synthesis gas to the atmosphere. In this configuration, an autothermal reformer is used to convert the light hydrocarbon gases produced during synthesis back into synthesis gas, which is then sent to the FT unit for further conversion into liquid fuels. The paraffinic diesel and the higher-range material made require additional refining to produce high-quality diesel and jet fuel. The naphtha-range material has a low octane number and thus requires substan-

---

[5]Key coal properties of Illinois no. 6 (as received) include 44.2 percent carbon, 9.7 percent ash, 11.1 percent moisture, 25.9 MJ/kg (lower heating value), 27.1 MJ/kg (higher heating value), and $1.71/GJ.

TABLE 4.1 Coal to Liquid Fuels by Fischer-Tropsch and Methanol to Gasoline Conversion Routes With and Without Carbon Capture and Storage

	CTL FT Recycling Without CCS	CTL FT Recycling With CCS	CTL MTG Recycling Without CCS	CTL MTG Recycling With CCS
Inputs:				
Coal, tons/day (as received)	26,700	26,700	22,900	23,200
Outputs:				
Diesel, bbl/d	28,700	28,700	0	0
Gasoline, bbl/d	21,290	21,290	50,000	50,000
Total liquid fuels, bbl/d	50,000	50,000	50,000	50,000
Efficiency, percent (low heating value)	49.1	47.6	54.2	52.9
Electricity, MW_e	427	317	145	111
CO_2 vented at the plant, tonnes/hr	1,427	209	1,200	230
CO_2 stored, tonnes/hr	0	1,217	0	970
Economics and metrics:				
Total plant cost (TPC), millions of dollars	4,880	4,950	3,940	4,020
Specific TPC, $/bbl per day	97,600	98,900	78,800	80,400
Total liquid fuels cost,[a] $/gal gasoline equivalent	1.50	1.64	1.47	1.57
Break-even oil price,[b] $/bbl	56	68	47	51
Life-cycle GHG emissions, kg CO_2 eq/GJ (low heating value)	205	98	192	109
FT liquids per petroleum-derived diesel emissions	2.23	1.07	2.09	1.18
Cost of avoided CO_2, $/tonne	Not applicable	11	Not applicable	10
Fuel cost:				
With $10/tonne CO_2, $/gal gasoline equivalent	1.71	1.74	1.69	1.69
With $50/tonne CO_2, $/gal gasoline equivalent	2.58	2.12	2.52	2.18
With $100/tonne CO_2, $/gal gasoline equivalent	3.67	2.60	3.66	2.79

Note: Details of models can be found in Kreutz et al. (2008) and Larson et al. (2008).

[a]For simplicity and consistency, the panel assumed that electricity was sold to the grid at the average 2007 generating price in the United States, which was $60/MW with $0/tonne of CO_2 charged. All table entries have that basis. If the value of the electricity is set at $80/MW, the total liquid-fuels cost decreases from $1.50/gal gasoline equivalent to $1.41/gal gasoline equivalent for CTL FT venting and from $1.64/gal gasoline equivalent to $1.58/gal gasoline equivalent for the CO_2 storage version. For $50/tonne of CO_2, the fuel cost decreases by $0.90 for venting and by $0.36 for CO_2 storage.

[b]The break-even crude-oil price is defined as the price of crude oil in dollars per barrel at which the wholesale prices of petroleum-derived products would equal (on a dollars-per-gigajoule basis) the calculated cost of production of the synthetic fuels. See Kreutz et al. (2008) for a detailed definition.

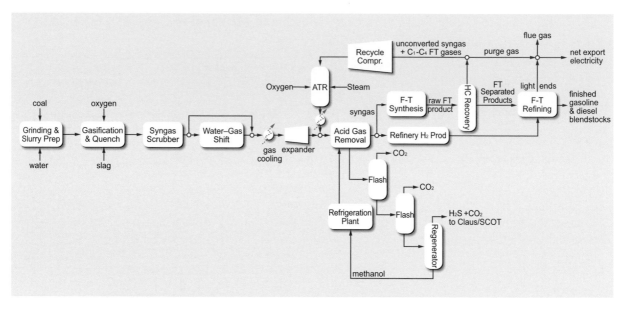

FIGURE 4.4 *Schematic of plant for production of diesel and gasoline from coal with FT synthesis, recycling of unconverted syngas, and reforming of light hydrocarbons; separated CO_2 is vented to the atmosphere.*

tial refining to produce high-octane gasoline. The estimates in Table 4.1 include the cost of upgrading to fuel products. Details of the Aspen Plus modeling and other aspects of the analysis are presented by Kreutz et al. (2008).

That commercial-scale conceptual plant produces gasoline and diesel at 50,000 bbl/d from 26,700 tons of as-received bituminous coal per day. That yields a ratio of 1.9 barrels (80 gal) per ton of coal and an overall plant efficiency of 49 percent (on the basis of the lower heating value [LHV]). The plant generates 874 MW of electric power; 447 MW are needed on site, and 427 MW are sold to the grid. In this configuration (Figure 4.4), the CO_2 produced during the conversion process, amounting to 1427 tonnes per hour, is vented to the atmosphere. The CTL plant with CCS takes advantage of the higher pressure of the CO_2 coming off the acid-gas removal flashes to minimize the compression-power requirements but still consumes more than 100 MW in compression-power consumption, reducing the plant export of power to the grid to 317 MW of electricity.

Figure 4.5 shows the schematic of a coal-to-gasoline plant that uses methanol synthesis followed by MTG. The plant uses the same equipment at the same size from coal storage up to the front of the synthesis loop as the FT plant. Because

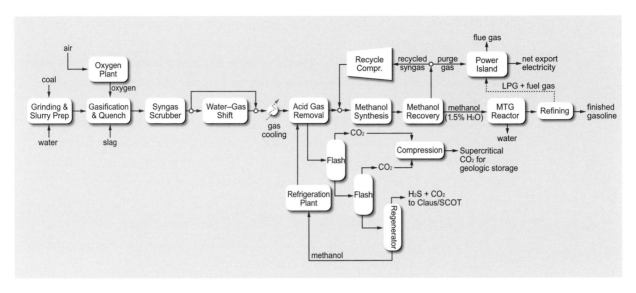

FIGURE 4.5 *Schematic of plant for production of gasoline from coal with methanol synthesis followed by MTG process with recycling and CCS.*

of the higher selectivity of the methanol synthesis and MTG conversion, the remainder of the plant is less complex than the FT plant. The plant vents the CO_2 separated from synthesis gas to the atmosphere and uses recycling of unconverted gases around the methanol-synthesis reactor. To be consistent with the FT plants producing only liquid transportation fuels and power, the LPG produced in the MTG process was burned to produce electricity in the power island. The plant produces gasoline at 50,000 bbl/d from 22,900 tons of as-received subbituminous coal per day. The MTG scheme yields 2.2 bbl of gasoline per ton of coal and an LHV plant efficiency of 54 percent. The somewhat higher liquid yields occur because the methanol syntheses and MTG are more selective in their conversion efficiency and gasoline is less dense than diesel fuel. The plant generates an estimated 440 MW_e and sells an estimated 145 MW_e to the grid. Good engineering data on the MTG portion of the plant are lacking, and the estimates of generated power need to be refined as better data become available. Higher plant efficiency occurs because the MTG plant produces less electricity, which has a lower efficiency of production. The plant vents CO_2 at about 1,200 tonnes/hr.

To estimate the total life-cycle greenhouse gas emissions from these processes (from coal mine to wheels), it is necessary to estimate the total emissions resulting from the mining and the transportation of the coal from the mine to the plant, including methane emissions from mining, emissions associated with fuel distribution from the conversion plant to the end user, emissions due to conversion processes at the plant, and emissions resulting from the combustion of the fuels produced.[6] Because the plants produce excess power, a greenhouse gas credit is given for power production on the basis of the greenhouse gas emissions associated with an IGCC plant that generates the same amount of power and has no CCS.

This carbon-accounting method estimates the life-cycle emission for the venting CTL FT case to be 205 kg CO_2 eq/GJ (LHV) of produced fuels (about 1 ton of CO_2 per barrel of product) and about 192 kg of CO_2 eq/GJ (LHV) of produced fuels for the coal-to-methanol-to-MTG case. For production of the fuels from conventional petroleum, the greenhouse gas life-cycle emission is estimated from Argonne National Laboratory's Greenhouse Gases, Regulated Emissions, and Energy Use in Transportation (GREET) model to be about 92 kg CO_2 eq/GJ. Therefore, the life-cycle emission is about 2.2 times that of fuels produced from petroleum.

The capital cost (total plant cost) of the FT plant without CCS (first column in Table 4.1) is estimated to be $4.9 billion. That is equivalent to a capital cost on a daily-barrel basis of $97,600. For the consistent economic parameters used in this report, a coal price of $42/ton ($1.71/GJ), and an electric-power value of $60/MWh, the resulting cost of the fuels would be $1.50/gal gasoline equivalent. In terms of a break-even oil price, that translates to $56/bbl (see Table 4.1 footnote for definition). If electricity is valued at $80/MW rather than $60/MW, the fuel-production cost decreases by $0.09/gal gasoline equivalent to $1.41/gal gasoline equivalent, and the decrease remains $0.09/gal gasoline equivalent for the several CO_2 cost entries in the table. The total plant cost of the MTG plant is estimated at $3.9 billion (third column of Table 4.1); on a daily-barrel basis, the capital cost is $78,800 per stream-day barrel (SDB). That is lower than the cost of the FT plant because of the somewhat higher complexity of the FT process and the larger refining requirement to produce fuels that meet the product specifications. The resulting cost of the high-octane gasoline produced is estimated at $1.47/gal, which equates to a break-even oil price of about $47/bbl. The impact of $80/MW

---

[6]Nonconversion-plant greenhouse gas emissions were estimated with the Argonne National Laboratory's Greenhouse Gases, Regulated Emissions, and Energy Use in Transportation (GREET) model version 1.8. Production, transportation, and refinery greenhouse gas emissions for petroleum-derived fuels were also estimated with GREET (Argonne National Laboratory, 2005).

versus $60/MW is a $0.04/gal gasoline equivalent reduction in fuel cost because an MTG plant sells less electricity than an FT plant does. For MTG, if LPG is sold at the current market price, the cost of fuel production decreases by about $0.20/gal gasoline equivalent to $1.26/gal gasoline equivalent. Those costs for the production of liquid transportation fuels from coal are comparable to the costs in a report by the RAND Corporation (Bartis et al., 2008).

The economic results shown above are for cases in which there is no tax on CO_2. If CO_2 were to be taxed in the future so that a plant operator had to pay to emit CO_2 to the atmosphere, the economic situation could change substantially. If the tax imposed on CO_2 were $100/tonne, the cost of fuel from this coal-based FT plant would increase from $1.50/gal gasoline equivalent to $3.67/gal gasoline equivalent. The MTG plant would see a similar impact on the cost of the gasoline produced and an increase from $1.47/gal gasoline equivalent to $3.66/gasoline equivalent.

The second and fourth columns in Table 4.1 summarize the results for the conceptual FT and methanol-to-MTG plants with recycling and with CCS. In this case for FT, bituminous coal at 26,700 tons/day produces liquid fuels at 50,000 bbl/d. Overall plant efficiency is reduced slightly, from 49 to 48 percent, in this case because of the need to compress and dry the captured CO_2 to 2,100 psi for pipelining and geologic storage. About 85 percent of the CO_2 produced during the conversion process is captured, and only 209 tonnes/hr are emitted to the atmosphere. Although the fuel output is the same with and without CCS, the net electric-power generation is reduced to power the compressors for the captured CO_2. A greater percentage of the CO_2 produced in the conversion process could be captured by changing the overall configuration to include, in addition to an autothermal reformer, additional water–gas shift and CO_2 capture facilities to produce more H_2, which could be used as fuel in the gas turbine in the combined-cycle power island. The same comments apply to the methanol-to-MTG case.

With the same method as in the previous case (except that the electric-power greenhouse gas credit is now based on an IGCC plant with CCS), the greenhouse gas life-cycle emission is estimated to be reduced to 98 kg of CO_2 eq/GJ for the FT unit producing liquid fuels. The ratio of the greenhouse gas life-cycle emissions for the FT liquids to that for petroleum-derived diesel is 1.1, which means FT liquids essentially have the same greenhouse gas life-cycle emissions as petroleum-derived fuels. The ratio must be interpreted carefully. The assignment of the greenhouse gas emissions associated with the generation of the excess electric power is somewhat arbitrary and depends on the power-generation technology that is displaced at the margin. It could be CO_2-free if power from nuclear energy were displaced, or the greenhouse gas emission credits could be high if power from conventional pulverized-coal plants were displaced. The method used in these analyses assumes that the reference plants are IGCC with no CCS for comparisons with venting cases and IGCC with CCS for cases with CCS. Thus, a consistent basis is used for assessing the greenhouse gas credit given to the excess power generated by these CTL plants. In addition, the assignment of the greenhouse gas life-cycle emissions for the production of low-sulfur diesel from conventional petroleum is arbitrary. There can be no single value for it. Crude oil varies in composition and in its ability to be refined, and refineries have different efficiencies and use a wide variety of refining processes. So, at best, the life-cycle emission can be only an approximate average. As a result of those uncertainties, a ratio of greenhouse gas emissions for coal-derived liquid fuels to emissions for petroleum-derived fuels of around 1 implies that the greenhouse gas emissions for the overall cycle can be about the same as or less than that for petroleum. In addition, if more CO_2 were captured in the coal-to-fuels conversion process by changing the process configuration, the life-cycle emission for the coal-to-fuels process could be further reduced to less than that of petroleum-derived fuels.

The capital cost of the FT plant with CCS is estimated to be about $5 billion, which is equivalent to a capital cost on a daily-barrel basis of just under $100,000. For the consistent economic parameters used in this analysis, the resulting cost of the fuels would be increased from $1.50/gal gasoline equivalent in the venting case to $1.64/gal gasoline equivalent in the CCS case. The cost of CO_2 avoided by this configuration is about $11/tonne. Those economic results are for a case in which there is no tax on CO_2 or equivalent shadow price reflecting cap-and-trade emission cost. If a carbon price of $100/tonne of CO_2 were imposed on fuels, the cost of fuel from this plant would increase from $1.64/gal gasoline equivalent to $2.60/gal gasoline equivalent, and the equivalent crude cost would be about $109/bbl. That is considerably less than for the case without CCS at that CO_2 price ($3.67/gal gasoline equivalent).

For the case of coal to gasoline via methanol to MTG with CCS, the total plant cost is estimated at $4 billion, and the cost per stream day barrel at $80,400. For consistent evaluation parameters, the fuel cost is $1.57/gal gasoline compared with $1.47 in the venting case. The equivalent crude cost for the CCS case is about $51/bbl. If the LPG is sold rather than used to produce power, the estimated fuel cost is reduced to $1.23/gal gasoline equivalent and $1.33/gal gasoline equivalent for MTG gasoline in the cases of CO_2 venting and geologic storage, respectively, about $0.20/gal gasoline equivalent less than for LPG use in power generation. The cost of implementing CCS at the plant level involves minimal changes. Essentially all that is needed is to add a compressor–dryer to compress the CO_2 stream that would otherwise be vented to the atmosphere because separation of the CO_2 is a required integral part of the overall process scheme. The cost of the avoided CO_2 is about $10/tonne, which includes the cost of CO_2 transport and geologic storage and is expressed as dollars per tonne of CO_2 equivalent avoided. The transport and storage costs used in the calculation were updated to 2007 by using recent reviews by McCollum and Ogden (2006) and Tarka (2008). Those cost estimates assume that 150 bar pressure CO_2 is transported 100 km and stored 2 km underground on the average.

Biomass to Liquid Fuels

The panel next considered biomass conversion to liquid fuels by thermochemical conversion to synthesis gas and then synthesis of the fuels. For the biomass case, a dry-feed gasifier is used for this system design because of handling problems. The biomass gasifier is a two-stage fluid-bed gasifier in which the second stage is at sufficiently high temperature to crack and react all the tars with steam (that is, to gasify) to produce syngas. The syngas is then filtered and undergoes cleanup and water–gas shift, with CO_2 removal, to produce the H_2:CO ratio desired for the synthesis reaction. Because of issues related to biomass availability, it is assumed that the maximum annual amount of biomass per plant available in a reasonable surrounding area would be 1.1 million dry tons. That equates to a biomass feed rate of 3940 tons/day. The plant size and design were based on that biomass feed rate. The biomass feedstock used for the design was switchgrass. In this case, only the design for FT synthesis of liquid fuels was analyzed, but the conclusions for methanol to MTG will be semiquantitatively similar. Table 4.2 summarizes the results.

The capital cost of this BTL plant with CO_2 venting is estimated to be $636 million. That is equivalent to a capital cost on a daily-barrel basis of $144,000, and the resulting cost of the fuels would be $3.05/gal gasoline equivalent, which converts to a break-even oil price of $127/bbl. Increasing the price of electricity sold from $60/MW to $80/MW decreases the fuel cost by about $0.08/gal gasoline equivalent. Those costs are higher primarily because of the smaller plant size, the diseconomies of scale, and the higher cost of biomass per unit of energy; if coal costs $42/ton, the cost of biomass is about $90/dry ton on an energy-equivalent basis. Larger plants would have lower unit costs, and the analysis of

TABLE 4.2 Thermochemical Conversion of Biomass (Switchgrass) to Liquid Fuels with Fischer-Tropsch Synthesis

	BTL FT Recycling Without CCS	BTL FT Recycling With CCS
Inputs:		
Biomass, tons/day	3,940	3,950
Biomass, millions of dry tons/year	1.1	1.1
Outputs:		
Total liquid fuels, bbl/d	4,410	4,410
Efficiency, percent (low heating value)	52	50
Electricity, MW_e	34.4	24.2
CO_2 vented, tonnes/hr	125	13
CO_2 stored, tonnes/hr	0	112
Economics and metrics:		
Total plant cost (TPC), millions of dollars	636	648
Specific TPC, $/bbl/d	144,000	147,000
Total liquid fuels cost, $/gal gasoline equivalent	3.05	3.32
Break-even oil price, $/bbl	127	139
Life-cycle GHG emission, kg of CO_2 eq/GJ (low heating value)	−8.3	−120
FT liquids per petroleum-derived diesel emissions	−0.09	−1.30
Cost of avoided CO_2, $/tonne	Not applicable	20
Fuel cost:		
Fuel cost (10$/mt CO_2), $/gal gasoline equivalent	3.02	3.15
Fuel cost (50$/mt CO_2), $/gal gasoline equivalent	2.87	2.50
Fuel cost (100$/mt CO_2), $/gal gasoline equivalent	2.69	1.69

Note: Details of models can be found in Kreutz et al. (2008) and Larson et al. (2008).

Figure 2.5 in Chapter 2 suggests that about 17 U.S. locations could have plants with twice that capacity using biomass delivered from within 40 miles. Figure 2.5 also suggests that about 80 locations are suitable for the plant size of Table 4.2. Other locations might face a cost of transporting biomass from much longer distances that outweighs the economies of scale gained for larger plants. This study did not assess the optimization of these issues. The above results represent a case in which the price of CO_2 is zero. For a CO_2 price of $100/tonne, the cost of fuel from this plant would decrease from $3.05/gal gasoline equivalent to $2.69/gal gasoline equivalent. This analysis placed a price on net greenhouse gas emissions from the production and use of the liquid fuel, including upstream and downstream emissions, all greenhouse gas emissions from the plant (including those associated with coproduct electricity), and CO_2 emissions from the combustion of the fuels. For the sake of simplicity, the CO_2 price was placed on the fuel produced at the plant gate and is thus included in the fuel price. For biomass-based fuels, the greenhouse gas emissions are the net value of total greenhouse gas emissions minus CO_2 capture by photosynthesis during biomass production. The analyses did not include any potential credit or losses due to soil carbon storage, because of its complexity and specificity. For biomass gasification, the greenhouse gas life-cycle emission is slightly negative because 10 percent of the carbon is assumed to be unconverted in the gasifier and to be permanently stored as carbon in the char. The char carbon storage provides a carbon credit, so the cost of the fuel decreases as the tax on CO_2 increases.

The second column of Table 4.2 summarizes the results for the conceptual BTL fuel plant with CCS. The same biomass feed is used as in the previous case. The energy penalty for capture is shown by the net power-production reduction to 24.2 MW_e. For this case, greenhouse gas life-cycle emission is estimated to be highly negative at -120 kg CO_2 eq/GJ of produced fuels. This illustrates the impact of the double benefit of using biomass with respect to greenhouse gas emissions when CCS is used. The biomass has already removed CO_2 from the atmosphere by photosynthesis during its growth, and then the CO_2 produced during the conversion process is captured and stored rather than allowed to be re-emitted to the atmosphere.

The capital cost of this plant is estimated to be about $147,000 on a daily-barrel basis, and the resulting cost of the fuels is 3.32/gal gasoline equivalent, corresponding to crude oil at about $139/bbl. Increasing the price of electricity sold from $60/MW to $80/MW decreases the cost of the fuel produced by $0.06/gal gasoline equivalent. The higher cost of biomass and the amount of biomass available affect the potential of thermochemical conversion of biomass to liquid fuels. The cost of CO_2 avoided by this configuration is about $20/tonne. If the price of CO_2 were $100/tonne, the cost of fuel from the biomass plant with CCS would decrease from $3.32/gal gasoline equivalent to $1.69/gal gasoline equivalent. The cost is decreased because of the carbon credit received by not emitting CO_2 to the atmosphere.

Coal and Biomass to Liquid Fuels

The benefit of producing liquid transportation fuels from biomass is that the greenhouse gas life-cycle emission is close to neutral (Bartis et al., 2008). With geologic storage of the captured CO_2, biomass-produced liquid fuels can have a large negative greenhouse gas impact. The main challenge is the higher cost due to the small plant size because of limitations on the local availability of biomass. Gasification of coal and biomass in the same plant allows the plant size to be increased without exceeding the local availability of biomass. The larger plant size would allow economies of scale to reduce the costs associated with the production of liquid transportation fuels. To assess the economics and greenhouse gas emissions of liquid fuels produced from coal and biomass, the panel evaluated a set of design cases in which the amount of biomass was fixed at 1.1 million tons/year (1.0 million tonnes/year) and coal was brought into the plant at a rate of 3030 tons/day. The coal represented 58 percent of the plant's energy input, and the biomass 42 percent. The plant size was more than doubled. Because of the different properties of coal and biomass, the plant was designed with two parallel gasification trains to accommodate them: an entrained-flow gasifier for coal and a two-stage fluid-bed gasifier for biomass. The syngas streams were combined to gain economies of scale for the remainder of the plant. A schematic of the plant is shown in Figure 4.6; the plant used recycling around the synthesis reactor, and the CO_2 removed from the synthesis gas stream was either vented to the atmosphere or captured and stored. Table 4.3 summarizes the results for the coal and biomass cases.

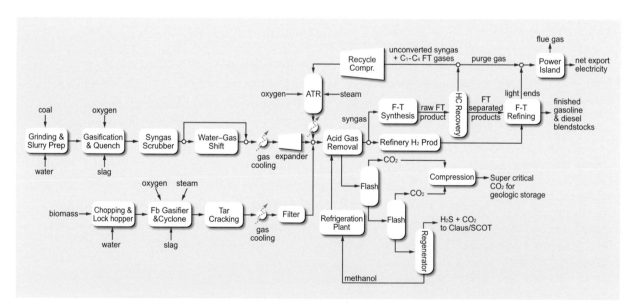

FIGURE 4.6 *Schematic of plant for gasification of coal and biomass with recycling around the synthesis loop and venting of the CO_2 removed from the synthesis gas stream.*

TABLE 4.3 Comparison of Coal-to-Liquid-Fuels Process With Coal-and-Biomass-to-Liquid-Fuels Process Using Fischer-Tropsch Synthesis

	CTL FT Recycling Without CCS	CTL FT Recycling With CCS	CBTL FT Recycling Without CCS	CBTL FT Recycling With CCS
Inputs:				
Coal, tons/day (as received)	26,700	26,700	3,030	3,030
Biomass, tons/day (dry)	0	0	3,950	3,950
Biomass, mass percent	0	0	57	57
Biomass energy, percent (low heating value)	0	0	42	42
Outputs:				
Gasoline, bbl/d	21,290	21,290	4,260	4,260
Diesel, bbl/d	28,700	28,700	5,740	5,750
Total liquid fuels, bbl/d[a]	50,000	50,000	10,000	10,000
Efficiency, percent (low heating value)	49.1	47.6	51.1	49.5
Electricity, MW$_e$	427	317	97	75
CO_2 vented, tonnes/hr	1,427	209	300	40
CO_2 stored, tonnes/hr	0	1217	0	262
Economics and metrics:				
Total plant cost (TPC), millions of dollars	4,880	4,950	1,320	1,340
Specific TPC, $/bbl/d	97,600	98,900	136,000	134,000
Total liquid fuels cost, $/gal gasoline equivalent	1.50	1.64	2.31	2.52
Break-even oil price, $/bbl	56	68	93	103
Greenhouse gas life-cycle emissions, kg of CO_2 eq/GJ (low heating value)	205	98	118	−2.3
FT liquids per petroleum-derived diesel emissions	2.23	1.07	1.28	−0.02
Cost of avoided CO_2, $/tonne	Not applicable	11	Not applicable	15
Fuel Cost:				
With $10/tonne CO_2, $/gal gasoline equivalent	1.71	1.74	2.42	2.50
With $50/tonne CO_2, $/gal gasoline equivalent	2.58	2.12	2.86	2.41
With $100/tonne CO_2, $/gal gasoline equivalent	3.67	2.60	3.40	2.29

Note: Details of models can be found in Kreutz et al. (2008) and Larson et al. (2008).

[a]For the CBTL cases, if the price of electricity is increased from $60/MW to $80/MW, the cost of transportation fuels decreases by about $0.10/gal gasoline equivalent and $0.08/gal gasoline equivalent in the venting and CCS cases, respectively.

The third data column of Table 4.3 summarizes the results for the conceptual CBTL plant with recycling and CO_2 venting. In this case, the plant gasifies both bituminous coal and biomass (switchgrass) to produce synthesis gas for conversion into FT liquid fuels. The plant consumes bituminous coal at 3,030 tons/day (dry) and switchgrass at 3,950 tons/day (dry) to produce liquid fuels at 10,000 bbl/d. Biomass is 58 percent of the total feed by mass and 42 percent of the total feed by energy; the remainder is coal. The CBTL plant is smaller than the CTL plant because of the limitation on availability of biomass in one location. In all these analyses, it is assumed that a maximum of 1.1 million dry tons of dry biomass per year can be supplied to the thermochemical conversion plants from the surrounding region.

In this case, the CO_2 produced during the conversion process is vented, and about 300 tonnes is emitted to the atmosphere per hour. To estimate the greenhouse gas life-cycle emission for CBTL plants, in addition to the greenhouse gas penalties from coal mining and transport, the greenhouse gas emissions associated with the production of biomass and its transport to the plant must also be accounted for. The GREET model was used to estimate these emissions. The carbon in the biomass was produced via photosynthesis by removing CO_2 from the atmosphere, so the biomass carbon is treated as a negative value in the carbon accounting. The credit for the excess power is estimated by assuming that an IGCC plant with no CCS was used to generate the power. The LCE is estimated to be 118 kg of CO_2 eq/GJ of produced fuels. This greenhouse gas life-cycle emission is slightly greater than that from the CTL plant with CCS.

The capital cost of this CBTL plant is estimated to be $1.3 billion, equivalent to a capital cost on a daily-barrel basis of $136,000. The resulting cost of the fuels is $2.31/gal gasoline equivalent, and the break-even crude-oil price is about $93/bbl. This case assumes a coal cost of $1.71/GJ and a biomass cost of $5/GJ and a zero carbon price. If the CO_2 price were $100/tonne, the cost of fuel from the plant would increase from $2.31/gal gasoline equivalent to $3.40/gal gasoline equivalent.

The fourth data column of Table 4.3 represents a conceptual CBTL plant with CCS. The same quantities of coal and biomass are used as in the CBTL plant with venting. The same quantity of liquid fuels is produced, but more electric power is needed to compress the captured CO_2. That reduces the net power production from 97 MW in the venting case to 75 MW. About 86 percent of the CO_2 produced during the conversion process is captured, and only 40 tonnes is emitted to the atmosphere per hour. The life-cycle emission is estimated to be reduced to the very low value of –2.3 kg of CO_2 eq/GJ. That means that the net greenhouse gas life-cycle emission from this CBTL plant configuration is essentially neutral: that is, the greenhouse gas emitted is balanced by the greenhouse gas avoided by photosynthesis and geologic storage of CO_2. The fuel produced is essentially a zero-carbon fuel.

The capital cost of this CBTL plant is estimated to be $1.3 billion, equivalent to a capital cost on a daily-barrel basis of $134,000. The resulting cost of the fuels would increase from $2.31/gal gasoline equivalent in the venting case to $2.52/gal gasoline equivalent with CCS. The cost of CO_2 avoided by this configuration would be about $15/tonne. If the CO_2 price were $100/tonne, the cost of fuel from the plant would decrease slightly from $2.52/gal gasoline equivalent to $2.29/gal gasoline equivalent; this is nearly 40 percent less than the cost of fuels from the CBTL case with venting at this carbon price.

Other Configurations: Polygeneration

Numerous cases for producing liquid transportation fuels from the thermochemical conversion of coal and biomass can be conceptualized. The cases evaluated above focused on configurations that maximized the amount of liquid fuels produced from a given amount of feedstock (Kreutz et al., 2008; Williams et al., 2008). Another approach is to consider process configurations that produce major quantities of different products; this is often referred to as polygeneration and can involve a number of options. To illustrate and evaluate this concept, the panel evaluated configurations that did not involve recycling around the synthesis reactor but used the unconverted synthesis gas and the nonfuel hydrocarbon fractions for power generation. Because the synthesis gas passed through the FT synthesis reactor only once, this case was referred to as once-through, or O-T. Four O-T cases involving coal and biomass feed and FT synthesis were evaluated and are briefly discussed below. Table 4.4 summarizes the O-T cases.

The first data column in Table 4.4 summarizes the results for the O-T case with venting of the CO_2 captured from syngas after the water–gas shift. The plant consumes biomass (as received) at 3940 tons/day and bituminous coal (as received) at 3760 tons/day. It produces liquid transportation fuels at 8100 bbl/d and power at 315 MW_e. On an energy basis, coal represents 63 percent of the feed to the system. For a carbon price of zero, the cost of the transportation fuels produced is $2.10/gal gasoline equivalent, and the break-even oil price is about $84/bbl when electricity is priced at $60/MWh. The ratio of the greenhouse gas

TABLE 4.4 Summary of Once-Through (O-T) Coal-and-Biomass-to-Liquid-Fuel Processes That Use Fischer-Tropsch Technology With and Without Carbon Capture and Storage

	O-T Without CCS	O-T With CCS	O-T With CCS and Root Carbon Credits[a]	O-T CCS Greenhouse Gas Equivalent Fuels
Inputs:				
Coal, tons/day (as received)	3,760	3,760	7,370	31,000
Biomass, tons/day (as received)	3,940	3,940	3,940	3,940
Biomass, mass percent (as received)	51	51	35	12.1
Biomass, energy percent (low heating value)	38.2	37	23	8.3
Outputs:				
Total FT fuels, bbl/d	8,100	8,100	13,000	46,200
Efficiency, percent (low heating value)	51	48	47	46.8
Electricity, MW_e	315	276	406	1,404
CO_2 vented, tonnes/hr	380	100	160	557
CO_2 stored, tonnes/hr	0	281	442	1,540
Economics and metrics:				
Total plant cost (TPC), millions of dollars	1,324	1,379	1,944	5,650
Specific TPC, $/bbl per day	163,000	170,000	149,000	122,000
Total FTL cost, $/gal gasoline equivalent	2.10	2.48	2.08	1.50
Break-even oil price, $/bbl	84	101	83	56
Greenhouse gas life-cycle emission for plant, kg of CO_2 eq/GJ (low heating value)	175	22	−17	110
FT liquids per petroleum-derived diesel emissions	1.90	0.24	−0.18	1.20
Cost of avoided CO_2, $/tonne	Not applicable	21	22	20

Note: Details of models can be found in Larson et al. (2008) and Williams et al. (2008).

[a]Accounts for carbon credit gained from CO_2 uptake in soil and roots assuming that the feedstock is mixed prairie grasses.

life-cycle emission for FT liquids and petroleum-derived diesel is 1.9. A greenhouse gas credit is taken for the coproduced electric power on the basis of emissions from an IGCC plant with CCS. However, if the greenhouse gas credit for the generated power is based on the much larger current grid-average greenhouse gas emissions instead of those of an IGCC-CCS plant, the greenhouse gas life-cycle emission for the liquid fuels would be about 72 percent of that for the same quantity of fuels produced from conventional petroleum. The greenhouse gas life-cycle emission for the complete fuel and power system would then be 13 percent lower than that for conventional petroleum and grid-based power generation.

If geologic storage of CO_2 is applied, the O-T CBTL plant would be expected to have the results summarized in the second data column of Table 4.4. The combination of coal and biomass with geologic storage of CO_2 produces liquid transportation fuels that are carbon-neutral or decarbonized over the life cycle. The electricity sold to the grid is effectively decarbonized also in that it has assumed carbon content equivalent to that of the greenhouse gas emissions from an IGCC plant with CCS. That means that the fuels are produced with no net greenhouse gas emissions, and that, in effect, so is the electric power. *Both transportation fuels and electric power could have absolutely zero greenhouse gas life-cycle emissions by increasing the fraction of biomass somewhat. This is a key observation and may represent a major opportunity to address emissions from both transportation and power production.* The liquid transportation fuels are available at about $2.48/gal gasoline equivalent, equivalent to a crude-oil price of about $101/bbl at zero CO_2 price. The transportation fuels produced with this approach become less expensive as the CO_2 price increases. The estimated cost of avoided CO_2 from the plant is about $21/tonne of CO_2. The cost of avoided CO_2 is low because separation and capture of CO_2 is an integral part of the synthesis process. The cost of separation and capture is included in the product cost, whether the captured CO_2 is transported and stored geologically or vented.

The third data column of Table 4.4 represents a scenario in which mixed prairie grasses grown on carbon-depleted soils is used as a feedstock and a carbon credit can be taken for soil or root sequestration for those grasses. The soil or root carbon credit is about 60 percent of the carbon in the harvested grasses. This case shows that because of the soil or root carbon credit, the same quantity of biomass—biomass (as received) at 3,940 tons/day or 1.1 million dry tons/year—can be mixed with more coal (7,370 tons/day) to produce 60 percent more fuel (13,000 bbl/d) and still attain a zero greenhouse gas life-cycle emission with indirect liquefaction.

The fourth data column in Table 4.4 shows the results for a large O-T CBTL plant with CCS in which biomass makes up only 8 percent of the feedstock on an energy basis and that uses coal at nearly 31,000 tons/day. The plant provides fuel at nearly 46,200 bbl/d and power at more than 1,000 MW_e. The fuel cost is estimated to be $1.50/gal gasoline equivalent, equivalent to a break-even oil price of $56/bbl when the electricity generated sells for $60/MWh. The ratio of FT liquid fuel to petroleum-derived fuel is about 1.2 for this plant option, which means that the greenhouse gas emissions from the fuels are equal to those from petroleum, but the electric power generated is decarbonized.

An example of the potential of polygeneration technology involves thermochemical conversion plants that use combined coal and biomass as feedstock and incorporate CCS. Such a plant consumes biomass at about 3,400 dry tons/day (1 million dry tons/year) and bituminous coal (as received) at about 4,000 tons/day to produce fuel at net output capacity of about 8,100 bbl/d and generate electric power at 280 MW. If three such plants start to be built in 2015 and the number of plants increases at 20 percent per year until 2035, there could be about 110 such polygeneration plants consuming biomass at 110 million dry tons/year and coal at 150 million tons/year in 2035. The plants would produce liquid transportation fuels at about 0.83 million bbl/d (13 billion gallons/year) and about 28 GW of continuous decarbonized electric power with a zero net greenhouse gas life-cycle emission. If 440 million dry tons of biomass and 600 million tons of coal (as received) were used in this configuration, 3.3 million barrels of liquid fuels and 100 GW of continuous power, both decarbonized, could be produced per year. Historically, petroleum companies have not been interested in power generation for sale (fuel was maximized, and power was sold to the extent that it was excess), and power companies were not interested in fuel production. That is an obvious barrier to making this approach viable.

Summary

Thermochemical conversion of coal with indirect CTL technologies could be used to produce clean, fungible transportation fuels for less than $2.00/gal gasoline equivalent. The technology used for the synthesis determines the products made but does not have a major effect on fuel cost. The FT process produces a mixture of gasoline and diesel and jet fuel. Methanol synthesis followed by MTG produces primarily high-octane gasoline. Because methanol synthesis is more selective, as is MTG, the yields are slightly higher and process simplicity results in slightly lower fuel costs than with FT. A combination of FT and MTG technologies could produce the desired mix of fuels required by the market. Without CCS, greenhouse gas life-cycle emission is estimated to be slightly more than twice that of producing and using liquid fuels from conventional petroleum. With CCS, however, greenhouse gas life-cycle emission is reduced to be about equal to or less than that of petroleum-derived fuels. Those results are comparable with results reported by other independent studies (Jaramillo et al., 2008; Bartis et al., 2008). By using a mixture of coal and biomass as a feedstock and storing the CO_2 captured in the process, essentially carbon-neutral fuels can be produced by using about 57 percent dry biomass by weight. The cost of fuel produced by such a plant configuration is estimated to be about $2.52/gal gasoline equivalent. For BTL plants, the greenhouse gas life-cycle emission is zero or negative because of the biomass-carbon photosynthesis credit, but the plants are necessarily small because of limitations of biomass availability. The small size of the plant and the high cost of biomass feedstock result in higher fuel costs of about $3/gal gasoline equivalent. A price on carbon would substantially reduce the costs, and for a CO_2 tax of $100/tonne, the fuel cost would be below $2/gal gasoline equivalent. The advantage of using the CBTL approach is that it allows for larger plants than biomass-only plants, and this can reduce capital and hence product costs. In addition, CBTL can reduce the greenhouse gas life-cycle emission compared with coal-only plants. Promising CBTL configurations are O-T plants that coproduce fuels and electric power. One of the conceptual O-T configurations discussed above could produce fuels at 46,200 bbl/d and electric power at 1,404 MW using only 8 percent biomass with greenhouse gas life-cycle emission of only 53 percent of that of an existing coal power and crude-oil products displaced.

Environmental Impacts

CTL plants can be configured to minimize their impact on the environment. Clean-coal technologies have been and are continuing to be developed in the United States and abroad. Many of the technologies are being developed for the electric-power industry, but they can also be used in CTL applications. For example, there is considerable similarity between an IGCC power plant and a CTL plant. Both plants need to produce clean synthesis gas from coal by using gasification and gas-cleaning technologies. The requirement for cleanness of the syngas is more stringent for CTL than for IGCC. CTL plants also need gas and steam turbines to produce their electric power. What has been learned in the power industry can be directly applied to a CTL industry. As a result, concerns over emissions of criteria pollutants and toxic chemicals—such as sulfur oxides, nitrogen oxides, particles, and mercury—would be minimal because CTL plants would use clean-coal technologies. Cleaner synthesis gas is needed in CTL technology than in power generation to avoid poisoning of the FT or MTG catalysts.

The sulfur compounds in the coal are converted into hydrogen sulfide and carbonyl sulfide, and these are fully recovered in the acid-gas treatment plant.

They are transformed into elemental sulfur that can be sold as a by-product. The ammonia in the synthesis gas resulting from the nitrogen in the coal is washed out in the water quench. The ammonia can be recovered and sold as a fertilizer or sent to wastewater treatment, where it is absorbed by bacteria. All the mercury, arsenic, and other heavy metals in the syngas are adsorbed on activated charcoal. The mineral matter (or ash) in the coal has been exposed to extremely high temperatures during gasification and has become vitrified into slag. The slag is non-leachable and finds use in cement or concrete for buildings, bridges, and roads. Nitrogen oxide emissions are reduced to about 3 ppm by using low-nitrogen-oxide burners in the gas turbines and selective-catalytic-reduction technology in the heat-recovery steam generators in the plant. The same or a similar pollution-control method would be used for CBTL and BTL plants. For BTL plants, additional syngas cleaning might be required (depending on the gasification technology used) for tar removal and removal of ash components that are not present so much in coal ash (for example, silica).

Water use in thermochemical-conversion plants depends primarily on the water-use approach used in designing the plants. In the conversion of coal and coal biomass to transportation fuels with all water streams recycled or reused, with or without CO_2 storage and with no power export, the major consumptive uses of water are for cooling, producing hydrogen, and solids handling. If water availability is unlimited because there is access to rivers, conventional forced- or natural-draft cooling towers would be used. In arid areas where water is limited, air cooling would be used as much as possible. Hybrid cooling systems that use both air and water cooling could also be used to limit overall water consumption. Depending on the magnitude of air cooling, water consumption could range from about 1 to 8 bbl of water per barrel of product. For CTL plants, environmental impacts will be associated with the mining of additional coal (NRC, 2002, 2007).

Product Characteristics

The low-temperature FT process produces about 10 percent hydrocarbon gases, 25 percent liquid naphtha, 22 percent distillate, and 46 percent wax or heavy-oil product. The wax can be selectively hydrocracked into distillate so that the overall product distribution can be skewed in favor of diesel. The clean fuels are recovered, and the wax is hydrocracked into more diesel fuel and naphtha. Any remaining synthesis gas is returned to the FT reactors for additional conversion to liquid fuels. The MTG process converts methanol to about 7 percent hydrocarbon gases,

82 percent liquid gasoline, and 11 percent butane, some of which can be added to the gasoline to give a product yield of around 88 percent regular unleaded gasoline.

DIRECT-LIQUEFACTION TECHNOLOGIES

Direct coal-liquefaction technologies are less developed than indirect-liquefaction technologies, and the uncertainties of capital costs and the refining necessary to produce fungible fuels make comparisons with indirect liquefaction difficult. More data may be available after the Chinese Shenhua plant reaches full operation (S. Tam, Headwaters, presentation to the panel on February 19, 2008). This section first discusses the history of direct liquefaction and then provides a technical overview, current status, technical challenges, process economics, potential environmental impacts, and product characteristics.

History

The pioneering developmental work in direct liquefaction is attributed to Friedrich Bergius and his colleagues dating from around the time of World War I. Commercial operation of direct-liquefaction plants began in Leuna, Germany, in 1927, under I.G. Farben. The first plant had a capacity of 100,000 tons/year. At about the same time, Imperial Chemical Industries (ICI) built a plant of similar capacity in Billingham, United Kingdom. In the late 1930s, the ICI plant was converted from direct liquefaction to produce aviation gasoline from creosote oil. However, direct liquefaction continued to be developed in Germany. The output of the plants had a substantial impact on liquid-fuel supply in Germany during World War II. Twelve plants collectively produced liquid fuels at about 4 million tons/year by 1944, after which production dropped dramatically because of the Allied bombing campaign. Many other countries were involved in direct liquefaction on a small scale during World War II; probably the most substantial effort was in Japan, in which four plants produced fuel at about 260,000 tons/year. During the 1950s, some modest attempts were made to continue direct liquefaction, including efforts in Germany and the United States. The American projects involved the Bureau of Mines, Consolidation Coal Company, and Union Carbide. All those efforts came to naught, primarily because they were not economically competitive with relatively inexpensive petroleum.

As a result of the oil embargo and price shocks of the 1970s, direct liquefaction underwent a major revival in the United States, Germany, and Japan. The U.S. work was supported by DOE and had the active involvement of numerous major oil companies, including Exxon, Hydrocarbon Research, Inc., and Gulf. Several large pilot plants, with nominal capacities for handling coal at up to about 250 tons/day, were constructed and operated with reasonable technical success in the late 1970s and into the 1980s. Those activities dwindled, one by one, during the 1980s as a result of changes in government policy and declining oil prices (Burke et al., 2001). The entire infrastructure (including pilot plants) of direct liquefaction in the United States was dismantled.

Technical Overview

The fundamental concept of direct liquefaction is simple. The intent is to convert coal into a petroleum-like liquid that can be refined into synthetic products that are comparable with current refinery products, such as gasoline, jet fuel, and diesel fuel. One can conceive of the empirical formula of a molecule of "petroleum" as $CH_{1.8}$ and that of "coal" as $CH_{0.8}$. Chemically, one can write

$$CH_{0.8} + \tfrac{1}{2} H_2 \rightarrow CH_{1.8}.$$

That simple chemical equation has proved to be difficult to reduce to successful engineering practice.

It is generally agreed that the hydrogenation of coal can proceed best when the coal is undergoing active thermal decomposition. For most coals, that means operating at 350°C or higher. Such temperatures are thought to be necessary to achieve adequate reaction rates. The reactions take place in a liquid medium, a process solvent in which primary reaction products from the coal dissolve. Because of the inverse dependence of the solubility of a gas (for example, H_2) on temperature, the liquefaction reactions have to take place at a high pressure of more than 134 bar at the reaction temperature.

Continuous feeding of a solid into a pressure vessel is a challenge. Therefore, virtually all direct-liquefaction process schemes rely on slurrying the coal in a liquid vehicle. The slurry is then pumped into the reactor. Various concepts for direct liquefaction used a process-derived recycling solvent as the slurry vehicle. That solvent might not be expected to participate actively in the chemical processes of liquefaction.

Two potential sources of hydrogen are considered. One approach is to use gaseous H_2. The use of gaseous H_2 in direct liquefaction would require the presence of an active hydrogenation catalyst. Iron compounds were favored as liquefaction catalysts because of their low cost, although other metals, such as molybdenum, are more active catalysts. The other approach is to use relatively hydrogen-rich compounds in the liquid to transfer hydrogen to molecular fragments liberated during the decomposition of the coal. The so-called hydrogen-donor compounds are exemplified by tetralin (1,2,3,4-tetrahydronaphthalene). Tetralin can transfer four of its hydrogen atoms to the coal fragments and be converted to naphthalene at the same time. Presumably, the "spent" hydrogen donors could be regenerated by hydrogenation during the liquefaction reaction or as a separate operation. Gaseous H_2 and a hydrogen-donor solvent can be used together.

Process concepts also differ in the number of reaction stages to be used. In principle, a multistage reaction offers an opportunity to optimize the process chemistry for the specific coal being liquefied. Stages can be operated at different temperatures and pressures; one could (conceptually) rely entirely on thermal processing in a donor solvent, a second could involve H_2 in the presence of a catalyst, and so on.

The numerous process concepts developed for direct liquefaction all represent approaches to adding hydrogen to coal to produce a petroleum-like liquid. The processes differ in the nature of the solvent to be used, how (if at all) spent solvent would be replenished, number of process stages, temperatures and pressures in each, residence time in each, hydrogenation catalysts to be used, and catalyst recovery and regeneration.

At the end of the last stage of liquefaction, the liquid products have to be separated from unconverted coal and mineral residue. The solid or liquid separation is a formidable operation, in part because the temperature of the liquid is dropping and, with pressure letdown, dissolved light molecules are probably flashing to vapor. Both effects raise the viscosity of the liquid, so the challenge is to separate finely divided solids from a highly viscous liquid. Centrifugation, solvent de-ashing, and pressure filtration appear to be the operations of choice.

The primary liquid will need further refining downstream to be converted to acceptable marketable products. The refining will probably include some combination of hydrotreating to remove heteroatoms, hydrogenation for further aromatic saturation, and hydrocracking to shift the products to lower-boiling-point materials. It has usually been presumed that the additional refining could be achieved in operations typical of oil refining.

A direct-liquefaction plant in Inner Mongolia, China, was in trial operation in December 2008. It ran for 300 hr during the trial. The plant, a $2 billion facility, will consume about 3.5 million tons of coal per year and produce 1.8 million tons of products, of which 70 percent is estimated to be clean diesel fuel. This project was initiated in 1996.

In 2006, the planning of another direct-liquefaction plant in Inner Mongolia, with Shell as a partner, was announced. The planned plant is estimated to have a capacity of 70,000 bbl/d, which is about 1 percent of Chinese petroleum consumption. The estimated cost of this plant is about $5–6 billion (although construction costs in China are not comparable with those in the United States). It is expected to come on line in 2012. Overall projections are for Chinese liquid-fuel production via direct liquefaction to reach 50 million tons/year by 2020. As far as is known, no other large-scale projects in direct liquefaction are under way elsewhere.

Technical Challenges

Downstream of the reactor, material selection for internals in pressure-letdown valves and selection of effective solid and liquid separation processes remain challenging. Not all coals are equally amenable to direct liquefaction. However, high-sulfur coals, which are undesirable for combustion, could be excellent liquefaction feedstocks because the pyrite in the coal serves as an in situ liquefaction catalyst. (In contrast, low-sulfur coals are preferred for indirect liquefaction because sulfur has to be removed from the syngas produced by coal gasification before synthesis.)

The optimal operating conditions for and the product yield slate from direct liquefaction are known to depend heavily on the specific coal feedstock being processed. It is questionable how far a universal approach could be used for the design and operation of plants if, for example, one used Powder River Basin coal, another Illinois Basin coal, and a third Appalachian coal.

Direct liquefaction requires substantial amounts of H_2. Although H_2 could come from a variety of sources, there would probably be a need to include a coal-gasification plant for H_2 production in or alongside the liquefaction plant.

One of the keys to future commercial development of direct liquefaction is to find low-severity process routes (for example, low temperature and low pressure) to obtain liquids from coal. That is likely to require a greater focus on fundamentals of coal chemistry than on process engineering.

Process Economics

A thorough and detailed economic analysis of direct liquefaction has not been done in almost 20 years. Numerous studies from the 1970s and 1980s are available. The numerical results of those studies need to be interpreted and used with caution. The panel estimated the costs of direct liquefaction on the basis of the DOE study *Direct Coal Liquefaction Baseline Design and System Analysis* (1993). Although the cost estimates are updated to reflect 2007 costs, they are not considered to be as accurate as or to be fully consistent with the estimates for indirect liquefaction.

The products of direct liquefaction are typically aromatic and contain large amounts of sulfur, nitrogen, and oxygen. Costs associated with the production of clean fuels that meet U.S. specifications have typically not been included in published estimates. For the panel's work, estimates were applied to include the cost of upgrading all product streams so that only clean transportation fuels are produced. Plant capital cost, including complete upgrading, is estimated at $5.5 billion, or about $115,000 per stream-day barrel. The overall thermal efficiency approaches 60 percent. The yield is below 2.5 bbl of liquid fuel products per ton of coal. Plant emissions are projected at 8.5 kg CO_2/gal product. The total plant CO_2 emissions, including fuel, are slightly less than those of an FT plant. The estimated cost of the liquids produced is about $0.20/gal higher than for a comparable CTL plant using FT. The overall greenhouse gas footprint of the venting plant is expected to be similar to or slightly better than that of the CTL plant using FT and venting CO_2. The direct-liquefaction plant with CCS is at a disadvantage relative to the indirect-liquefaction plant because it has more flue-gas CO_2 to be recovered. The recovery of CO_2 from several flue-gas streams in a direct-liquefaction plant needs additional equipment and is much more expensive than CO_2 recovery in an indirect-liquefaction plant. That disadvantage could be eliminated through engineering modification of the plant design, but such changes would come at a cost.

The performance estimate is consistent with data on the Chinese plant under construction. The product quality of the Chinese plant might meet the quality of the Chinese transportation-fuel system, but transportation-fuel blending stocks in the United States essentially have to meet the quality of petroleum blending stocks because of tight specifications for final fuel. Either indirect or direct coal liquefaction requires about $5 billion in capital for a commercial-scale plant. Raising such capital may require substantial government intervention in the form, for example, of loan guarantees, incentive programs to offset capital and operations and maintenance costs, or guaranteed purchases of products to get the industry started. A government–private sector partnership might be necessary for the setup of the first few direct- or indirect-liquefaction plants.

Environmental Impact

Because coal's hydrogen:carbon ratio is lower than that of petroleum, transportation fuels produced from direct liquefaction of coal would have much higher greenhouse gas emissions than gasoline has. If nonfossil sources of energy were used for hydrogen production and process heat for the conversion processes, the net effect of coal-based fuels would be about the same as that of fuels from petroleum (NRC, 1990). As discussed earlier, using biomass–coal mixtures in indirect-liquefaction plants could result in substantial reductions in greenhouse gas life-cycle emission. That strategy has not been tested for direct liquefaction but should be investigated for potentially comparable reductions of greenhouse gas emissions.

"The conversion of coal into synthetic fuels can embrace practically any potential form of pollution and health hazard which can be associated with coal, including combustion products and ash, phenolic liquors and coal liquids which are exceptionally rich in known or suspected carcinogens" (Grainger and Gibson, 1981).

Data on water use, especially in the last few years, seem to be sparse. One estimate suggests water consumption of about 200 million gallons per year for operation of a plant with a coal capacity of 2000 tons/day (Comolli et al., 1993). The estimate of about 2 gal of water per gallon of product is consistent with water needs for indirect liquefaction.

Product Characteristics

Finished products from direct liquefaction are intended to be fully fungible with respect to comparable petroleum products, but that has not been adequately demonstrated. Direct liquefaction produces low-cetane fuel (cetane index, about 45) (Mzinyati, 2007). As a replacement for fuel oils, coal liquids are considered to be more difficult to store, to have higher concentrations of potential carcinogens, to produce higher quantities of nitrogen oxides, and to have a greater soot-forming tendency. Blends of coal products with petroleum might form precipitates. Production of lighter transportation fuels appears to be accompanied by high rates of catalyst deactivation and to require high hydrogen consumption.

FINDINGS AND RECOMMENDATIONS

Gasoline and diesel can be produced from the abundant U.S. coal reserves to have greenhouse gas life-cycle emissions similar to or less than those of petroleum-based fuels in 2020 or sooner if existing thermochemical technology is combined with geologic storage of CO_2. Widespread deployment of such facilities will require major increases in coal mining and transportation infrastructure either for moving coal to the plants or moving fuel from the plants to the market.

Finding 4.1

Despite the vast coal resource in the United States, it is not a forgone conclusion that adequate coal will be mined and be available to meet the needs of a growing coal-to-fuels industry and the needs of the power industry.

Recommendation 4.1

The U.S. coal industry, the U.S. Environmental Protection Agency, the U.S. Department of Energy, and the U.S. Department of Transportation should assess the potential for a rapid expansion of the U.S. coal-supply industry and delineate the critical barriers to growth, environmental effects, and their effects on coal cost. The analysis should include several scenarios, one of which assumes that the United States will move rapidly toward increasing use of coal-based liquid fuels for transportation to improve energy security. An improved understanding of the immediate and long-term environmental effects of increased mining, transportation, and use of coal would be an important goal of the analysis.

Geologic storage of CO_2, however, would have to be demonstrated at commercial scale and implemented by then. Without CCS, the greenhouse gas life-cycle emission will be more than twice those from petroleum-based fuels. Coal can be combined with biomass at a ratio of 60:40 (on an energy basis) to produce liquid fuels that have greenhouse gas emissions comparable with those from petroleum-based fuels if CCS is not implemented. With CCS, fuels produced from coal and biomass would have a slightly negative to roughly zero carbon balance. Cellulosic dry biomass also can be converted thermochemically to synthetic gasoline and diesel without coal. The greenhouse gas life-cycle emissions from those fuels should be close to zero without CCS and highly negative with CCS, but the cost of fuel products will be higher than the cost of those produced from coal or combined coal and biomass.

Finding 4.2

Technologies for the indirect liquefaction of coal to transportation fuels are commercially deployable today; but without geologic storage of the CO_2 produced in the conversion, greenhouse gas life-cycle emissions will be about twice those of petroleum-based fuels. With geologic storage of CO_2, CTL transportation fuels could have greenhouse gas life-cycle emissions equivalent to those of equivalent petroleum-derived fuels.

Finding 4.3

Indirect liquefaction of combined coal and biomass to transportation fuels is close to being commercially deployable today. Coal can be combined with biomass at a ratio of 60:40 (on an energy basis) to produce liquid fuels that have greenhouse gas life-cycle emissions comparable with those of petroleum-based fuels if CCS is not implemented. With CCS, production of fuels from coal and biomass would have a carbon balance of about zero to slightly negative.

Finding 4.4

Geologic storage of CO_2 on a commercial scale is critical for producing liquid transportation fuels from coal without a large adverse greenhouse gas impact. This is similar to the situation for producing power from coal.

Recommendation 4.2

The federal government should continue to partner with industry and independent researchers in an aggressive program to determine the operational procedures, monitoring, safety, and effectiveness of commercial-scale technology for geologic storage of CO_2. Three to five commercial-scale demonstrations (each with about 1 million tonnes of CO_2 per year and operated for several years) should be set up within the next 3–5 years in areas of several geologic types.

The demonstrations should focus on site choice, permitting, monitoring, operation, closure, and legal procedures needed to support the broad-scale application of geologic storage of CO_2. The development of needed engineering data and determination of the full costs of geologic storage of CO_2—including engineering, monitoring, and other costs on the basis of data developed from continuing demonstration projects—should have high priority.

The configuration of the thermochemical conversion plants produces a concentrated stream of CO_2 that must be removed before the fuel-synthesis step, even in noncapture designs. Thus, the requirement for geologic storage has only a small effect on cost and efficiency. On a plant basis, the engineering cost of CO_2 avoided is about $10–15/tonne, but the cost is based on a "bottom-up" engineering estimate of expenses for drying, compression, transport, land purchase, drilling wells and injecting CO_2, monitoring, and capping wells. Experience with a variety of energy technologies suggests that the full cost of geologic storage cannot be captured by such an approach, because some implementation barriers increase costs and are difficult to quantify in advance. Accordingly, the numerical geologic cost used in this report, which is based on factors quantified by an engineering analysis, and life-cycle costs for fuels that entail carbon storage may constitute a lower bound on future costs.

Finding 4.5

There do not appear to be any technical issues that cannot be resolved or any cost showstoppers associated with geologic storage of CO_2. There is, however, much to be developed in siting, permitting, monitoring, and site closure; it is essential that public and political uncertainty be resolved and that costs be better defined. Uncertainty among the general public and policy makers about the efficacy and regulatory environment has the potential to raise storage cost. Ultimately, the requirements for siting, design, operation, monitoring, carbon-accounting procedures, liability, and the associated regulatory frameworks need to be developed to avoid unanticipated delays in initiating demonstration projects and, later, in permitting and licensing of individual commercial-scale projects. Extensive experience with storage in deep saline aquifers has yet to be gained and evaluated. A full assessment of the future cost of CCS should emphasize, at least qualitatively, the uncertainty arising from such factors.

Recommendation 4.3

The government-sponsored geologic CO_2 storage projects need to address issues related to the concerns of the general public and policy makers about geologic CO_2 storage through rigorous scientific and policy analyses. As the work on geologic storage progresses, any factors that might result in public concerns and uncertainty in the regulatory environment should be evaluated and built into the project decision-making process because they could raise storage cost and slow projects.

The key technologies required to convert coal and cofed coal and biomass to liquid transportation fuels have been commercially demonstrated and are ready for commercial deployment. With geologic storage of CO_2, coal can be used to produce liquid transportation fuels that have greenhouse gas life-cycle emission that is equivalent to that of petroleum-derived fuels. Cofed biomass and coal can be used to produce liquid transportation fuels that are equivalent to those produced from petroleum with respect to greenhouse gas life-cycle emission without geologic storage of CO_2 and fuels that have lower greenhouse gas life-cycle emission with geologic CO_2 storage. Technology for producing liquid transportation fuels with biomass only (BTL) has been demonstrated but requires additional development to be ready for commercial deployment. It can produce carbon-neutral fuels; with geologic CO_2 storage, liquid transportation fuels so produced can have negative greenhouse gas life-cycle emission. Carbon storage in soils by the biomass crops can enhance the favorable effect of biomass conversion to fuels but is hard to project because it depends on many situational and agricultural factors. Liquid transportation fuels produced from biomass alone would be more expensive than CTL fuels because of the high cost of biomass and the diseconomies of scale for plants that are small because of limited regional biomass availability. Using both coal and biomass (CBTL) allows larger plants that can benefit from economies of scale, that have lower capital costs and use cheaper coal, and that therefore have lower production costs.

Finding 4.6

The advanced technologies for gasification, syngas cleanup, and Fischer-Tropsch synthesis have been demonstrated on a commercial scale. Their integration on the scale required to have a substantial impact on fuel production has not been dem-

onstrated but is not considered a major issue. For first-mover projects to produce liquid transportation fuels from coal on the scale of a large plant poses a degree of technical risk; in addition, the risk of price and cost volatility that energy markets have shown recently has to be considered. The risk greatly increases the difficulty of developing and funding first-mover projects.

Finding 4.7

Technologies for the indirect liquefaction of coal to produce liquid transportation fuels with greenhouse gas life-cycle emissions equivalent to those of petroleum-based fuels can be commercially deployed before 2020 only if several first-mover plants are started up soon and if the safety and long-term viability of geologic storage of CO_2 is demonstrated in the next 5-6 years.

Recommendation 4.4

A program of aggressive support for first-mover commercial plants that produce coal-to-liquid transportation fuels and coal-and-biomass-to-liquid transportation fuels with integrated geologic storage of CO_2 should be undertaken immediately to address U.S. energy security and to provide fuels with greenhouse gas emissions similar to or less than those of petroleum-based fuels. The demonstration and deployment of "first-mover" coal or coal-and-biomass plants should be encouraged on the basis of the primary technologies, including CCS to demonstrate the technological viability of CTL and CBTL fuels and to reduce the technical and investment risks associated with funding of future plants. If decisions to proceed with commercial demonstrations are made soon so that the plants could start up in 4–5 years and if CCS is demonstrated to be safe and viable, those technologies would be commercially deployable by 2020.

Recommendation 4.5

The first-mover coal or coal–biomass plants recommended above should be sited so that they provide CO_2 for several of the sponsored geologic CO_2-storage projects, and their progress should be expedited to facilitate the geologic CO_2-storage projects and the further development of conversion technologies. To the extent possible, the conversion plants and geologic storage should be implemented as a package. As a first step, a few CTL plants and CBTL plants could serve as sources of CO_2 for a small number of CCS demonstration projects. However, so-called capture-ready plants that vent CO_2 would create liquid fuels with higher CO_2 emissions per unit of usable energy than those from petroleum-based fuels; their commercialization should not be encouraged before commercially available CCS is proved to be safe and sustainable.

Finding 4.8

The technology for producing liquid transportation fuels from biomass or from combined biomass and coal via thermochemical conversion has been demonstrated but requires additional development to be ready for commercial deployment.

Recommendation 4.6

Key technologies should be demonstrated for biomass gasification on an intermediate scale, alone and in combination with coal, to obtain the engineering and operating data required to design commercial-scale synthesis gas-production units.

Finding 4.9

Conversion plants that use 60 percent coal and 40 percent biomass as feedstock can be configured to eliminate recycling of unconverted synthesis gas and thereby generate a substantial amount of additional electric power. If the CO_2 captured from such a plant is stored geologically, both the liquid transportation fuels and the electric power produced for sale to the grid could have zero greenhouse gas life-cycle emissions. That approach might present a key opportunity to address emissions from both transportation and power.

Recommendation 4.7

A thorough systems analysis should be developed for process configurations of coal-and-biomass-to-liquids plants that eliminate recycling of unconverted synthesis gas and generate substantial additional electric power. The plants' fuel cost and power costs, potential to address greenhouse gas emissions, and potential impact on U.S. oil consumption should be assessed thoroughly.

Finding 4.10

Technologies for direct liquefaction of coal are less well developed, and the uncertainties of capital costs and of the refining necessary to produce high-quality transportation fuels are substantial. The uncertainties will be reduced after the Chinese Shenhua plant reaches full operation if adequate data are made available.

Recommendation 4.8

The performance, product spectrum, and projected economics of direct and indirect coal liquefaction should be evaluated and reviewed on the basis of commercial demonstrations in China and other countries.

REFERENCES

Anderson, S., and R. Newell. 2004. Prospects for carbon capture and storage technologies. Annual Review of Environment and Resources 29:109-142.

Argonne National Laboratory. 2005. The Greenhouse Gases, Regulated Emissions, and Energy Use in Transportation (GREET) Model. UChicago Argonne, LLC. Available at http://www.transportation.anl.gov/modeling_simulation/GREET/index.html. Accessed January 30, 2009.

Bartis, J.T., F. Camm, and D.S. Ortiz. 2008. Producing Liquid Fuels from Coal: Prospects and Issues. Santa Monica, Calif.: RAND Corporation.

Burke, F.P., S.D. Brandes, D.C. McCoy, R.A. Winschel, D. Gray, and G. Tomlinson. 2001. Summary Report of the DOE Direct Liquefaction Campaign of the Late Twentieth Century: Topical Report. Washington, D.C.: U.S. Department of Energy.

Chu, W., R. Kieffer, A. Kiennemann, and J.P. Hindermann. 1995. Conversion of syngas to C1-C6 alcohol mixtures on promoted CuLa2Zr2O7 catalysts. Applied Catalysis A: General 121:95-111.

Clifford, C.B., and H.H. Schobert. 2007. Development of coal-based jet fuel. In The 234th ACS National Meeting, Boston, Mass.

Cobb, J.T., Jr. 2007. Survey of commercial biomass gasifiers. Paper read at American Institute of Chemical Engineers Annual Meeting, Salt Lake City, Utah, November 4-9, 2007.

Comolli, A.G., E.S. Johanson, W.F. Karolkiewicz, L.K. Lee, G.A. Popper, R.H. Stalzer, and T.O. Smith. 1993. Close-Coupled Catalytic Two-Stage Liquefaction Process Bench Studies. Washington, D.C.: U.S. Department of Energy.

EIA (Energy Information Agency). 2009. Coal resources, current and back issues. Available at www.eia.doe.gov/cneaf/coal/reserves/reserves.html#_ftp/. Accessed July 30, 2009.

Fan, L.S., and F.X. Li. 2007. Clean coal. Physics World 20:37-41.

Fan, L.S., and M. Iyer. 2006. Coal cleans up its act. The Chemical Engineers October:36-38.

Furimsky, E. 1998. Gasification of oil sand coke: Review. Fuel Processing Technology 56:263-290.

Grainger, L., and J. Gibson. 1981. Coal Utilisation: Technology, Economics and Policy. London: Graham and Trotman.

Gupta, P., L.G. Velazquez-Vargas, and L.S. Fan. 2007. Syngas Redox (SGR) process to produce hydrogen from coal derived syngas. Energy and Fuels 21:2900-2908.

Herman, R.G. 2000. Advances in catalytic synthesis and utilization of higher alcohols. Catalysis Today 55:233-245.

IEA (International Energy Agency). 2007. Thermal gasification of biomass: Highlights of current gasification activities in member countries: Task 33. Technical Presentations from Second Semi-Annual Task Meeting, Berrgen & Petten, Netherlands, October 24, 2007.

IRGC (International Risk Governance Council). 2008. Regulation of Carbon Capture and Storage. Available at http://www.irgc.org/IMG/pdf/Policy_Brief_CCS.pdf. Accessed January 14, 2009.

Jaramillo, P., M. Griffin, and H.S. Matthews. 2008. Comparative analysis of the production costs and life-cycle GHG emissions of FT liquid fuels from coal and natural gas. Environmental Science and Technology 42:7559-7565.

Knoef, H.A.M. 2005. Handbook Biomass Gasification. Enschede, Netherlands: BTG.

Kreutz, T.G., E.D. Larson, G. Liu, and R.H. Williams. 2008. Fischer-Tropsch fuels from coal and biomass. In 25th Annual International Pittsburgh Coal Conference, Pittsburgh, Pa.

Kung, H.H. 1980. Methanol synthesis. Catalysis Reviews 22:235-259.

Larson, E.D., G. Fiorese, G. Liu, R.H. Williams, T.G. Kreutz, and S. Consonni. 2008. Coproduction of synthetic fuels and electricity from coal + biomass with zero carbon emissions: An Illinois case study. Energy Procedia 1:4371-4378.

Li, D., C. Yang, W. Li, Y. Sun, and B. Zhong. 2005. Ni/ADM: A high activity and selectivity to C2 + OH catalyst for catalytic conversion of synthesis gas to C1-C5 mixed alcohols. Topics in Catalysis 32:233-239.

Lovell, J. 2008. Britain seeks to set pace in carbon capture quest. Reuters News Service, July 1. Available at http://www.reuters.com/article/environmentNews/idUSL302998920080630. Accessed October 12, 2008.

McCollum, D.L., and J.M. Ogden. 2006. Techno-Economic Models for Carbon Dioxide Compression, Transport, and Storage. Davis: University of California, Davis.

MIT (Massachusetts Institute of Technology). 2007. The Future of Coal. Cambridge: Massachusetts Institute of Technology.

Mzinyati, A.B. 2007. Fuel-blending stocks from the hydrotreatment of a distillate formed by direct coal liquefaction. Energy and Fuels 21:2751-2762.

NRC (National Research Council). 1990. Fuels to Drive Our Future. Washington, D.C.: National Academy Press.

NRC. 2002. Evolutionary and Revolutionary Technologies for Mining. Washington, D.C.: National Academy Press.

NRC. 2007. Coal Research and Development to Support National Energy Policy. Washington, D.C.: The National Academies Press.

Palmgren, C., M.G. Morgan, W. Bruine de Bruin, and D.W. Keith. 2004. Initial public perceptions of deep geological and oceanic disposal of carbon dioxide. Environmental Science and Technology 38:6441-6450.

Socolow, R.H. 2005. Can we bury global warming? Scientific American (July):49-55.

Tarka, T. 2008. Systems Analysis Technical Note, CO_2 Transport and Storage Costs. Pittsburgh, Pa.: U.S. Department of Energy.

Tomlinson, G., A. El Sawy, and D. Gray. 1989. Economics of advanced indirect liquefaction processes. In Contracts Review Meeting. CONF-891131—DE90 008422. Springfield, Va.: National Technical Information Service.

Tomlinson, G., and D. Gray. 2007. Chemical Looping Process in a Coal-to-Liquids Configuration. Washington, D.C.: U.S. Department of Energy.

Williams, R.H., E.D. Larson, G. Lui, and T.G. Kreutz. 2008. Fischer-Tropsch fuels from coal and biomass: Strategic advantages of once-through (polygeneration) configurations. Energy Procedia 1(1):4379-4386.

Wilson, E.J., S.J. Friedmann, and M.F. Pollak. 2007. Risk, regulation and liability for carbon capture and sequestration. Environmental Science and Technology 41:5945-5952.

5

Distribution

The final stage in supplying any fuel is distribution to users. Gasoline and diesel fuels benefit from a well-established distribution system that cost-effectively makes them available for customers to purchase individually at service stations and for vehicle fleets at refueling stations. In 2008, about 160,000 refueling stations in the United States supplied several grades of gasoline and sometimes diesel. The distribution system meets the demand for gasoline with varied octane (antiknock) ratings, whose composition varies seasonally to compensate for changes in ambient temperature and because of fuel requirements related to air-quality control. Diesel fuel also varies seasonally in composition because of cold-weather requirements.

The biofuel that is most available and most used today is ethanol, which accounts for nearly 6 percent of U.S. gasoline use by volume and 4 percent by energy content. The supply of ethanol is expected to increase steadily over the next decade and beyond as a result of the Renewable Fuel Standard as amended in the 2007 U.S. Energy Independence and Security Act (EISA). One of the important challenges to widespread use of ethanol as a transportation fuel is that it cannot be transported and delivered in existing petroleum-delivery systems (for example, pipelines) because of the incompatibility of materials and water absorption by ethanol in the pipelines. Therefore, this chapter focuses on the transportation and distribution of ethanol. Synthetic gasoline and diesel fuels produced from biomass and coal and future synthetic biofuels, such as biobutanol, are expected to be compatible with the existing infrastructure for petroleum products; distribution costs for these synthetic fuels are expected to be similar to those for petroleum-based fuels. Although their volume is expected to grow over the next decade or two, the increasing volume could be accommodated incrementally in the existing infrastructure, so it is not explicitly discussed here.

In considering ethanol as a transportation fuel, one needs to be aware that ethanol is not a one-to-one replacement for its petroleum-based counterparts. Ethanol contains only two-thirds of the energy of the same volume of gasoline. The corn grain used and the cellulosic biomass to be used to produce ethanol also vary in density and other physical characteristics that affect their costs of transportation from field to conversion plants, as discussed in Chapter 2. Most of the biomass will be produced in the interior of the United States (or from wood in the Northwest). The economics of transporting biomass feedstock versus finished transportation fuel favor biorefineries in the Midwest, but 80 percent of the U.S. population—representing the largest current and future transportation-fuel markets—lives along the coasts. Figure 5.1 maps existing biorefineries and those

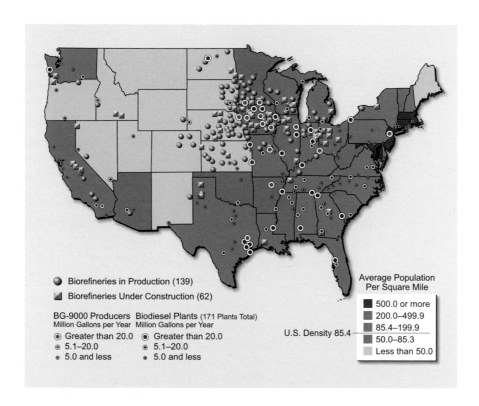

FIGURE 5.1 *U.S. ethanol and biodiesel plant locations compared with state population density as of July 1, 2007.*
Source: Adapted from NBB (2007), U.S. Census Bureau (2007), and RFA (2008).

under construction that produce ethanol and biodiesel relative to U.S. state population densities. As discussed later in this chapter, there is an ethanol-compatible infrastructure for transporting ethanol to markets today for the relatively small volumes produced. However, if the ethanol markets increase rapidly over the next decade, the lack of delivery capacity and the complexity of constructing it will challenge the industry.

As of 2008, most of the ethanol produced is blended in gasoline at up to 10 percent by volume; such ethanol-containing gasoline is designated E10. There are few outlets for higher ethanol blends, such as E85; close to 15 percent gasoline (by volume) is blended with pure ethanol for several reasons, for example, to improve vehicle cold-starting. If E10 fuels were sold at every refueling station in the United States, about 15 billion gallons of ethanol would be consumed each year. If the United States plans to produce about 40 billion gallons of cellulosic ethanol each year to improve energy security and to reduce carbon emissions, the number of E85 refueling stations will have to be increased to more than the 1900 stations that exist in 2008.

ETHANOL TRANSPORTATION

More than two-thirds of the quantity of U.S. petroleum products is shipped via pipeline, and the rest via barge (27 percent), truck (3 percent), or rail (2 percent) (Booz Allen Hamilton, 2007). Ethanol is not compatible with existing petroleum pipelines; it can damage pipeline seals and other equipment and even induce cracking in pipeline steel (Farrell et al., 2007). The pipeline industry, however, is considering dedicated pipeline for ethanol as an option (as discussed later in this chapter).

A typical ethanol-distribution system is shown in Figure 5.2. Typical ethanol transportation fuels are E10 and E85. As shown in the figure, truck transportation is the critical last step in the system (transportation from blender to fueling station). It is unlikely that any other mode of transportation will replace trucks at this stage in the distribution system, because trucks are the most economical mode of short-range transportation.

If larger volumes are to be carried between biorefineries and blending stations, however, there could be several competing modes. A next-generation ethanol plant taking in 4 million tons of biomass per year would produce 30,000 bbl of ethanol per day. One inland barge would transport about 30,000 bbl (1.3 million gallons) of denatured ethanol, one railroad car about 750 bbl (33,000 gal),

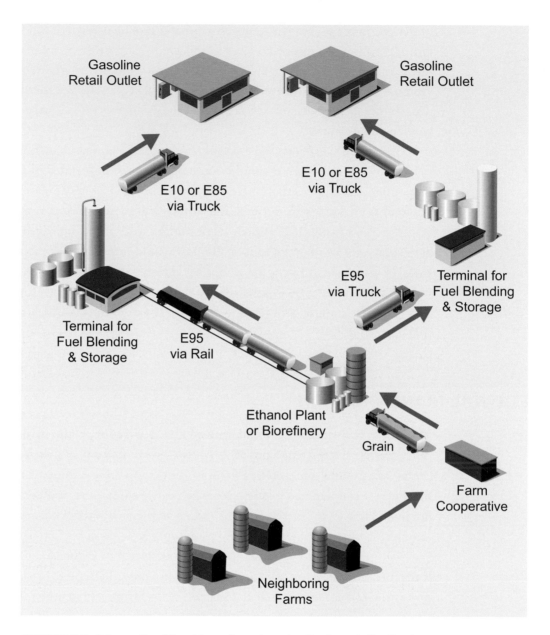

FIGURE 5.2 *Schematic of land-based truck and rail ethanol-distribution system. Source: USDA-AMS, 2007.*

TABLE 5.1 Ethanol-Transportation Costs, by Mode of Transportation

	Mode of Transportation		
	Truck	Rail	Barge
Loading and unloading	$0.02/gal	$0.015/gal	$0.015/gal
Time-dependent	$32/h per truckload	Not applicable	Not applicable
Fixed cost	Not applicable	$8.80/100 gal	$1.40/100 gal
Distance-dependent	$1.3/mile per truckload	$0.0075/mile per 100 gal	$0.015/mile per 100 gal
Truck capacity	8,000 gal	33,000 gal	1.26 million gallons

Source: Adapted from Jenkins et al., 2008.

and one truck about 200 bbl (8,000 gal) (USDA-AMS, 2007). For comparison, a 12-inch pipeline could move up to 100,000 bbl (4.2 million gallons) per day.

The cost of transportation and distribution is a substantial component of the total cost of ethanol. In the United States, ethanol is carried from biorefineries to staging[1] and blending[2] terminals and then to fueling stations by trains, trucks, and barges. Table 5.1 lists estimated costs associated with shipping ethanol via truck, rail, or barge (Jenkins et al., 2008). The cost of transporting petroleum fuels from refineries to fueling stations is about $0.03-0.05/gal, whereas the combined cost of transporting ethanol from production plants to fueling stations is estimated to be $0.13–0.18/gal (Morrow et al., 2006; GAO, 2007). However, as discussed later, the costs of ethanol transportation could be considerably higher if the delivery system is not optimized.

Trucks

Trucking of fuel ethanol is the most efficient and cost-effective transportation mode for distances up to about 300 miles (Reynolds, 2002). Transportation costs for longer distances are much higher, $0.20/gal. A typical transport truck can carry about 8,000 gal/load (Reynolds, 2002).

Increasing the amount of E10 or E85 fuel distributed would require increases in the number of transport trucks. Today, for every 10 trucks transporting E10 fuel from a blending terminal to a fueling station, there is one truck carrying dena-

---

[1]Terminal where smaller shipments of ethanol are received and held until there is sufficient fuel to transport.

[2]Terminal where fuel-grade ethanol is blended with gasoline. Typical blends are E10 and E85.

tured ethanol to a blending terminal. Expanded E85 distribution would require 8.5 truckloads of fuel ethanol for every 10 trucks leaving the blending terminal (API, 2008). The most pressing constraint on increasing ethanol transportation is the industry-wide shortage of drivers of long-haul, heavy-duty trucks, especially drivers with HAZMAT certification. The truck-transportation industry is already experiencing a shortage of drivers, and the imbalance between driver demand and working truck drivers is predicted to grow to more than 100,000 by 2014 (American Trucking Associations, 2005).

Barges

Ethanol transportation via barge or ship is limited to locations near large waterways, so only about 10 percent of the ethanol produced in the United States is transported by barges (USDA-AMS, 2007). Barges can, however, transport a large volume of ethanol in a single shipment. Inland tank barges, for example, can transport 1 million gallons of ethanol (USDA-AMS, 2007). Larger, ocean-faring ships and barges can transport about 1–12 million gallons of ethanol, depending on the size of the vessel and the final destination. Inland barges transport ethanol down the upper Mississippi River to staging terminals below the river's navigation locks. Larger barges or ships that transport ethanol between staging facilities on the lower Mississippi and blending terminals on the West Coast require passage through the Panama Canal and take about 34 days to make the trip. Transportation times to East Coast locations take about 24 days (Reynolds, 2002). Shipments to the densely populated Northeast coast add $0.10–0.12/gal[3] to the price of ethanol (Reynolds, 2002).

Funding to increase the size of the 1930s-built upper Mississippi navigation-lock system was approved in November 2007. The expansion from 600-ft locks to 1200-ft locks will open the upper Mississippi to larger modern barges. Increased barge and ship transportation will also require more and larger multipurpose staging facilities along waterways. Such intermodal facilities will be able to accept and quickly unload ethanol arriving by barge, rail, and truck from the Midwest and then distribute it by the most economical mode of transportation (truck, train, or barge or ship).

---

[3]$0.02–0.04/gal for ship or ocean barge shipments, and $0.08–0.16/gal for inland barge shipments.

TABLE 5.2 Costs of Ethanol Transportation Between Southwest Iowa to Illinois and from Southwest Iowa to California or the Louisiana Basin via Unit Train, Gathered Train, or Single Car

Route	Unit Train[a]	Gathered Train[b]	Single Car
		$/car	
Southwest Iowa to Illinois	2100	2500	2900
Southwest Iowa to California or Louisiana Basin	3900	4400	5300
		$/gal	
Southwest Iowa to Illinois	0.07	0.09	0.10
Southwest Iowa to California or Louisiana Basin	0.13	0.15	0.18

[a]95-car ethanol train originating at one plant.
[b]Ethanol train originating at two or three plants.
Source: BNSF Railway Company, 2007.

Rail

Trains currently transport most fuel ethanol from biorefineries to blending terminals throughout the United States. Unit trains[4] are the most economical and efficient mode of transportation (see Table 5.2). Typical unit-train turnaround time for delivery from the Midwest to coastal blending terminals is about 6 wk. However, if a biorefinery has insufficient capacity to fill a unit train, a single-commodity train can gather ethanol from several biorefineries, or it can be shipped in a single car as part of a multicargo train. In those cases, assembly and disassembly increase the delivery time and cost. Because the costs and time for transporting ethanol via rail and barge are similar (Table 5.1), the main benefit of marine cargo transport is the ease of unloading at the destination because of its high volume (Reynolds, 2002).

As of January 1, 2007, 41,000 rail tank cars capable of shipping ethanol were in use (USDA-AMS, 2007). Existing rail capacity can accommodate current ethanol transportation demand, but increases in ethanol production will lead to a possible railcar shortage. Railcar manufacturers have a 1.5-yr backlog of tank-car orders, and shortages are expected to continue for the next few years because of the steep rise in ethanol production (Crooks, 2006).

---

[4]A single-commodity train shuttling between a sole point of origin and a sole destination.

Pipelines

Pipelines are the most economical method for transporting large quantities of liquids over long distances. Shipping fuel through pipelines costs $0.025/gal or less per 1,000 miles (Curley, 2008). Gasoline-transportation costs by the various modes were estimated as follows (Curley, 2008):

- *Pipeline:* $0.015–0.025/gal per 1000 miles.
- *Barge:* $0.04–0.05/gal per 1000 miles.
- *Train:* $0.075–0.125/gal per 1000 miles.
- *Truck:* $0.30–0.40/gal per 1000 miles.

Pipelines are not available for commercial biofuel transportation in the United States, although they might have much lower costs. The combination of increased ethanol production and demand with the time, volume, and cost benefits associated with pipeline transportation is spurring interest in overcoming the operational, technical, and economic issues associated with biofuel pipeline transportation. The primary issues associated with such pipeline transportation are these:

- Ethanol has a greater affinity for water than does gasoline.
- Ethanol is a much better solvent than gasoline.
- Ethanol is more corrosive than petroleum products.
- Existing pipelines are not near biorefineries.
- The practical bounds on the "transmix" (a term that the industry uses for a mixture of immiscible or otherwise incompatible fluids that need to be separated) have yet to be established.

Because of its affinity for water, ethanol is hydrated by ambient water from other transported fuels, terminals, and tank roofs as it flows through a multiuse, multipipeline network. Pipelines for transporting traditional petroleum fuels are not airtight, so moisture can get into the pipeline; the small amount of water that enters, however, does not mix with the gasoline and can be easily drained off. Ethanol, in contrast, has a higher affinity for water than does gasoline. Water contamination picked up during ethanol transportation will increase the fraction of water above allowable fuel-ethanol specifications, and the fuel will not be able to be sold to consumers. In a blended ethanol-gasoline fuel, once the ethanol absorbs enough water in a pipeline system, the fuel does not stay blended and separates into an aqueous phase that contains ethanol and water and a gasoline-rich phase. In both cases, ethanol can be recovered from the aqueous phase only by distillation. Fuel ethanol is routinely shipped via pipeline in Brazil, where phase separation is mitigated by first shipping hydrous ethanol and then anhydrous ethanol (Hammel-Smith et al., 2002). Nonetheless, pipeline shipment of ethanol in the United States will require capital investment and involve additional maintenance costs.

Because ethanol is a better solvent than petroleum products, transporting ethanol through existing multiuse pipelines dissolves many common polymers in the pipelines and thus contaminates the fuel ethanol. According to ethanol-pipeline testing conducted by Buckeye and Williams Energy Services, frequent dewatering of pipelines,[5] closed floater storage tanks, dry storage tanks, inline corrosion monitoring, and filtration systems would be required to transport ethanol through multiuse pipelines on a regular basis. Ethanol-only pipelines constructed of ethanol-compatible materials would avoid many contamination issues. However, construction of new pipelines would cost $1–2 million per mile, depending on, for example, right-of-way issues and material and labor costs (GAO, 2007). Costs could be even higher in a tight construction market as existed in the middle of 2008.

Another issue associated with the solvent properties of ethanol is an increase in stress-corrosion cracking (SCC) in pipelines in high-stress locations. Cases of SCC have been reported in pipelines, storage tanks, and associated handling equipment at distribution and blending terminals. No cases of SCC have been reported at biorefineries or after blending or in trucks, railcars, or barges (Kane and Maldonado, 2003). Thus, it might be possible to design new pipelines that minimize stress to reduce the possibility of SCC.

The pipeline industry throughout the world is seeking solutions to those issues. As Curley (2008) observed in his review of the problem,

> Since increased ethanol usage is being mandated in autos by the federal government, cheaper ways to transport ethanol are needed. Using an existing pipeline to transport ethanol is likely not practical because all the valves, gaskets, and tank seals on floating roofs would need to be checked to see if the construction materials are compatible with ethanol. However, a new multi-products pipeline could easily be designed with ethanol compatible polymers in valves, gaskets, and seals. The steel for the pipeline could be specified to minimize the possibility of stress corrosion cracking (SCC).

---

[5]To remove water from pipelines.

The main unresolved issue is how to handle the ethanol transmix in a cost-effective manner. If this issue could be resolved, transporting ethanol in a multi-products pipeline could occur. Nevertheless, a small diameter dedicated ethanol pipeline may be the best alternative since there are no transmix or product quality issues with this alternative. A 4- or 6-inch dedicated pipeline could be placed in the same trench as a new gasoline/diesel pipeline at a relatively low cost.

The Brazilians are studying running a small diameter (about 12-inch) carbon steel pipeline from the interior (ethanol production areas) of the country to the east coast of Brazil.

Kinder Morgan Energy Partners started up a converted 106-mile-long etha-nol-pipeline test in late 2008. The conversion and cleaning of the petroleum-fuel pipeline included replacing gaskets and rotating element pumps with ones that are compatible with ethanol. Extra pipe scrubbers were sent to remove excess buildup to prevent ethanol from picking up contaminants along the way. The pipeline was used to send pure ethanol that was batched side by side with gasoline going through the pipe. Corrosion inhibitor was to be injected into the entire batch (Gunter, 2008a). In October 2008, Kinder Morgan said that the pipeline test was a success and that clients could start shipping ethanol in the middle of November 2008 (Gunter, 2008b).

Retrofitting all existing multipurpose pipelines is likely not practical; however, new multipurpose pipelines could be designed with ethanol-compatible polymers in valves, gaskets, and seals and designed to minimize SCC. Laying ethanol-only pipelines next to existing multipurpose pipelines would be more cost-effective than purchasing rights-of-way for new routes.

Integration of ethanol into the commercial fuel-shipment schedule with other fuels and fuel grades requires that economical trailback tests[6] and transmix-handling procedures be developed. Transporting a biofuel, such as ethanol, directly before or after a conventional petroleum fuel will result in a mixture of ethanol and hydrocarbons that is beyond the acceptable specification for either fuel. The points where the transmix starts and ends are related to the acceptability of trace amounts of fuel contamination (Curley, 2008). The trailback threshold depends on whether low levels of biofuels affect the performance of other fuels, such as aviation fuel, and whether the biofuels meet acceptable specification after shipment.

---

[6]Trailback refers to the contamination of products in a multipurpose pipeline by additives or residues left on the pipeline walls by ethanol products that were shipped previously. Trailback tests assess the level of contamination in products shipped.

Expansion of the Delivery System

As of 2008, delivery of ethanol from plant to refueling station takes place in an existing system of trucks, barges, rail, and central blending plants. The system has expanded incrementally as ethanol production has increased. However, if the large production volumes envisioned by the EISA occur and the majority of ethanol is produced in the Midwest and consumed on the coasts, the delivery infrastructure will have to be expanded considerably. The alternative would be to use ethanol close to where it is produced.

Although the cost of delivery is a small portion of the overall ethanol-fuel cost, the logistics and capital requirements for widespread expansion could be substantial hurdles if they are not planned well. Morrow et al. (2006) provided a thorough analysis of the complexities and costs of widespread fuel ethanol expansion in the United States. If planning is suitable and fuel ethanol's cost is competitive in the fuel market, the panel believes that the ethanol-delivery infrastructure would be expanded to meet demand. Brazil, where almost all new vehicles sold are capable of using fuel from E20 to E95, provides a good example of how a distribution system for ethanol, the retailing of fuel, and the production of flexible-fuel vehicles could work smoothly without being expensive.

The panel cautions that biofuel technologies will probably evolve from ethanol to biofuels that are more compatible with the existing petroleum infrastructure. Thus, planning for biofuel expansion should consider when biofuels other than ethanol might come onto the market and ensure that the ethanol-delivery infrastructure is not overbuilt and underused.

THE MARKET FOR BIOFUELS

Consumer acceptance of biofuels will be determined by a combination of favorable prices relative to those of conventional petroleum products; subsidies and mandates for biofuel; the prevalence of E85 and biodiesel fueling stations (Figure 5.3); and the availability and affordability of flexible-fuel vehicles.

Distribution Infrastructure

Refueling availability is fundamental to the widespread adoption of alternative fuels. Biofuel-refueling stations must grow beyond niche markets to a density sufficient for supporting alternative-fuel vehicles. The number of fueling stations

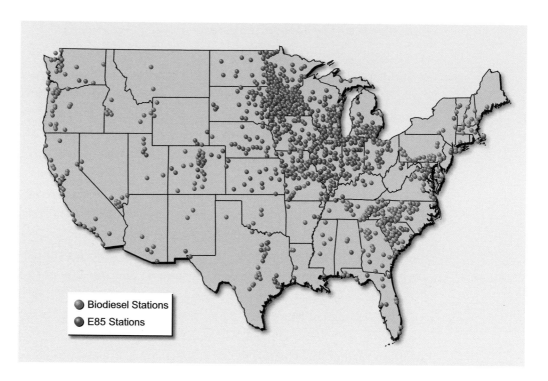

FIGURE 5.3 *U.S. biodiesel and E85 fueling-station locations.*

required to service a community adequately is unknown. However, a recent assessment based on urban population density relative to existing petroleum fueling-station density estimated that about 50,000 stations would provide sufficient coverage for the general public while providing retail competition between stations. That translates to about 0.5 station per square mile in a city with a population of 500,000 and a population density of 2,000 per square mile. The estimated upper and lower bounds on sufficient urban station coverage are shown in Figure 5.4 (Melaina and Bremson, 2008).

Retrofitting existing fueling stations with storage and dispensing equipment compatible with the chemical properties of E85 fuels is often expensive, and some station owners are averse to carrying these biofuels. To retrofit existing fueling stations, underground storage-tank systems, pumps, and dispensers must be converted to be compatible with the higher-ethanol blends. Several issues associated with retrofitting existing fueling stations are similar to those associated with pipeline transportation of ethanol and ethanol blends: phase separation as a result

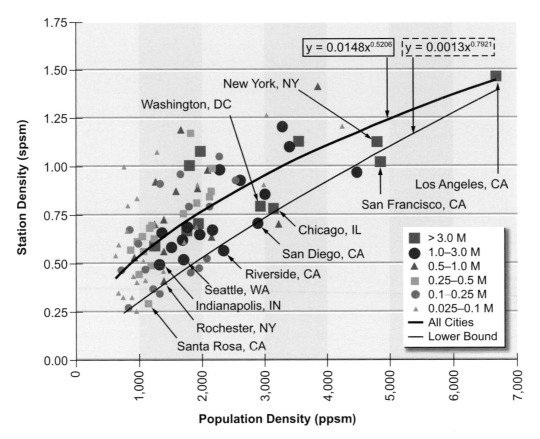

FIGURE 5.4 *Estimated sufficient alternative fueling-station coverage relative to urban population density. Upper and lower boundaries are shown.*
Note: ppsm = persons per square mile; spsm = stations per square mile.
Source: Reprinted from Melaina and Bremson, 2008. Copyright 2008, with permission from Elsevier.

of hydration, SCC, and contamination as a result of incompatible materials commonly found throughout conventional fueling stations.

Another hurdle in adding E85 capability to fueling stations is the restrictions placed on branded stations and the ability to obtain insurance (Johnson and Melendez, 2007). Many refining companies that own major gasoline brands do not allow E85 dispensers to be placed under the same canopy as their branded gasoline dispensers. If a branded fueling station wishes to add E85 capacity, a separate canopy must be added. That requirement increases the total cost of an E85 retrofitting project (Johnson and Melendez, 2007).

TABLE 5.3 Generalized Costs for Installing New E85 Equipment or Retrofitting Existing Petroleum Equipment to Be Ethanol-Compatible

Scenario	Estimated Cost	Description	Major Variables Affecting Cost
New tank, new or retrofit dispensers	$50,000–200,000	Includes new underground storage tank, pump, dispensers, piping, electric, excavation, concrete work	Dispenser needs, concrete work, excavation, sell-backs, canopy, tank size, location, labor cost, permitting requirements
Converting existing tank, new or retrofit dispensers	$2,500–30,000	Tank cleaning, replacement of incompatible components in piping and dispensers	Dispenser needs, number of incompatible components, location, labor cost, permitting requirements

Source: NREL, 2008, and references cited therein.

The National Renewable Energy Laboratory conducted a survey and literature search on the cost of adding E85-fueling capacity to existing gasoline stations (NREL, 2008). The survey included the costs incurred for 120 E85 stations, 84 of which would have new tanks installed and the remainder would convert existing tanks. Replacing one gasoline dispenser and retrofitting existing storage tanks to carry E85 at an older existing fueling station would cost $1,736–68,000. The installation cost for a new E85-compatible fueling station would be $7,559–247,000. The generalized costs for retrofitting or building a new E85-compatible fueling station collected from the literature are listed in Table 5.3 (NREL, 2008). The installation or retrofitting costs might not be considered high for some, but they might not be feasible for some individual station owners.

Fuel Use by Flexible-Fuel Vehicles

As of 2007, there were about 5 million flexible-fuel cars and light trucks in the United States—about 2 percent of the total U.S. fleet. Since their introduction, the number of flex-fuel vehicles[7] available has steadily increased. The increase, however, is largely a result of U.S. corporate average fuel economy (CAFE) mile-

---

[7]Vehicles are designed to run on ethanol–gasoline blends from E10 up to E85 fuel.

age rating requirements as opposed to the availability of E85 fuel or consumer demand for alternative-fuel vehicles.

Building a vehicle capable of running on E85 fuel adds about $100 to its production cost. Flexible-fuel vehicles have upgraded fuel systems with large fuel pumps and injectors that enable them to accommodate the greater fuel volume required for the same energy content as gasoline (Hammel-Smith et al., 2002). The fuel tanks and lines are composed of ethanol-compatible materials. Operating on E85 involves an increase of a few percent in horsepower as a result of extra cooling of the in-cylinder air and the higher fuel octane rating, and maximum vehicle acceleration is higher. Because of the lower energy of ethanol relative to gasoline, fuel consumption (in gallons per 100 miles) with ethanol is about 25 percent higher, although overall vehicle efficiency can be up to 5 percent better (West et al., 2007). The attractive antiknock characteristics of ethanol could be used to improve the efficiency with which engines use gasoline in a dual-fuel engine setup (Edmunds, 2008).

Encouraging the use of flexible-fuel vehicles and the use of gasoline with a high proportion of ethanol is a complex issue. As discussed earlier in this chapter, getting large quantities of those fuels to where the vehicles are can be challenging. In the past, the U.S. Congress mandated that federal agencies gradually increase the number of flexible-fuel vehicles in their fleets, but most purchased vehicles are used in places where flexible fuel is not readily available (Kindy and Keating, 2008). To complicate the issue, many of those flexible-fuel vehicles are large sedans and sport utility vehicles. Because of their large flexible-fuel engines and low fuel efficiency, those vehicles used more gasoline than smaller and more fuel-efficient vehicles (Kindy and Keating, 2008). The simultaneous implementation and market penetration of E85 fuel and flexible-fuel vehicles is an important practical consideration.

FINDINGS AND RECOMMENDATIONS

Finding 5.1

The need to expand the delivery infrastructure to meet a high volume of ethanol deployment could delay and limit the penetration of ethanol into the U.S. transportation-fuels market. Replacing a substantial proportion of transportation gasoline with ethanol will require a new infrastructure for its transport and distribution. Although the cost of delivery is a small fraction of the overall fuel-ethanol cost, the logistics and capital requirements for widespread expansion could present many hurdles if they are not planned for well.

Recommendation 5.1

The U.S. Department of Energy and the biofuels industry should conduct a comprehensive joint study to identify the infrastructure system requirements of, research and development needs in, and challenges facing the expanding biofuels industry. Consideration should be given to the long-term potential of truck or barge delivery versus the potential of pipeline delivery that is needed to accommodate increasing volumes of ethanol. The timing and role of advanced biofuels that are compatible with the existing gasoline infrastructure should be factored into the analysis.

Finding 5.2

Expansion of the flexible-fuel vehicle fleet needs to be complemented by the presence of ethanol stations close to where the vehicles are used. Past policy that mandated the increased use of alternative-fuel vehicles did not result in reduced gasoline consumption, because ethanol pumps were not readily available in many areas where flexible-fuel vehicles were used. The close coupling of alternative fuels and alternative-fuel vehicles is an important practical consideration. Future policy measures need to take into account implementation of alternative-fuel vehicles, availability of alternative fuels, and proximity of vehicles to fueling stations to ensure an effective vehicle and fuel transition.

REFERENCES

American Trucking Associations. 2005. The U.S. truck driver shortage: Analysis and forecasts. Available at http://www.truckline.com/StateIndustry/Pages/DriverShortageReport.aspx. Accessed October 19, 2008.

API (American Petroleum Institute). 2008. Shipping Ethanol Through Pipelines. Available at http://www.api.org/aboutoilgas/sectors/pipeline/upload/pipelineethanolshipmentfinal-3.doc. Accessed October 19, 2008.

BNSF Railway Company. 2007. Pricing Update: October 2007 Ethanol Rate Adjustment. Available at http://www.bnsf.com/markets/agricultural/ag_news/year2007/pricing07/p08-23-07a.html. Accessed October 13, 2008.

Booz Allen Hamilton. 2007. Rationalizing the Regulatory Environment for Renewable Energy: Overcoming Constraints for Rapid Growth in the Biofuels Industry. McLean, Va.: Booz Allen Hamilton.

Crooks, A. 2006. From grass to gas: On the road to energy independence, how soon will cellulosic ethanol be a factor? Rural Cooperatives (Ethanol Issue, September/October):16-18.

Curley, M. 2008. Can ethanol be transported in a multi-product pipelines? Pipeline and Gas Journal 235:34.

Edmunds, Inc. 2008. Edmunds.com. Available at http://www.edmunds.com. Accessed October 19, 2008.

Farrell, A., D. Sperling, S.M. Arons, A.R. Brandt, M.A. Delucchi, A. Eggert, B.K. Haya, J. Hughes, B.M. Jenkins, A.D. Jones, D.M. Kammen, S.R. Kaffka, C.R. Knittel, D.M. Lemoine, E.W. Martin, M.W. Melaina, J. Ogden, R.J. Plevin, B.T. Turner, R.B. Williams, and C. Yang. 2007. A Low-Carbon Fuel Standard for California. Part 1: Technical Analysis. Davis: University of California, Davis.

Hammel-Smith, C., J. Fang, M. Powders, and J. Aabakken. 2002. Issues Associated with the Use of Higher Ethanol Blends (E17-E24). Golden, Colo.: National Renewable Energy Laboratory.

GAO (U.S. Government Accountability Office). 2007. Biofuels: DOE Lacks a Strategic Approach to Coordinate Increasing Production with Infrastructure Development and Vehicle Needs. Washington, D.C.: GAO.

Gunter, F. 2008a. Kinder Morgan makes waves with Florida ethanol line. Houston Business Journal, March 28. Available at http://houston.bizjournals.com/houston/stories/2008/03/31/story4.html. Accessed November 11, 2008.

Gunter, F. 2008b. Ethanol pipeline test a success, says Kinder Morgan. Houston Business Journal, October 16. Available at http://houston.bizjournals.com/houston/stories/2008/10/13/daily42.html. Accessed February 11, 2009.

Jenkins, B., N. Parker, P. Tittman, Q. Hart, J. Cunningham, and M. Lay. 2008. Strategic Development of Bioenergy in the Western States. Available at http://www.westgov.org/wga/initiatives/transfuels/reports/Biomass%20.ppt#256,1. Accessed October 13, 2008.

Johnson, C., and M. Melendez. 2007. E85 Retail Business Case: When and Why to Sell E85. Golden, Colo.: National Renewable Energy Laboratory.

Kane, R.D., and J.G. Maldonado. 2003. Stress Corrosion Cracking of Carbon Steel in Fuel Grade Ethanol: Review and Survey. Washington, D.C.: American Petroleum Institute.

Kindy, K., and D. Keating. 2008. Problems plague U.S. flex-fuel fleet: Most government-bought vehicles still use standard gas. Washington Post, November 23. Available at http://www.washingtonpost.com/wpdyn/content/article/2008/11/22/AR2008112200886_pf.html. Accessed February 9, 2009.

Melaina, M., and J. Bremson. 2008. Refueling availability for alternative fuel vehicle markets: Sufficient urban station coverage. Energy Policy 36:3223-3231.

Morrow, W.R., W. Michael Griffin, and H. Scott Matthews. 2006. Modeling switchgrass derived cellulosic ethanol distribution in the United States. Environmental Science Technology 40:2877-2886.

NBB (National Biodiesel Board). 2007. National Biodiesel Board Annual Report 2007. Jefferson City, Mo.: NBB.

NREL (National Renewable Energy Laboratory). 2008. Cost of Adding E85 Fueling Capacity to Existing Gasoline Stations: NREL Survey and Literature Search. Golden, Colo.: NREL.

Reynolds, R.E. 2002. Infrastructure Requirements for an Expanded Fuel Ethanol Industry. South Bend, Ind.: Downstream Alternatives.

RFA (Renewable Fuels Association). 2008. Changing the Climate: 2008 Ethanol Industry Outlook. Washington, D.C.: RFA.

U.S. Census Bureau. 2007. Population Profile of the United States. Available at http://www.census.gov/population/www/popprofile/profiledynamic.html. Accessed October 10, 2008.

USDA-AMS (U.S. Department of Agriculture, Agricultural Marketing Service). 2007. USDA Ethanol Transportation Backgrounder. Available at http://www.ams.usda.gov/AMSv1.0/getfile?dDocName=STELPRDC5063605&acct=atpub. Accessed October 13, 2008.

West, B.H., A. Lopez, T. Theiss, R. Graves, J. Storey, and S.A. Lewis, Sr. 2007. Fuel economy and emissions of the ethanol optimized Saab 9-5 biopower. In Society of Automotive Engineers Powertrain and Fluids Meeting, Chicago, Ill.

6

Comparison of Options and Market Penetration

Chapters 2, 3, and 4 provide estimates of costs of fuel products and life-cycle carbon dioxide (CO_2) emission from liquid transportation fuels produced from biomass, coal, and coal and biomass via different conversion pathways.[1] This chapter compares the life-cycle costs, CO_2 emission, and potential supply of the alternative fuel options by analyzing the supply chain beginning with the biomass (Chapter 2) and coal feedstocks, ending with conversion to alternative liquid fuels (Chapters 3 and 4), including carbon balances. The result of the panel's analysis is a potential supply curve related to alternative liquid fuels that use biomass, coal, or combined coal and biomass as feedstocks. However, the actual supply in 2020 could well be smaller than the potential supply because there are important lags in decisions to construct new conversion plants and in construction. In addition, some of the biomass supply that appears to be economical might not be made available for conversion to alternative fuels because of logistical, infrastructure, and agricultural-organization issues. The analysis shows how the potential supply curve might change with alternative CO_2 prices and alternative capital costs. The comparisons in this chapter are based on a point-in-time estimate of costs and the panel's judgment of technological advancement in the next 10–15 years. The conclusions are drawn from consistent comparisons among alternative liquid-fuel options, but they are not predictions of what the fuel costs or market penetration would be in 2020 or 2035 inasmuch as such factors as

---

[1]This chapter assesses only CO_2 emission because the panel was not able to determine changes in other greenhouse gases throughout the life cycle of fuel production. Changes in greenhouse gases other than CO_2 are likely to be small or nonexistent.

technological changes, policies that encourage development of one option rather than another, and market forces could alter the conclusions.

COMPARISON OF COSTS, GREENHOUSE GAS EMISSIONS, AND POTENTIAL FUEL SUPPLY

To examine the potential supply of liquid transportation fuels from nonpetroleum sources, the panel developed estimates of the unit costs and quantities of various cellulosic biomass sources that could be produced sustainably as discussed in Chapter 2. The panel's analysis was based on land that is not now used for growing foods although the panel cannot ensure that none of that land will be used for food production in the future. The estimates of biomass supply were combined with the amount of corn grain that would probably be used to produce fuels to satisfy the current legislative requirement to produce 15 billion gallons of ethanol per year. The panel's analysis allowed it to estimate a supply function for biomass that shows the quantities of cellulosic biomass feedstocks that would potentially be available at the various unit costs. The panel assumed that coal would not be limiting in that it would be available in sufficient quantities at a constant unit cost if used with biomass in thermochemical conversion processes. The panel developed quantitative comparative analyses of alternative pathways to convert biomass, coal, or combinations of coal and biomass to liquid fuels (either ethanol or synthetic diesel and gasoline). Pathways, in principle, could include any combination of the various biomass feedstocks and coal and could include either thermochemical or biochemical conversion processes.[2] However, rather than treating all possible combinations, the panel first examined the cost of and the CO_2 emissions associated with each of the various thermochemical and biochemical conversion processes that would use one biomass feedstock and then examined the costs, supplies, and CO_2 emissions associated with one thermochemical conversion process and one biochemical conversion process that would use each of the biomass feedstocks.

The first set of analyses compared the costs and greenhouse gas emissions from fuels produced by biochemical and thermochemical conversion. The panel

---

[2]The panel also included biochemical conversion of corn grain to ethanol but did not focus the quantitative analysis on this process.

recognizes that the cost of fuel and the greenhouse gas emissions from biofuels vary with feedstock. Because the purpose of the first set of analyses was to compare biochemical and thermochemical conversion, using one biomass feedstock in the analyses would better illustrate the differences between the conversion processes. *Miscanthus*, a high-yield perennial grass, was the biomass feedstock used for each conversion process (except those using only coal) because its cost and chemical composition are about the medians of the estimated costs and chemical composition of different cellulosic feedstocks. That analysis allowed the panel to estimate unit costs of each of the thermochemical and biochemical conversion processes on the assumption that *Miscanthus* was the biomass feedstock used for each process.

For the second set of comparisons, the panel chose two generic conversion processes—conversion of each of the lignocellulosic biomass feedstocks to produce ethanol, and thermochemical conversion of a combination of coal with each of the lignocellulosic biomass feedstocks (in a coal:biomass ratio of 60:40 on an energy basis) to produce synthetic diesel and gasoline. The estimated supply function for biomass provided information about feedstock quantities and costs. That information was combined with information about conversion costs to obtain supply functions for alternative fuels produced via either thermochemical or biochemical conversion and the assumed corn grain ethanol.

In its analyses, the panel made the following assumptions. Changes in the assumptions would normally change the estimated potential supply function. And because uncertainty is associated with each of the assumptions, the collection of uncertainties translates to important uncertainties in the potential supply curve.

- All available land discussed in Chapter 2 will be made available for growing biomass for liquid fuels; none will be used for stand-alone electricity production. This assumption implies that renewable portfolio standards for electricity production will not result in the use of biomass to satisfy the requirements for renewable supplies of electricity.
- Prices of biomass correspond to the costs of producing the biomass, including the opportunity cost of land. (See Chapter 2 for cost estimation.) All available biomass will be priced at those costs. As in Chapter 4, a coal price of $42/ton was used.
- Conversion plants that use biomass as feedstock will have the capacity of using it at about 4000 dry tons per day, and all plants will run at 90 percent of capacity.

- Biochemical conversion plants use 0.45–0.51 dry tons of biomass for each barrel of ethanol produced, with variations among different feedstocks based on their chemical compositions.
- Capital costs for all investment are based on a 7 percent pretax, no-subsidy real discount rate. Possible variations in discount rate are ignored.
- Where specified, carbon capture and storage (CCS) will be used to dispose of CO_2 permanently. The CCS costs represent estimates of engineering costs to implement CCS. Although there is considerable uncertainty in CCS costs because of potential social, legal, and political issues, these issues are not included in the analyses. Thus, the full cost of CCS could be higher than that used in the analyses and will not be known until CCS is implemented on a commercial scale. (See Chapter 4 and Appendix K.)
- If a greenhouse gas price is imposed, it applies to the entire life-cycle CO_2 net emission, including emission released in growing biomass, in the conversion processes, and in the ultimate combustion of the liquid fuels, minus CO_2 removed from the atmosphere in growing the biomass.[3] A process that removes more CO_2 from the atmosphere than it produces would receive a net payment for CO_2.
- The panel cannot project the carbon price. When a carbon price is included, it is assumed to be $50/tonne of CO_2 in 2020 and in the years shortly thereafter. The actual carbon price could be larger or smaller than that.
- To be consistent with the analysis in Chapter 2, these analyses assume that no indirect greenhouse gas emissions result from land-use changes in the growing and harvesting of cellulosic biomass. All biomass volumes in Chapter 2 were estimated under the constraint that they could

---

[3]Emissions released in growing biomass included estimates of petroleum, natural gas, and fertilizer used for growing, harvesting, and transporting the biomass. Increases in carbon in soil were subtracted. For waste, there is no such reduction for growing biomass, because any such reductions would be independent of whether waste was used as feedstock or permanently stored in landfill. Carbon emissions of the conversion process included total carbon inputs—biomass, coal, and electricity—minus carbon remaining in the fuel. For processes that generated electricity, electricity input was a negative number that reduced the calculated carbon release. This carbon credit for electricity generation was based on 0.61 tonne of CO_2 per megawatt-hour of electricity generated by the process. It was assumed that on combustion all carbon remaining in the fuel would be released into the atmosphere as CO_2.

be grown and harvested without creating indirect greenhouse gas emissions.

- Production of corn grain has indirect greenhouse gas emissions, but the panel's cost analyses assume that a U.S. carbon price will not be imposed on such indirect emissions.
- Electricity produced as a coproduct has a value of $80/MWh[4] in the absence of any price placed on greenhouse gases. If a greenhouse gas price is imposed, the value of coproduct electricity includes, in addition to $80/MWh, the cost of the CO_2 emission for electricity generation on the basis of the average of all U.S. electricity generation.
- The biomass and cofed coal and biomass conversion plants are sized for biomass feed rates of about 4,000 dry tons per day.
- The high-yield perennial grass is *Miscanthus* at $101 per dry ton.

Chapter 2 discusses the projected costs and availability of the various biomass feedstocks in 2020. The data from Chapter 2 have been combined to estimate a supply function for biomass to show the quantities of biomass feedstocks available at the various unit costs. That supply function is shown in Figure 6.1. As discussed in Chapter 2, the unit costs of most of the feedstocks—straw, woody biomass, corn stover, *Miscanthus*, native and mixed grasses, and switchgrass—are built up from estimates of the various costs of growing and transporting them. The costs of two feedstocks—corn grain and hay—are based on recent market prices. In particular, the panel assumed that by 2020 the corn price will have dropped sharply from the 2008 high of $7.88/bushel to $3.17/bushel, corresponding to $130 per dry ton, a price more consistent with historical prices. The panel assumed that the price of dryland or field-run hay will be $110/ton, which is similar to historical prices. Finally, the cost of using wastes is based on a rough estimate of the costs of gathering, transporting, and storing municipal waste. Such costs can be expected to be highly variable, but the panel assumed that gathering, transporting, and storing will add up to $51 per dry ton.

The costs of producing alternative liquid fuels via the various pathways were estimated on the basis of the costs of feedstocks, capital costs, operating costs, conversion efficiencies, and the assumptions outlined above. Scaling factors

----

[4]This is the value at the busbar. $80/MWh is the assumed wholesale price of electricity in 2020 in the absence of any carbon prices. The panel did not estimate the feedback from changes in policy options on that electricity price, other than the effects of including carbon prices.

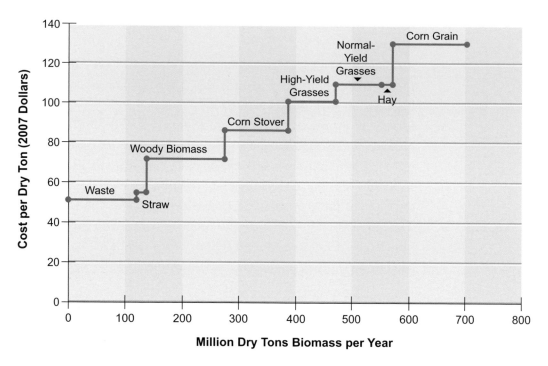

FIGURE 6.1 *Supply function for biomass feedstocks in 2020. High-yield grasses include* Miscanthus *and normal-yield grasses include switchgrass and prairie grasses.*

for capital costs of biochemical and thermochemical conversion plants were derived from two independent analyses and so might not be directly comparable. A factor of 0.70 was used for biochemical conversion plants, and a factor of 0.90 was used for thermochemical conversion plants. Figure 6.2 shows the estimate of the gasoline-equivalent[5] cost of alternative liquid fuels, without a CO_2 price, produced from coal, biomass, or combined coal and biomass. As indicated above, liquid fuels would be produced by using biochemical conversion of *Miscanthus* to ethanol (biochemical ethanol) or by using thermochemical conversion via the Fischer-Tropsch (FT) process or a methanol-to-gasoline (MTG) process. For thermochemical conversion, FT and MTG are shown both with and without CCS. As discussed in Chapter 4, the cost of CCS was based on engineering estimates of expenses for transport, land purchase, permitting, drilling, all required capital

---

[5]Costs per barrel of ethanol are divided by 0.67 to put ethanol costs on an energy-equivalent basis with gasoline. For Fischer-Tropsch liquids, the conversion factor is 1.0.

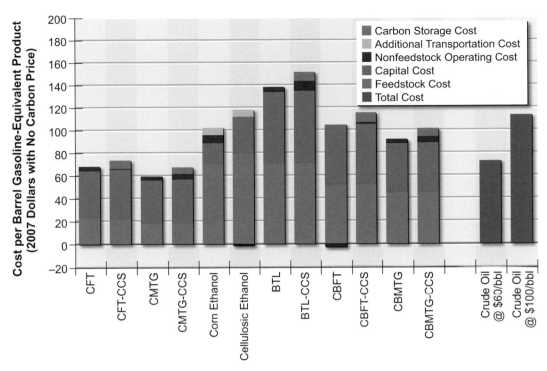

FIGURE 6.2 *Costs of alternative liquid fuels produced from coal, biomass, or coal and biomass with zero carbon price. Costs are given per barrel of gasoline equivalent. Note: BTL = biomass-to-liquid fuel; CBFT = coal-and-biomass-to-liquid fuel, Fischer-Tropsch; CBMTG = coal-and-biomass-to-liquid fuel, methanol-to-gasoline; CCS = carbon capture and storage; CFT = coal-to-liquid fuel, Fischer-Tropsch; CMTG = coal-to-liquid fuel, methanol-to-gasoline.*

equipment, storing, capping wells, and monitoring for an additional 50 years. The full cost of CCS could be higher as a result of uncertainty about the regulatory environment of CO_2 storage. The supply of ethanol produced from corn grain is also included in the figure. For comparison, costs of gasoline are shown in Figure 6.2 for two different crude oil prices: $60 per barrel and $100 per barrel.

Figure 6.3 shows the net CO_2 emission per gasoline-equivalent barrel produced by various production pathways. Figure 6.4 shows the detailed flows of CO_2 underlying the net flows in Figure 6.3. The CO_2 released on combustion is similar among the various pathways, with ethanol releasing less CO_2 on combustion than either gasoline or synthetic diesel and gasoline do. The large variation in net releases is the result of the large variations in the CO_2 taken from the atmosphere in growing biomass and the large variations in the CO_2 released into the atmosphere in the conversion process.

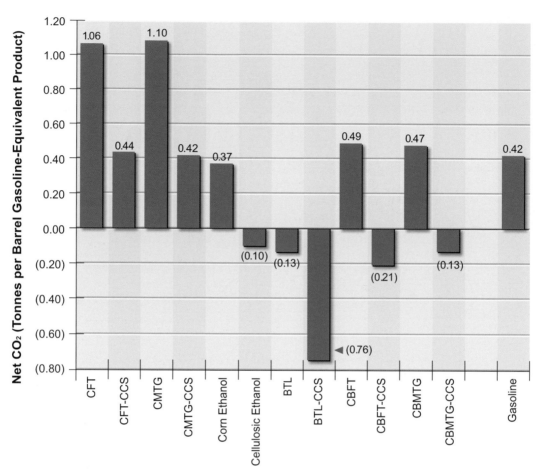

FIGURE 6.3 *Estimated CO₂ emission over the life cycle of alternative-fuel production from the mining and harvesting of resources to the conversion to and consumption of fuel. CO₂ is expressed as tonnes of CO₂ per barrel of gasoline-equivalent liquid fuels; a barrel of ethanol is assumed to have 67 percent as much energy as a barrel of gaso-line. The life-cycle CO₂ emission from biofuels includes a CO₂ credit from photosynthetic uptake by plants, but indirect greenhouse gas emissions, if any, as a result of land-use changes are not included.*
Note: BTL = biomass-to-liquid fuel; CBFT = coal-and-biomass-to-liquid fuel, Fischer-Tropsch; CBMTG = coal-and-biomass-to-liquid fuel, methanol-to-gasoline; CCS = carbon capture and storage; CFT = coal-to-liquid fuel, Fischer-Tropsch; CMTG = coal-to-liquid fuel, methanol-to-gasoline.

The results in Figure 6.2 show that FT and MTG coal-to-liquid (CTL) fuel products with and without CCS are cost-competitive at crude prices of about $60/bbl, but Figure 6.3 shows that without CCS the process vents a large amount of CO₂, almost twice that of petroleum gasoline on a life-cycle basis. With CCS, the CO₂ life-cycle emission is about the same as that of petroleum gasoline. The biochemical conversion of biomass produces fuels that are more expensive than

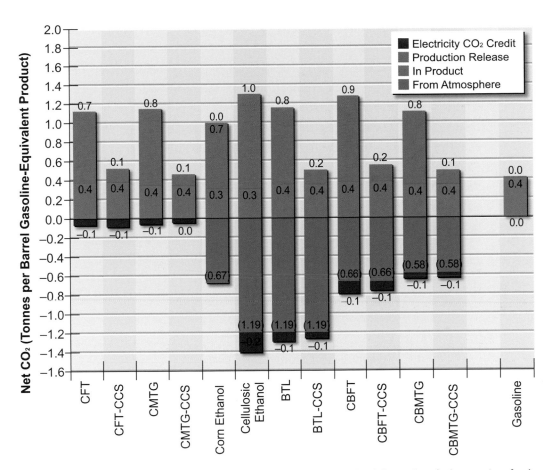

FIGURE 6.4 *Detailed flows of CO₂ emission over the life cycle of alternative-fuel production from the mining and harvesting of resources to the conversion to and consumption of fuel. CO₂ emission is expressed as tonnes of CO₂ per barrel of gasoline equivalent.*
Note: BTL = biomass-to-liquid fuel; CBFT = coal-and-biomass-to-liquid fuel, Fischer-Tropsch; CBMTG = coal-and-biomass-to-liquid fuel, methanol-to-gasoline; CCS = carbon capture and storage; CFT = coal-to-liquid fuel, Fischer-Tropsch; CMTG = coal-to-liquid fuel, methanol-to-gasoline.

CTL fuels because the conversion plants are smaller and the feedstock more expensive: biomass costs 3–4 times as much as coal on an energy-equivalent basis. Because of the lower capital cost of the biochemical conversion plants, even the smaller plant produces cellulosic ethanol competitively, at about $115/bbl of gasoline equivalent. CO_2 emission from the corn grain ethanol is slightly lower than that from gasoline. In contrast, CO_2 emission from cellulosic ethanol without CCS is close to zero.

The cost of liquid fuel from thermochemical conversion of biomass, with CO_2 venting and without coal, is about $140 and is higher than that from biochemical conversion. Most of the difference in cost results from the greater electricity sales to the grid in connection with the biochemical conversion process. Thermochemical conversion of biomass has the potential of large negative net releases of CO_2 with CCS; that is, the process leads to a net removal of CO_2 from the atmosphere. Particularly interesting is the results from the relatively small (8,000 tons/day total feed) cofed coal and biomass plant with CCS. The fuel costs are about $110/bbl of gasoline equivalent, and CO_2 atmospheric releases from plants with CCS are negative. Those results point to the importance of that option in the U.S. energy strategy.

The important influence of a carbon price on fuel price is shown in Figure 6.5 and Table 6.1. The figure and table show that a $50/tonne CO_2 price increases the costs of the fossil-fuel options, including the costs of petroleum-based gasoline,

TABLE 6.1 Comparison of Costs of Alternative Liquid Fuels Produced from Coal, Biomass, or Coal and Biomass With and Without a $50/tonne CO_2 Price

		Cost of Fuel ($/bbl of gasoline equivalent)				
		Thermochemical Conversion Without CCS		Thermochemical Conversion With CCS		Biochemical Conversion Without CCS
Carbon Price ($/tonne of CO_2 equivalent)	Feedstock	FT	MTG	FT	MTG	
0	Coal	68	59	74	67	Not applicable
0	Coal and biomass	101	92	115	102	Not applicable
0	Biomass	138	Not estimated	151	Not estimated	117
0	Crude oil	At crude-oil price of $60, cost of gasoline = $73/bbl				
		At crude-oil price of $100, cost of gasoline = $113/bbl				
50	Coal	121	115	95	88	Not applicable
50	Coal and biomass	126	116	105	95	Not applicable
50	Biomass	132	Not estimated	114	Not estimated	111
50	Crude oil	At crude-oil price of $60, cost of gasoline = $94/bbl				
		At crude-oil price of $100, cost of gasoline = $134/bbl				

substantially. The carbon price brings the cost of biochemical conversion options to $110/bbl of gasoline equivalent. The large amount of CO_2 vented in the CTL process almost doubles the cost of product once the carbon price of $50/tonne of CO_2 is imposed.

Inclusion of a carbon price does not increase the total costs for all pathways (Table 6.1). For example, thermochemical conversion of biomass costs about $140/bbl of gasoline equivalent without CCS, but the produced fuels with the carbon price and CCS are competitive with petroleum-based fuels in the range of $115/bbl of gasoline equivalent (or a crude oil price of $100/bbl). In general, if any pathway takes more CO_2 from the atmosphere than it releases in other parts of its life cycle, the inclusion of a carbon price reduces the total cost of producing liquid fuel via that pathway.

In reading the graphs, it is important to note that Figures 6.5, 6.6, and 6.7 show the breakdown of all costs, including negative costs, such as credit from electricity generation or from carbon uptake. The negative costs must be subtracted from the positive costs to obtain the actual costs. For example, BTL/CCS cost is $151/bbl – $37/bbl = $114/bbl.

Those estimates are all based on costs of small gasification units operating with a feed rate of 4,000 dry tons per day. Each unit is capital-intensive. Therefore, larger units can be expected to be deployed in regions where potential biomass availability is large—for example, 10,000 dry tons per day. Such units could result in much lower costs.

The panel also conducted a sensitivity analysis to assess the effect of uncertainty in capital costs on the cost of fuel products. A variation of a 30 percent increase to a 20 percent decrease in capital costs was evaluated. Results are shown in Figures 6.6 and 6.7. The capital-cost variations affect fuel costs of the capital-intensive gasification processes more than those of the biochemical conversion processes, but the variations do not have a major effect on the costs of fuel products relative to each other, particularly in light of the wide swings in crude-oil price in 2008. Although it is not shown in the figures, another less-developed concept is biochemical conversion with CCS. The panel made a rough first-pass estimate of the cost reduction in biochemical conversion ($125/bbl of gasoline equivalent) and found that with a CO_2 price of $50/tonne, cost could be reduced substantially through CCS. That cost, however, was not fully quantified.

As noted previously, the cost estimates for biochemical conversion and thermochemical conversion are based on only one biomass feedstock, *Miscanthus*. Figures 6.5 through 6.7 do not show how much fuel could be produced at the

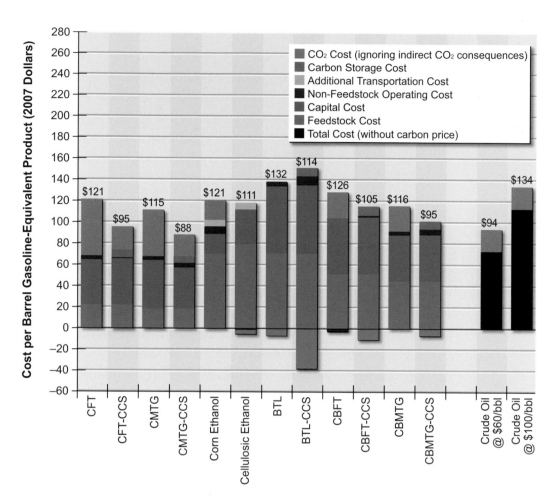

FIGURE 6.5 *Cost of alternative liquid fuels produced from coal, biomass, or coal and biomass with a CO_2-equivalent price of $50/tonne. The negative costs must be subtracted from the positive costs to obtain the actual costs; for example, BTL-CCS cost is $151/bbl – $37/bbl = $114/bbl.*
Note: BTL = biomass-to-liquid fuel; CBFT = coal-and-biomass-to-liquid fuel, Fischer-Tropsch; CBMTG = coal-and-biomass-to-liquid fuel, methanol-to-gasoline; CCS = carbon capture and storage; CFT = coal-to-liquid fuel, Fischer-Tropsch; CMTG = coal-to-liquid fuel, methanol-to-gasoline.

estimated costs. To provide a complete supply function for alternative liquid fuels, the supply function from Figure 6.1 for all biomass feedstocks has been combined with the conversion cost estimates. The results are shown in Figures 6.8 through 6.10. Figure 6.8 shows the potential gasoline-equivalent supply of ethanol from biochemical conversion of lignocellulosic biomass with 2020 deployable technology. It shows potential supply, not the panel's projected penetration of cellulosic

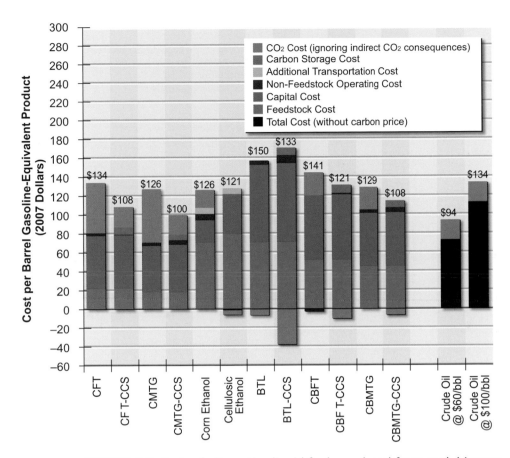

FIGURE 6.6 *Costs of alternative liquid fuels produced from coal, biomass, or coal and biomass with a CO₂ price of $50/tonne when capital costs are 30 percent higher than the panel's estimates.*
Note: BTL = biomass-to-liquid fuel; CBFT = coal-and-biomass-to-liquid fuel, Fischer-Tropsch; CBMTG = coal-and-biomass-to-liquid fuel, methanol-to-gasoline; CCS = carbon capture and storage; CFT = coal-to-liquid fuel, Fischer-Tropsch; CMTG = coal-to-liquid fuel, methanol-to-gasoline.

ethanol in 2020, because it does not incorporate lags in implementation of the technology that will result from the need to permit and build the infrastructure to produce and transport the alternative fuels. Figure 6.9 shows the potential gasoline-equivalent supply of ethanol from biochemical conversion of both ligno-cellulosic biomass and ethanol distilled from corn, again with 2020 deployable technology. The estimated supply of synthetic gasoline and diesel derived from coal and biomass as feedstocks is shown in Figure 6.10. Two supply functions are shown: one with CCS and the other without CCS. The comparison shows that if

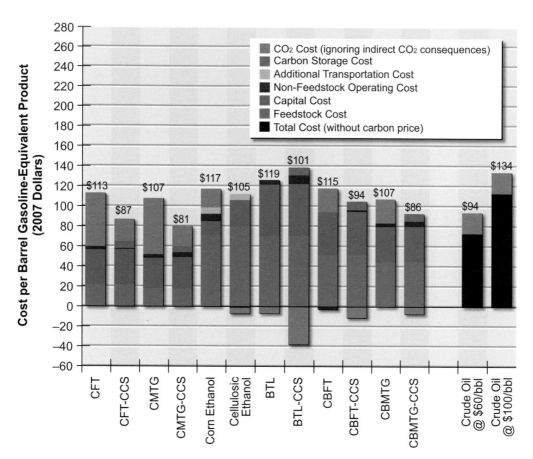

FIGURE 6.7 *Costs of alternative liquid fuels produced from coal, biomass, or coal and biomass with a CO₂ price of $50/tonne when capital costs are 20 percent lower than the panel's estimates.*
Note: BTL = biomass-to-liquid fuel; CBFT = coal-and-biomass-to-liquid fuel, Fischer-Tropsch; CBMTG = coal-and-biomass-to-liquid fuel, methanol-to-gasoline; CCS = carbon capture and storage; CFT = coal-to-liquid fuel, Fischer-Tropsch; CMTG = coal-to-liquid fuel, methanol-to-gasoline.

the CCS technologies are viable and if a CO₂ price of $50/tonne is implemented, for each feedstock it will cost less to use CCS than to release the CO₂ into the atmosphere.

Either of the production processes underlying Figure 6.8 or Figure 6.9 and Figure 6.10 would use the same supplies of biomass. Therefore, the quantities cannot be added. If all the production (in addition to ethanol produced from corn grain) is based on cellulosic conversion, Figure 6.8 would be potentially applicable. If all production is based on thermochemical conversion, the quanti-

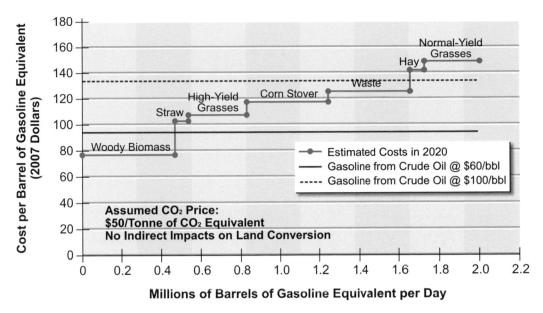

FIGURE 6.8 *Estimated supply of cellulosic ethanol at different price points in 2020. Because the costs of feedstocks vary, ethanol costs will vary with the feedstock used. The cost estimates include a $50 tax for each tonne of CO_2 released on a well-to-wheel basis. The red solid and dotted lines show the supply of crude oil at $60/bbl and $100/bbl for comparison.*

ties in Figure 6.10 would be potentially applicable. Most likely, some production would be based on biochemical processes and some on thermochemical processes, so the actual potential supply function would lie between the supply functions of Figures 6.8 and 6.10. In addition, ethanol would be produced from corn grain at roughly 0.67 million bbl/d of gasoline equivalent.

To put the results into perspective, light-duty vehicle gasoline and diesel use in the United States in 2008 is estimated to be about 9 million bbl of oil equivalent per day (EIA, 2008); 1 bbl of crude oil produces about 0.85 bbl of gasoline and diesel. Total oil used in the United States was 21 million bbl/d, of which 14 million bbl was used for transportation and 12 million bbl was imported (EIA, 2008). Thus, 2 million bbl of gasoline-equivalent ethanol produced from cellulosic biomass and 0.7 million bbl of gasoline-equivalent ethanol produced from corn grain have the potential to replace about 30 percent of the petroleum-based fuel consumed in the United States by light-duty vehicles or 20 percent of all transportation fuels. The difference between current technology and 2020 technology shows the importance of reductions in feedstock costs and increase in yield to achieving the supply target.

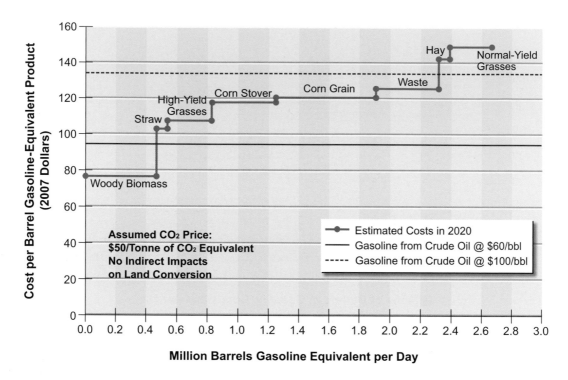

FIGURE 6.9 *Estimated supply of cellulosic ethanol plus ethanol produced from corn grain at different price points in 2020. Cost estimates include a $50 tax for each tonne of CO_2 released on a well-to-wheel basis. The red solid and dotted lines show the supply of crude oil at $60/bbl and $100/bbl for comparison.*

The potential supply of gasoline or diesel from thermochemical conversion of a combination of coal and biomass (with CCS) is greater than that from biochemical conversion that uses only biomass. The thermochemical costs are similar to or smaller than the biochemical conversion costs. The costs differ because coal costs less than biomass. In addition, using a combination of coal and biomass allows a larger plant to be built and reduces capital costs per volume of product.

The combination of coal and biomass allows more alternative fuel to be produced than would be possible with biomass alone. The quantity of biomass limits the overall production in either case. Thus, the addition of coal increases the total amount of liquids that could be produced from a given quantity of biomass. Using the combination of coal and biomass, oil potentially can be displaced from transportation at almost 4 million bbl/d (40 percent of gasoline and diesel used

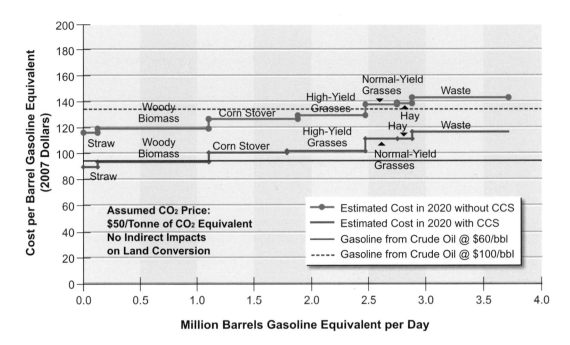

FIGURE 6.10 *Estimated supply of gasoline and diesel produced by thermochemical conversion of combined coal and biomass via FT with and without CCS at different price points in 2020. The cost estimates include a $50 tax for each tonne of CO_2 released on a well-to-wheel basis. The red solid and dotted lines show the supply of crude oil at $60/bbl and $100/bbl for comparison.*

by light-duty vehicles in 2008). As noted above, this analysis assumes that all cellulosic biomass sustainably grown for fuel will be used for liquid transportation fuel. See Box 6.1 for further discussion.

MARKET PENETRATION

The discussion above focuses on biomass supply and fuel technology deployable by 2020, but a potential supply of alternative liquid fuel does not translate to the supply that would actually be available in 2020. The following section discusses issues that might limit the rate of market penetration. For biochemical conversion, two scenarios of potential biochemical penetration are presented. The actual penetration could be slower or faster, depending on crude-oil price, expectations of future prices, federal and state policy, the U.S. construction industry, and other variables.

BOX 6.1 *Preferential Use of Biomass—Power Generation or Liquid Transportation Fuels*

A number of factors can be expected to influence the use of biomass to support U.S. energy requirements. The major options are use of biomass to generate power and to produce liquid transportation fuels. Biomass can be expected to be used for both options according to policies that mandate a minimum requirement for renewable energy and fuels. Those include minimum requirements for renewable power generation and coal power-plant permits that mandate that a given percentage of biomass be fed with coal. Mandating minimum requirements for renewable transportation fuels will drive the use of biomass to produce fuels. Other factors will also be influential in determining the use of biomass.

First, the lack of feedstock options other than biomass for producing liquid transportation fuels with reduced CO_2 emission means that biomass will have to be a component. The use of coal with CCS can provide liquid transportation fuels and move the United States away from reliance on petroleum, but it does not reduce CO_2 emission from the transportation sector. At its best, it is neutral relative to conventional gasoline from the point of view of climate change. Power generation has a number of options other than biomass that can provide electricity with reduced CO_2 emission. From a renewables point of view, there are wind and solar sources. Nuclear power also has low CO_2 emission. Furthermore, the use of coal with CCS can produce electricity with marked reductions in CO_2 emission—by, say, 80 or 90 percent—and in mercury and sulfur emissions. Thus, power generation truly has options other than biomass to address greenhouse gas and other environmental issues. That points to the use of biomass for liquid transportation fuels as an essential component in any greenhouse gas management program. In addition, biomass for liquid transportation fuels provides diversity of supply and enhances energy security.

If biomass is to be used as a component in a CO_2-management approach, it should be used in a way that provides the lowest-cost CO_2 reduction in terms of dollars per tonne of CO_2 avoided. The avoided cost of CO_2 is projected to be much lower when biomass is used to produce liquid fuels than when it is used to produce power.

That leads to the conclusion that the use of biomass to produce liquid transportation fuels has more societal advantages than its use to generate electricity, because the use of biomass is an effective route to reducing CO_2 emission from the transportation sector where few other options exist and it does so at a much lower cost per tonne of CO_2 avoided.

Biochemical Conversion

Production of ethanol from grain is fully commercial. U.S. production capacity grew from 0.28 million bbl/d at the end of 2004 to 0.38 million bbl/d by the end of 2006 and to about 0.46 million bbl/d by the end of 2007. (Those figures correspond to 4.3, 5.9, and about 7 billion gallons per year by the end of 2007.[6])

The capacity-build rate of grain ethanol averaged 25 percent per year over a 6-year period. At the maximum build rate, 1–2 billion gallons of annual ethanol-production capacity was added per year or an annual addition of 0.065–0.13 million bbl/d; at an average plant size of 3300 bbl of ethanol per day or 50 million gallons per year, that means 20–40 plants/year at the maximum. Considering current plant construction that was under way, ethanol-production capacity would have been about 0.5 million bbl/d by the end of 2008. However, 12–15 billion gallons of grain ethanol per year (0.8–1.0 million bbl/d) is probably the limit with respect to corn availability, assuming that corn yields and acreage increase modestly.

Production of ethanol from cellulose has yet to be demonstrated on a commercial scale, and there remain questions about the economic and commercial viability of the technology. Within the next 3–5 years, five or six technology-demonstration plants (on a noncommercial scale) are expected. The plants will provide valuable information on cost, engineering design, technology robustness, and particularly commercial viability on the scale required to warrant large-scale cellulosic-ethanol production. That information should be available by 2012. The commercial and economic issues potentially will be gradually resolved as cellulosic-ethanol production technology matures and development of new strains of organisms and manufacturing methods reach commercial implementation. As commercially proven technology for cellulosic-ethanol production evolves in scale and efficiency, growth in cellulosic-ethanol production capacity could approach or even exceed the growth experienced in grain ethanol. Cellulosic-ethanol plants are similar to grain-ethanol plants but somewhat more complex; and because of the dispersed nature of biomass, they might be comparable in size to, or even up to twice as large as, typical grain-ethanol plants. For the rest of this discussion, it is assumed that cellulosic ethanol will be commercially demonstrated by 2012 and that it will be either economically competitive with petroleum-based fuels or made

---

[6]In oil-equivalent figures, these rates—adjusted for energy content—correspond to 0.19, 0.26, and 0.31 bbl of oil equivalent per day.

competitive through the use of subsidies or policy so that capacity will be built with private funds. The U.S. Department of Energy roadmap for cellulosic ethanol proposes "to accelerate cellulosic ethanol research, helping to make biofuels practical and cost-competitive by 2012."

Here, cellulosic-ethanol plants with a collective capacity of 1 billion gallons per year are assumed to be in operation by 2015 as a result of overall commercial development and demonstration activities and that the capacity-build beyond 2015 will track one of two scenarios based on the capacity-build experienced by grain ethanol (1–2 billion gallons of new capacity per year) (Figure 6.11). One scenario tracks the maximum capacity-build experienced for grain ethanol, and the second scenario is more aggressive and reaches about twice the capacity achieved for grain ethanol. The two scenarios project 7–12 billion gallons of cellulosic ethanol per year by 2020 (0.5–0.8 million bbl/d). Continued aggressive capacity-building could achieve the renewable fuel standard (RFS) mandate capacity of 16 billion gallons of cellulosic biofuel per year by 2022, but it would be a stretch. The RFS was created by the 2005 U.S. Energy Policy Act. However, the 2007 U.S. Energy Independence and Security Act amended the RFS to set forth "a phase-in for renewable fuel volumes beginning with 9 billion gallons in 2008 and ending at 36 billion gallons in 2022" (0.6 and 2.4 million bbl/d, respectively). If the more aggressive scenario plays out, capacity-building could yield 1.5–2 million bbl of cellulosic ethanol per day by 2030 and up to 2.6 million bbl/d shortly thereafter, consuming about 440 million dry tons of biomass per year. However, it should be stressed that whether the production capacity expands more rapidly or less rapidly will depend heavily on economic incentives and policies and on the actual and projected prices of crude oil.

Thermochemical Conversion

For coal plants, the gasification, FT, and MTG technologies are developed. However, there is no experience with integrated plants that would use all the technologies combined with CCS. To have CTL ready to supply fuels in the shortest time possible to improve energy security, an immediate start on the design and construction of commercial demonstration plants with CCS is critical. CO_2 capture is built into the FT and MTG processes, but learning from demonstration-plant operations is critical for decreasing cost and improving performance. CO_2 storage will require adding compressors to the plants and locating the demonstrations close to CO_2 repositories (for example, saline aquifers, geological formations, or

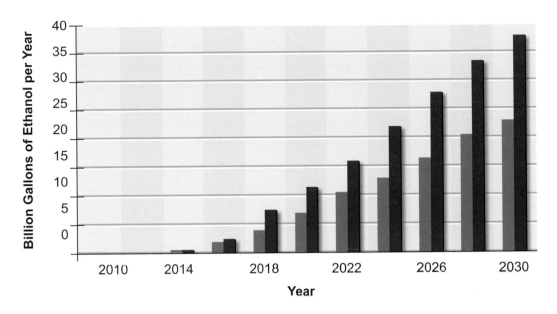

FIGURE 6.11 *Cellulosic-ethanol capacity-building scenarios starting with commercial demonstration plants in 2009 with first commercial-scale plants following thereafter, building to 1 billion gallons of cellulosic ethanol per year in 2015. Capacity-building beyond 2015 is in accordance with the maximum capacity build achieved for grain ethanol (blue bars), and a more aggressive capacity build of about twice that achieved for grain ethanol. The maximum build rate could achieve the 2022 Renewable Fuel Standard mandate of 16 billion gallons of cellulosic ethanol per year, but it would be a stretch.*

sites of enhanced oil-recovery opportunities). Experience from the demonstrations is also needed to resolve scientific and regulatory issues to make CCS viable. If the demonstrations are started immediately and CCS is proved viable and safe by 2015, economically viable commercial plants could be starting up before 2020.

For thermochemical processing of biomass and cofed coal and biomass plants, a timeline similar to that for CTL applies. CCS is not necessary for biomass-to-liquid fuel plants to produce carbon-neutral fuels, and commercial demonstration can start immediately if society places a high enough value on carbon-neutral fuels (fuels with zero greenhouse gas life-cycle emissions). Although CCS may not be required in coal-and-biomass-to-liquid (CBTL) fuel plants if the proportion of biomass to coal is high in the feedstock, such plants will have to deal with the problems of feeding biomass to gasifiers and locating the plants in a region that could supply sufficient biomass (about 4000 dry tons of biomass per day) and have access to sufficient coal (about 3000 tons/day as received). For CBTL plants, the technology is close to developed, and several commercial demonstration plants are in operation or being built with and without CCS. However, gaining operational experience in the plants with CCS is critical because cost reductions will result from the experience. Because CTL fuel has twice the CO_2 life-cycle emission of gasoline unless it uses CCS, CCS will probably be required. Penetration rates of the CBTL plants could be expected to be similar to or slightly less than that of the cellulosic-ethanol build-out case that follows the experience of grain ethanol discussed earlier. Penetration rates for biomass plants can be expected to be similar to that for the cellulosic-ethanol case, but both plants depend heavily on the ability to reduce fuel-production costs and on the presence of a substantial carbon policy. The biomass-gasification penetration rate will depend heavily on getting the biomass supply up to about a million dry tons per year per site or higher. Cellulosic ethanol could be applied on a smaller scale of biomass availability.

To get some perspectives on capacity growth for CBTL and CTL plants, the panel presents the following analysis. The capacity growth rates could be higher or lower, depending on such factors as government policy, oil prices, carbon price, and the labor and commodity markets.

Consider a CBTL plant integrated with CCS that uses about 40 percent biomass and 60 percent coal on an energy basis. Such a plant produces liquid transportation fuels that are essentially carbon-free and, to the extent that it produces electricity for the grid, the electricity is also carbon-free. In the recycle case designed to maximize the liquid-fuels production with CCS, the plant produces about 10,000 bbl of liquid hydrocarbon transportation fuels per day. The size of the plant considered in this case is 4,000 dry tons of biomass per day. The CBTL plant is more complex and its capital cost is substantially higher on the basis of a barrel of fuel produced than is a biochemical conversion plant of comparable biomass feed capacity. When the difference is put on the basis of energy-equivalent fuels, it is reduced but is still important. As mentioned above, the build-out of grain ethanol for fuel capacity averaged 25 percent per year over a 6-year period and was the basis of the estimation of the build-out rate of cellulosic ethanol at sites of 1.1 million dry tons per year. The cellulosic-ethanol build-out had 225 plants producing 1.5 million bbl of ethanol per day (1 million bbl of gasoline equivalent per day) in 2030 at a total running sum cost of about $100 billion for the base case and 370 plants producing 2.4 million bbl/d (1.6 million bbl of gasoline equivalent per day) and consuming about 440 million dry tons of biomass per year.

For CBTL plants, the panel used a slightly lower build-out rate because of issues of accessing sites with about 1.1 million dry tons of biomass per year and a similar availability of coal. In this case, a total of 200 plants were in place by 2030 and producing 2 million bbl of gasoline equivalent per day at a running sum cost of about $260 billion. It was assumed that three plants were commissioned in 2015 and that growth is expanded as capacity to build them increases; this largely follows the cellulosic-ethanol projection to achieve the numbers summarized above. That would consume about 220 million dry tons of biomass and about 200 million tons of coal per year. If that growth rate could be continued to 2035, an estimated 2.5 million bbl of gasoline-equivalent fuels could be produced, consuming less than the projected biomass availability; but siting plants to access both biomass and coal is probably the limiting factor for CBTL plants. The analysis shows that the capacity growth rates would have to exceed historical rates considerably if 550 million dry tons of biomass per year is to be converted to liquid fuels in 2030.

For CTL plants with CCS, consider a plant build-out rate of two to three plants per year each with 50,000-bbl/day capacity for 20 years starting in 2015 (when the first plants are commissioned). This scenario would reduce dependence on imported oil, but it would not reduce CO_2 emission from transportation. At a build-out rate of two plants per year, 2 million bbl of liquid fuels per day would be produced from 390 million tons of coal per year by 2035 at a total cost of about $200 billion for all the plants built. At a build-out rate of three plants per year, 3 million bbl of liquid fuels per day would be produced from about 580 million tons of coal per year. The latter case would replace about one-third of U.S. oil use in light-duty transportation and increase U.S. coal production by 50 percent.

FINDINGS AND RECOMMENDATIONS

Finding 6.1

Alternative liquid transportation fuels from coal and biomass have the potential to play an important role in helping the United States to address issues of energy security, supply diversification, and greenhouse gas emissions with technologies that are commercially deployable by 2020.

- **With CO_2 emissions similar to those from petroleum-based fuels,** a substantial supply of alternative liquid transportation fuels can be produced with thermochemical conversion of coal with geologic storage of CO_2 at a gasoline-equivalent cost of $70/bbl.

- **With CO$_2$ emissions substantially lower than those from petroleum-based fuels,** up to 2 million bbl/d of gasoline-equivalent fuel can technically be produced with biochemical or thermochemical conversion of the estimated 550 million dry tons of biomass available in 2020 at a gasoline-equivalent cost of about $115–140/bbl. Up to 4 million bbl/d of gasoline-equivalent fuel can be technically produced if the same amount of biomass is combined with coal (60 percent coal and 40 percent biomass on an energy basis) at a gasoline-equivalent cost of about $95–110/bbl. However, the technically feasible supply does not equal the actual supply inasmuch as many factors influence the market penetration of fuels.

Finding 6.2

If commercial demonstration of cellulosic-ethanol plants is successful and commercial deployment begins in 2015 and if it is assumed that capacity will grow by 50 percent each year, cellulosic ethanol with low CO$_2$ life-cycle emissions can replace up to 0.5 million barrels of gasoline equivalent per day by 2020 and 1.7 million bbl/d by 2035.

Finding 6.3

If commercial demonstration of coal-and-biomass-to-liquid plants with carbon capture and storage is successful and the first commercial plants start up in 2020 and if it is assumed that capacity will grow by 20 percent each year, coal-and-biomass-to-liquid fuels with low CO$_2$ life-cycle emissions can replace up to 2.5 million barrels of gasoline equivalent per day by 2035.

Finding 6.4

If commercial demonstration of coal-to-liquid plants with carbon capture and storage is successful and the first commercial plants start up in 2020 and if it is assumed that capacity will grow by two to three plants each year, coal-to-liquid fuels with CO$_2$ life-cycle emissions similar to those of petroleum-based fuels can replace up to 3 million barrels of gasoline equivalent per day by 2035. That option would require an increase in U.S. coal production by 50 percent.

Recommendation 6.1

Detailed scenarios of market penetration rates of biofuels, coal-to-liquid fuels, and associated biomass and coal supply options should be developed to clarify hurdles and challenges to achieving substantial effects on U.S. oil use and CO_2 emissions. The analysis will provide policy makers and business leaders with the information needed to establish enduring policies and investment plans for accelerating the development and penetration of alternative-fuels technologies.

REFERENCE

EIA (Energy Information Administration). 2008. Annual Energy Outlook 2008 with Projections to 2030. Washington, D.C.: U.S. Department of Energy, Energy Information Administration.

7 Overall Findings and Recommendations

Alternative liquid transportation fuels from coal and cellulosic biomass have the potential to play an important role in helping the United States to address a variety of issues—including energy security, supply diversification, and greenhouse gas emissions—with technologies that could be commercially deployable by 2020. Several options are available for increasing domestic fuel supply while using either thermochemical conversion of coal, biomass, or both or using biochemical conversion of biomass. Different options have different potential supplies and greenhouse gas effects; the choice will most likely depend on U.S. carbon policy.

- *Biomass supply*—The panel projects the amount of cellulosic biomass that can technically be produced and harvested sustainably for biochemical or thermochemical conversion (or other energy uses) to be 550 million dry tons per year by 2020.
- *Coal-to-liquid fuels by thermochemical conversion*—At an estimated cost of about $70/bbl of gasoline equivalent (that is, less than $60/bbl of oil equivalent), gasoline and diesel can be produced from the abundant U.S. coal reserves to have life-cycle carbon dioxide (CO_2) emission similar to that of petroleum-based gasoline in 2020 or sooner if existing thermochemical technology is combined with carbon capture and storage (CCS). CCS, however, would have to be demonstrated on a commercial scale and implemented by then. The supply will be limited by the amount of coal that can mined to meet the needs of a growing coal-to-liquid fuels industry.

- *Biomass-to-liquid fuels by thermochemical conversion*—The estimated 550 million tons of dry biomass can be converted by thermochemical conversion to up to about 30 billion gallons of synthetic gasoline and diesel at an estimated cost of about $140/bbl of gasoline equivalent. The CO_2 life-cycle emission will be close to zero without CCS.
- *Biomass-to-liquid fuels by biochemical conversion*—The estimated 550 million tons of dry biomass can be converted by biochemical conversion to up to about 45 billion gallons of ethanol (equivalent on an energy basis to about 30 billion gallons of gasoline), at about $115/bbl of gasoline equivalent. The CO_2 life-cycle emission will be close to zero.
- *Coal-and-biomass-to-liquid fuels by thermochemical conversion*—The estimated 550 million tons of biomass can be combined with coal at a ratio of 40:60 (on an energy basis) to produce up to 60 billion gallons of liquid fuels per year on a gasoline-equivalent basis by thermochemical conversion at an average estimated cost of about $95/bbl gasoline equivalent without CCS and $110/bbl of gasoline equivalent with CCS. The CO_2 life-cycle emissions of the fuels produced without CCS would be comparable with those of petroleum-based fuels without CCS and zero or slightly negative with CCS.

Although alternative liquid fuel technology can be deployable and supply a substantial volume of clean fuels for U.S. transportation at a reasonable cost, it will take more than a decade for the fuels to reach full market penetration. The supply of 30–60 billion gallons of clean fuels per year will require the design, permitting, and construction of hundreds of conversion plants and associated fuel transportation and delivery infrastructure.

Recommendation 7.1

Detailed scenarios of market penetration rates of biofuels, coal-to-liquid fuels, and associated biomass and coal supply options should be developed to clarify hurdles and challenges to achieving substantial effects on U.S. oil use and CO_2 emissions. The analysis will provide policy makers and business leaders with the information needed to establish enduring policies and investment plans for accelerating the development and penetration of alternative-fuels technologies.

In thermochemical conversion of coal or combined coal and biomass to produce transportation fuels, CCS is critical for reducing CO_2 emission. The $10–15 estimated cost of CCS used in this study's analyses represents preliminary engineering costs. Ultimate requirements for design, monitoring, carbon-accounting procedures, liability, and associated regulatory frameworks, are yet to be developed, and there is potential for unanticipated delay in initiating demonstration projects and, later, in licensing individual commercial-scale projects. Uncertainty about the regulatory environment arising from concerns of the general public and policy makers have the potential to raise storage costs. Hence, the full cost of CCS is difficult to determine without some commercial-scale experience with geologic CO_2 storage. Large-scale demonstration and establishment of procedures for long-term monitoring of CCS have to be pursued aggressively in the next few years if thermochemical conversion of biomass and coal with CCS is to be ready for commercial deployment by 2020.

Recommendation 7.2

The federal government should continue to partner with industry and independent researchers in an aggressive program to determine the operational procedures, monitoring, safety, and effectiveness of commercial-scale technology for geologic storage of CO_2. Three to five commercial-scale demonstrations (each with about 1 million tonnes CO_2 per year and operated for several years) should be set up within the next 3–5 years in areas of several geologic types.

The demonstrations should focus on site choice, permitting, monitoring, operation, closure, and legal procedures needed to support the broad-scale application of geologic storage of CO_2. The development of needed engineering data and determination of the full costs of geologic storage of CO_2—including engineering, monitoring, and other costs based on data developed from continuing demonstration projects—should have high priority.

Recommendation 7.3

The government-sponsored geologic CO_2 storage projects need to address issues related to the concerns of the general public and policy-makers about geologic CO_2 storage through rigorous scientific and policy analyses. As the work on geological storage progresses, any factors that might result in public concerns and uncertainty in the regulatory environment should be evaluated and built into the project decision-making process because they could raise storage cost and slow projects.

The amount of cellulosic biomass that could potentially be produced sustainably with today's technologies and management practices is estimated to be about 400 million dry tons per year. Production could potentially be increased to about 550 million dry tons by 2020. The panel believes that that quantity of biomass can be produced from dedicated energy crops, agricultural and forestry residues, and municipal solid wastes without affecting U.S. food and fiber production or having adverse environmental effects. The supply of cellulosic biomass is limited by the amount that can be grown and harvested in a sustainable manner on marginal lands or agriculturally degraded lands. Improved agricultural practices and improved plant species and cultivars will be required to increase the sustainable production of cellulosic biomass and to achieve the full potential of biomass-based fuels. A sustained research and development (R&D) effort in increasing productivity, improving stress tolerance, managing diseases and weeds, and improving the efficiency of nutrient use would help to improve biomass yields. To use biomass as a resource for energy in a sustainable manner requires that the effects of biomass production or harvesting on a range of factors—soil, water, and air quality; food, feed, and fiber production; carbon sequestration; wildlife habitat and biodiversity; rural development—and other issues and the resulting supply of energy be assessed in a holistic way so that multiple public and private concerns are addressed simultaneously. Incentives and best agricultural practices will probably be needed to encourage sustainable production of biomass for biofuel production. Producers need to grow biofuel feedstocks on degraded agricultural land to avoid direct and indirect competition with the food supply, and they need to minimize land-use practices that result in substantial net greenhouse gas emissions.

Recommendation 7.4

The federal government should support focused research and development programs to provide the technical bases of improving agricultural practices and biomass growth to achieve the desired increase in sustainable production of cellulosic biomass. Focused attention should be directed toward plant breeding, agronomy, ecology, weed and pest science, disease management, hydrology, soil physics, agricultural engineering, economics, regional planning, field-to-wheel biofuel systems analysis, and related public policy.

Cellulosic ethanol is in the early stages of commercial development; a few commercial plants are expected to begin operations in the next several years.

Over the next decade, process improvements in this generation of technology are expected to come from evolutionary developments and knowledge gained through commercial experience and increases in scale of operation. Incremental improvements in biochemical conversion technologies can be expected to reduce nonfeedstock process costs by about 25 percent by 2020 and 40 percent by 2035. Because of lack of commercial experience, costs might be higher than estimated during initial commercialization but decrease thereafter as experience is gained. Future improvements in cellulosic technology that entail invention of biocatalysts and biological processes could produce fuels that supplement ethanol production in the next 15 years. In addition to ethanol, advanced biofuels (for example, lipids, higher alcohols, hydrocarbons, or other products that are easier to separate than ethanol) should be investigated because they could have higher energy content, would be less hygroscopic than ethanol, and therefore could fit more smoothly into the current petroleum infrastructure than ethanol.

Recommendation 7.5

The federal government should ensure that there is adequate research support to focus advances in bioengineering and the expanding biotechnologies on developing advanced biofuels. The research should focus on advanced biosciences—genomics, molecular biology, and genetics—and biotechnologies that could convert biomass directly to produce lipids, higher alcohols, and hydrocarbons fuels that can be directly integrated into the existing transportation infrastructure. The translation of those technologies into large-scale commercial practice poses many challenges that need to be resolved by R&D and demonstration if major effects on production of alternative liquid fuels from renewable resources are to be realized.

Without CO_2 sequestration, technologies for the indirect liquefaction of coal to transportation fuels are commercially deployable today and can produce gasoline and diesel at an estimated cost of about $65/bbl of gasoline equivalent, but life-cycle CO_2 emission will be more than twice that of petroleum-based fuels. The coal-to-liquid plant configuration produces a concentrated stream of CO_2 that has to be removed before the fuel-synthesis step even in nonsequestration plants. Requiring carbon storage would have a relatively small effect on cost and efficiency. Thus, with CCS, indirect liquefaction processes can have essentially the same CO_2 life-cycle emission as petroleum-based liquid fuels, or less, and still produce fuels at an estimated cost of about $70/bbl of gasoline equivalent.

Cogasification of biomass and coal to produce liquid fuels would have similar CO_2 life-cycle emissions as processing of the same amount of biomass and coal separately for liquid fuels. Cogasification, however, allows a larger scale of operation than would be possible with biomass only and reduces costs per unit capacity. However, penalties associated with the preprocessing of the biomass and the technical problems in feeding biomass to high-pressure gasification systems have to be taken into account. Successful feeding of raw biomass to high-pressure gasification systems could pose a challenge because biomass, unlike coal, is soft and fibrous and therefore difficult to reduce to the sizes necessary for efficient gasification. CCS has yet to be demonstrated and implemented for this alternative.

To have thermochemical conversion of coal or coal and biomass to liquid fuels ready for deployment by 2020, the development of coal or coal and biomass gasification technology combined with fuel synthesis and CCS technology would have to be accelerated and proceed simultaneously so that the technologies can be implemented as a package. As a first step, a few coal-to-liquid plants and coal-and-biomass-to-liquid plants could serve as sources of CO_2 for a small number of CCS demonstration projects. However, so-called capture-ready plants that vent CO_2 would create liquid fuels with higher CO_2 emission per unit usable energy than petroleum-based fuels; their commercialization should not be encouraged unless those plants are integrated with CCS at their start-up. It is critical for construction of demonstration plants integrated with CCS to start as soon as possible so that commercial-plant and CCS design data can be collected.

Thermochemical and biochemical conversion approaches for the production of clean fuels both entail practical and technical challenges. The supply of biomass could limit plant size and influence the cost of fuel products from any plant that uses it as a feedstock irrespective of the conversion approach. The supply of available biomass will probably be limited to within 40 miles of the conversion plant because biomass is bulky, expensive, and difficult to transport. The density of biomass (quantity per acre) will vary considerably from region to region across the country, ranging from a supply of less than 1,000 tons/day to 10,000 tons/day. Technologies that increase the density of biomass in the field to decrease transportation cost and logistic issues should be developed. The density associated with such technologies as field-scale pyrolysis could facilitate its transportation to larger-scale regional conversion facilities. Thermochemical conversion plants require larger capital investment than do biochemical conversion plants, so the former benefit to a greater extent than the latter from economies of scale.

Finding 7.1

A potential optimal strategy for producing biofuels in the United States could be to locate thermochemical conversion plants that use coal and biomass as a combined feedstock in regions where biomass is abundant and locate biochemical conversion plants in regions where biomass is less concentrated. Thermochemical plants require larger capital investment per barrel of product than do biochemical conversion plants and thus benefit to a greater extent from economies of scale. This strategy could maximize the use of cellulosic biomass and minimize the costs of fuel products.

Recommendation 7.6

The U.S. Department of Energy and the U.S. Department of Agriculture should determine the spatial distribution of potential U.S. biomass supply to provide better information on the potential size, location, and costs of conversion plants. The information would allow determination of the optimal size of conversion plants for particular locations in relation to the road network and the costs and greenhouse gas effects of feedstock transport. The information should also be combined with the logistics of coal delivery to such plants to develop an optimal strategy for using U.S. biomass and coal resources for producing sustainable biofuels.

Because ethanol cannot be transported in pipelines used for petroleum transport, an expanded infrastructure will be required to replace gasoline with a larger proportion of ethanol produced via biochemical conversion. Ethanol is currently transported by rail or barges and not by pipelines, because it is corrosive in the existing infrastructure and can damage seals, gaskets, and other equipment and induce stress-corrosion cracking in high-stress areas. If ethanol is to be used in fuel at concentrations higher than 20 percent (for example, E85, which is a blend of 85 percent ethanol and 15 percent gasoline), the number of refueling stations will have to be increased to support alternative-fuel vehicles. The transport and distribution of synthetic diesel and gasoline produced via thermochemical conversion will be less challenging because they are compatible with the existing infrastructure for petroleum-based fuels.

Recommendation 7.7

The U.S. Department of Energy and the biofuels industry should conduct a comprehensive joint study to identify the infrastructure system requirements of, research and development needs in, and challenges facing the expanding biofuels industry. Consideration should be given to the long-term potential of truck or barge delivery versus the potential of pipeline delivery that is needed to accommodate increasing volumes of ethanol. The timing and role of advanced biofuels that are compatible with the existing gasoline infrastructure should be factored into the analysis.

Finding 7.2

The deployment of alternative liquid transportation fuels aimed at diversifying the energy portfolio, improving energy security, and reducing the environmental footprint by 2035 would require aggressive large-scale demonstration in the next few years and strategic planning to optimize the use of coal and biomass to produce fuels and to integrate them into the transportation system. Given the magnitude of U.S. liquid-fuel consumption (14 million barrels of crude oil per day in the transportation sector) and the scale of current petroleum imports (about 56 percent of the petroleum used in the United States is imported), a business-as-usual approach is insufficient to address the need to find alternative liquid transportation fuels, particularly because development and demonstration of technology, construction of plants, and implementation of infrastructure require 10–20 years per cycle.

Recommendation 7.8

The U.S. Department of Energy should partner with industry in the aggressive development and demonstration of cellulosic-biofuel and thermochemical-conversion technologies with carbon capture and storage to advance technology and to address challenges identified in the commercial demonstration programs. The current government and industry programs should be evaluated to determine their adequacy to meet the commercialization timeline required to reduce U.S. oil use and CO_2 emissions over the next decade.

8

Key Challenges to Commercial Deployment

This chapter summarizes the challenges to commercial deployment of facilities for biochemical conversion of cellulosic biomass to ethanol and for thermochemical conversion of coal, biomass, or combined coal and biomass to liquid fuels.

CHALLENGE 1

Several technological and sociological issues pose a serious challenge to the development of the biomass-supply industry for the production of cellulosic biofuels:

- Developing a systems approach through which farmers, biomass integrators, and those operating biofuel-conversion facilities can build a well-organized and sustainable cellulosic-ethanol industry that will address the relevant issues such as biofuel; soil, water, and air quality; carbon sequestration; wildlife habitat; rural development; and rural infrastructure—without creating unintended consequences through piecemeal development efforts.
- Determining the full life-cycle greenhouse gas emissions of various biofuel crops.
- Certifying the greenhouse gas benefits of different potential biofuel scenarios.
- Overcoming the perception that crop residues and similar materials are literally "trash" or waste products and therefore have little or no value for farmers.

Those issues, although formidable, can be overcome by developing a systems approach that has multiple end points and that collectively can provide a variety of credits or incentives—such as carbon sequestration, water quality, soil quality, wildlife, and rural development—and thus strengthen the U.S. agricultural industry. Failure to link the various critical environmental, economic, and social needs and to address them as an integrated system could reduce the availability of biomass to amounts substantially below the 550 million tons technically deployable in 2020.

CHALLENGE 2

If thermochemical conversion of coal or combined coal and biomass is to be important in reducing U.S. reliance on crude oil and reducing CO_2 emission in the next 20–30 years, CCS will have to be shown to be safe and economically and politically viable. The capture of CO_2 is proved, and commercial-scale demonstration plants are needed to measure and improve cost and performance. Separate large-scale programs will be required to resolve storage and regulatory issues associated with geologic CO_2 storage approaching an annual rate of gigatonnes. The analyses presented in this report assume that the viability of CCS will be demonstrated by 2015 so that integrated coal-to-liquid plants can start up by 2020. In that scenario, the first coal or coal-and-biomass gasification plant would not be in operation until 2020. The assumption of CCS demonstration by 2015 is ambitious and will require focused and aggressive government action to realize it. Uncertainty about the regulatory environment arising from concerns of the general public and policy makers have the potential to raise storage costs above the costs assumed in this report. Ultimate requirements for selection, design, monitoring, carbon-accounting procedures, liability, and associated regulatory frameworks are yet to be developed, and there is a potential for unanticipated delays in initiating demonstration projects and later in licensing individual commercial-scale projects. Large-scale demonstrations and establishment of procedures for operation and long-term monitoring of CCS projects have to be pursued aggressively in the next few years if thermochemical conversion of biomass and coal with CCS is to be ready for commercial deployment by 2020.

CHALLENGE 3

Cellulosic ethanol is in the early stages of commercial development; a few commercial demonstration plants are expected to begin operations in the next several years. Over the next decade, process improvements in this generation of technology are expected to come from evolutionary developments and knowledge gained through commercial experience and increases in scale of operation. Incremental improvements in biochemical conversion technologies can be expected to reduce nonfeedstock process costs by about 25 percent by 2020 and 40 percent by 2035. It will take focused and sustained industrial and government action to achieve those cost reductions.

The key technical issues to be resolved to achieve cost reductions are these:

- More efficient pretreatment to free up celluloses and hemicelluloses and to enable more efficient downstream conversion. Improved pretreatment is not likely to reduce product cost substantially, because pretreatment cost is small relative to other costs.
- Better enzymes that are not subject to end-product inhibition to improve the conversion process.
- Maximizing of solids loading in the reactors.
- Engineering of organisms that can ferment sugars in a toxic biomass hydrolysate and produce high concentrations of the final biofuel. Improving microorganism tolerance of toxicity is a key issue.

CHALLENGE 4

An expanded ethanol transportation and distribution infrastructure will be required if ethanol is to be used in much greater amounts than now in light-duty vehicles. Ethanol cannot be transported in pipelines that are used for petroleum transport. It is currently transported by rail or barges, not by pipelines, because it is corrosive in the existing infrastructure and can damage seals, gaskets, and other equipment and induce stress-corrosion cracking in high-stress areas. If ethanol is to be used in fuel at concentrations higher than 20 percent (for example, E85, which is a blend of 85 percent ethanol and 15 percent gasoline), the number of refueling stations offering these options to alternative-fuel vehicles will have to be increased. To enable widespread availability of ethanol in the fuel system, the challenge of fuel distriubtion must be addressed. However, if cellulosic biomass were dedicated to thermochemical conversion with a Fischer-Tropsch or methanol-to-gasoline process, the resulting fuels would be chemically equivalent to conventional gasoline and diesel, and the infrastructure challenge posed by the use of ethanol would be minimized.

CHALLENGE 5

The panel's analyses provide a snapshot of the potential costs of liquid fuels produced by biochemical or thermochemical conversion of biomass and thermochemical conversion of biomass and coal. Fuel costs are dynamic and fluctuate as a result of other externalities. With the wide variation in the prices of most commodities, especially oil, investors will have to have confidence that such mandates as carbon caps, carbon tax, or tariffs on imported oil will ensure that alternative liquid transportation fuels can compete with fuels refined from crude oil. The price of carbon emission or the existence of fuel standards that require specified reductions in greenhouse gas life-cycle emissions from fuel will affect economic choices.

Other economic issues are specific to particular types of plants. For biochemical conversion and thermochemical conversion plants that use biomass as feedstock, the volatility of feedstock costs is a concern: the supply and costs of feedstock can be affected dramatically by weather. For thermochemical conversion plants, the investment risk is considerable because of the high capital expenditure. Because a 50,000-bbl/d plant could cost $4–5 billion, the plants could be expected to approach a cost of $100,000 per daily barrel, which is about 6 times the capital investment cost for crude oil in deepwater Gulf of Mexico.

9

Other Alternative Fuel Options

This report has focused so far on the major part of the panel's statement of task, namely, liquid fuels for transportation that use biomass and coal as feedstocks. To address the requirement of the statement of task regarding competitive fuels, the panel reviewed other potential fuels that could be available over the next 25 years. This final chapter briefly discusses other fuel technologies and their advantages and disadvantages. Compressed natural gas (CNG) is reviewed first, and then liquid fuels that can be produced from syngas, including gas-to-liquid (GTL) diesel, dimethyl ether, and methanol. The chapter discusses technology implications of using hydrogen in fuel-cell-powered vehicles.

Chapter 4 discussed how coal or coal and biomass gasification produces syngas, which can be converted to diesel and gasoline or to methanol, which can be converted to gasoline. Syngas can also be produced by reforming natural gas. Only if large supplies of inexpensive domestic natural gas were available—for example, from natural-gas hydrates—would the United States be likely to use natural gas as feedstock for transportation-fuel production. Chapter 4 discussed how methanol can be produced from coal synthesis gas, but the panel believes that the best approach is to convert synthesis gas to methanol and use methanol-to-gasoline technology to produce gasoline, which fits directly into the existing U.S. fuel-delivery infrastructure. Hydrogen has the potential to reduce U.S. greenhouse gas emissions and oil use, as discussed in two recent National Research Council reports, *Transitions to Alternative Transportation Technologies—A Focus on Hydrogen* (NRC, 2008) and *The Hydrogen Economy: Opportunities, Costs, Barriers, and R&D Needs* (NRC, 2004); but it is a long-term option.

COMPRESSED NATURAL GAS

In 2007, the main U.S. uses for natural gas were apportioned as follows: electric-power generation, 30 percent; industrial use, 29 percent; residential use, 20 percent; and commercial use, 13 percent (EIA, 2008). Only 0.11 percent was used as fuel in transportation vehicles. It is the primary feedstock for fertilizers and petrochemicals. The cleanest and most efficient hydrocarbon fuel, natural gas is environmentally superior to coal for electric-power generation, and for similar reasons it could be a sound choice for transportation fuels.

The chapter on fossil fuel of the report *America's Energy Future: Technology and Transformation* (NAS-NAE-NRC, 2009) provides estimates of U.S. natural gas resources. Current natural-gas consumption needs are met mainly by domestic production. A switch to natural gas for a large segment of U.S. transportation use would probably trigger increased importation of natural gas or fuels produced from natural gas.

Technologies for producing transportation fuels from natural gas are ready for deployment by 2020. If natural gas were used for transportation instead of for electricity, there would be a potential to supply roughly one-fifth to one-fourth of transportation needs from North American natural-gas reserves, but only with investment in distribution infrastructure. Supplying more would require importing natural gas.

Compressed natural gas fuels natural-gas vehicles (NGVs). Natural gas is not a liquid fuel and it must be compressed to supply sufficient fuel for a vehicle. In 2008, there were more than 150,000 NGVs and 1,500 NGV fueling stations in the United States. Natural gas is sold in gallons of gasoline equivalent; a gallon of gasoline equivalent has the same energy content (124,800 Btu) as a gallon of gasoline. NGVs are more expensive than hybrid or gasoline vehicles. The Civic GX NGV has a manufacturer's suggested retail price of $24,590 compared with $22,600 for the hybrid sedan and $15,010 for the regular sedan (Rock, 2008).

Of all the fossil fuels, natural gas produces the least carbon dioxide (CO_2) when burned because it contains the lowest carbon:hydrogen ratio. It also releases smaller amounts of criteria air pollutants. NGVs emit unburned methane (which has a higher greenhouse forcing potential than CO_2), but this may be compensated for by the substantial reduction in CO_2 emission. Dedicated NGVs emit less carbon monoxide (CO), nonmethane organic gas, nitrogen oxides (NO_x), and CO_2 than do gasoline vehicles.

Natural-gas engines are more fuel-efficient than gasoline engines. The main benefit of CNG in the past was its low price (about 80 percent that of gasoline on the gallons of gasoline equivalent basis). Transport and distribution of natural gas are relatively inexpensive because the infrastructure for industrial and household use already exists (Yborra, 2006).

Despite a possibly advantageous fuel-supply situation, NGVs still have a lot of hurdles to overcome. The two main challenges faced by NGVs are insufficient refueling stations and inconvenient on-board CNG tanks that take up most of the trunk space. An NGV market can be analyzed by using the vehicle-to-refueling-station index, defined as the ratio of the number of NGVs (in thousands) to the number of natural-gas refueling stations. According to Yeh (2007), "using techniques including consumer preference surveys and travel time/distance simulations, it has been found out that the sustainable growth of alternative fuel vehicles . . . during the transition from initial market development to a mature market requires the number of alternative-fuel refueling stations be a minimum of 10 to 20 percent of the number available for conventional gasoline stations." A thriving NGV market tends to have an index of 1; this gives rise to a problem: new stations are not being opened because of the lack of users, but few people use NGVs because of the lack of refueling stations.

A key disadvantage of NGVs is their low range. The average range of a gasoline or diesel vehicle is 400 miles, and the range of an NGV is only 100–150 miles, depending on the natural-gas compression. Because of the dearth of natural-gas refueling stations, the current prevalent choice is to use a bifuel NGV that can run on both natural gas and gasoline. The problems associated with bifuel engines include slightly less acceleration and about 10 percent less power than a dedicated NGV because bifuel engines are not optimized to work on natural gas. Furthermore, warranties on new gasoline vehicles are severely reduced if they are converted to bifuel NGVs. The most important barriers for NGVs might be a public perception that CNG is a dangerous explosive to have on one's vehicle and a perception that self-service refueling with a high-pressure gas is too risky to offer to the general public.

About 22 percent of all new public-transit bus orders are for NGVs. Buses and corporate-fleet cars that stay in town have been the main market for NGVs, and both uses are mainly in response to the Clean-Fuel Fleet Program set up by the Environmental Protection Agency to reduce air pollution.

ALTERNATIVE DIESEL

Syngas-production technology has been discussed in the context of coal-to-liquid (CTL) fuels. The GTL process for producing diesel is similar to the indirect liquefaction of coal. Instead of producing syngas via gasification of coal or coal and biomass, the syngas is produced by steam reforming of natural gas. As with CTL, synthesis gas can be converted to a distillate and wax with a catalytic modification of the Fischer-Tropsch process[1] discussed in Chapter 4. The distillate and wax are hydrocracked to produce high-quality diesel and naphtha, as well as other streams that form the basis of such specialty products as synthetic lubricants. Although it is technically difficult, the naphtha can also be upgraded to gasoline.

Naphtha is an ideal feedstock for the manufacture of chemical building blocks (for example, ethylene), and GTL diesel is a high-quality automotive fuel or blending stock (Johnson-Matthey, 2006). GTL is an option for producing diesel from "stranded" natural gas like that which exits in the Middle East and Russia. However, a couple of those plants would begin to swamp the chemical naphtha market with material.

Hypothetically, converting natural gas to GTL diesel has several advantages over the use of CNG. All diesel vehicles can run on GTL diesel, and this gives gas producers access to new market opportunities. The range of diesel vehicles is much higher than that of NGVs because of diesel's higher fuel density. Engine efficiency and performance are not compromised by the adjustment for GTL diesel. GTL diesel can be shipped in normal tankers and unloaded at ordinary ports (*The Economist*, 2006).

There are several commercial GTL plants, including those of Sasol in Nigeria and Qatar and Shell in Malaysia and Qatar that produce GTL diesel; and a number of companies, including World GTL and ConocoPhillips, have plans to build GTL plants in the next several years. The economics of GTL plants are closely tied to the price of natural gas, and their viability depends on inexpensive stranded gas. GTL diesel is viewed mainly as an alternative to liquefied natural gas for monetizing associated natural gas or large natural-gas accumulations like the one in Qatar. The high cost of producing GTL makes it unlikely that GTL processes

---

[1]Most evaluations of CTL assume the use of iron-based Fischer-Tropsch catalysts largely because of impurities. GTL typically uses rhodium-based catalysts that do not have the poor selectivity of the iron-based catalysts and do not produce olefinic stocks.

will be developed in the United States unless an abundant and inexpensive source of natural gas, such as natural-gas hydrates, is found.

METHANOL

Methanol, an alcohol, is a liquid that can be used in internal-combustion engines to power vehicles. During the late 1980s, it was seen as a route to diversifying the fuels for the U.S. transportation system by converting natural gas from remote fields around the world to methanol and transporting the methanol to the United States to be used in the transportation system. That strategy was seen by energy planners as a way to convert what was cheap remote natural gas (around $1.00 per thousand cubic feet) to a marketable product. Today, it is used mainly as a commodity chemical and is produced primarily from natural gas.

Methanol has a higher octane rating than gasoline and is therefore a suitable neat fuel (that is, 100 percent methanol) for internal-combustion engines (for example, in racing cars). In practical terms, the penetration of methanol into a transportation system for light-duty vehicles that are fueled mainly by gasoline would require the construction of a distribution system and the use of flexible-fuel vehicles that could run on a mixture of gasoline and methanol. The use of a mixture of 85 percent methanol and gasoline (M85) would avoid the cold-start problem caused by methanol's low volatility, but methanol has about half the energy density of gasoline, and this affects the range that a vehicle can achieve on a full tank of fuel. Other drawbacks of methanol include its corrosivity, hydrophilicity, and toxicity. Methanol can cause various harmful effects to human health, including blindness and death if ingested, absorbed through the skin, or inhaled (Fisher Scientific, 2008). It would present substantial environmental, safety, health, and liability issues for station owners if it were introduced on a wide scale. One means of avoiding the infrastructure would be to convert the methanol to gasoline.

DIMETHYL ETHER

Dimethyl ether (DME) is a liquid fuel, at low pressure, with properties similar to those of liquefied petroleum gas. When burned, it produces less CO and CO_2 than gasoline and diesel do because of its lower carbon:hydrogen ratio. DME contains oxygen, so it requires a lower air:fuel ratio than gasoline and diesel do. DME has a higher thermal efficiency than diesel fuel, so it could enable higher-efficiency engine design. The presence of oxygen in the structure of DME minimizes soot formation (Arcoumanis et al., 2008). Other exhaust emissions—such as unburned hydrocarbons, NO_x, and particulate matter—are also reduced. The California Air Resources Board emission standards for automotive fuel are surpassed by DME; it is an ultraclean fuel.

At present, the preferred route and most cost-effective method for producing DME is through the dehydrogenation of methanol from synthesis gas, which is a mixture of CO and hydrogen. The basic steps for producing DME are as follows:

1. Syngas production by steam reforming of natural gas or by partial oxidation of coal, oil residue, biomass, or a combination of those.
2. Methanol synthesis with the use of copper-based or zinc oxide catalysts.
3. Methanol dehydrogenation to DME with the use of a zeolite-based catalyst.

The DME fuel produced is unsuitable for spark-ignition engines because of its high cetane number, but it can fuel a diesel engine with little modification. DME has properties similar to those of GTL diesel, including (Yao et al., 2006; Arcoumanis et al., 2008; Kim et al., 2008) good cold-flow properties, low sulfate content, and low combustion noise.

The principal advantage of using DME as an automotive fuel is that it is a clean-burning fuel that is easy to handle and store. It has thermal efficiency and ignitability similar to those of conventional diesel. As in the case of other potential alternative fuels, the primary challenge to the use of DME as an automotive fuel is the need for an infrastructure for its distribution. Disadvantages of using DME include low viscosity, poor lubricity, a propensity to swell rubber and cause leaks, and a heating value lower than that of conventional diesel.

HYDROGEN

Hydrogen, like electricity, is an energy carrier that can be generated from a wide variety of sources, including nuclear energy, renewable energy, and fossil fuels. Hydrogen also can be made from water via the process of electrolysis, although this appears to be more expensive than reforming natural gas. Used in vehicles, both hydrogen and electricity make efficient use of energy compared with liquid-fuel options on a well-to-wheel basis. As generally envisioned, hydrogen would generate electricity in a fuel cell, and the vehicle would be powered by an electric motor.[2] Developments in battery technology that may make plug-in hybrid electric and all-electric vehicles feasible will be discussed in several forthcoming National Research Council reports.

Hydrogen fuel-cell vehicle (HFCV) technology has progressed rapidly over the last several years, and large numbers of such vehicles could be introduced by 2015. Current HFCVs are very expensive because they are largely hand-built. For example, in 2008, Honda released a small number of HFCVs named FCX Clarity which cost several hundred thousands of dollars to produce (Fackler, 2008). However, technological improvements and economies of scale brought about by mass production should greatly reduce costs.

This section provides a synopsis of the National Research Council report *Transitions to Alternative Transportation Technologies—A Focus on Hydrogen* (NRC, 2008), by the Committee on Assessment of Resource Needs for Fuel Cell and Hydrogen Technologies. That committee concluded that the maximum practical number of HFCVs that could be operating in 2020 would be about 2 million, among 280 million light-duty vehicles in the United States. By about 2023, as costs of the vehicles and hydrogen drop, HFCVs could become competitive on a life-cycle basis. Their number could grow rapidly thereafter to about 25 million by 2030, and by 2050 they could account for more than 80 percent of new vehicles entering the U.S. light-duty vehicle market. Those numbers are not predictions by that committee but rather a scenario based on an estimate of the maximum penetration rate if it is assumed that technical goals are met, that consumers readily accept HFCVs, and that policy instruments are in place to drive the introduction of hydrogen fuel and HFCVs through the market transition period.

The scenario would require that automobile manufacturers increase production of HFCVs even while they cost much more than conventional vehicles and that investments be made to build and operate hydrogen fueling stations even while the market for hydrogen is very small. Substantial government actions and assistance would be needed to support such a transition to HFCVs by 2020

---

[2]Hydrogen also can be burned in an internal combustion engine (ICE), but the overall efficiency is much lower than that with a combination of fuel cells and a motor. It would be difficult to store enough hydrogen onboard to give an all-hydrogen ICE vehicle an acceptable range. The BMW hydrogen ICE also can use gasoline.

even with continued technical progress in fuel-cell and hydrogen-production technologies.

A large per-vehicle subsidy would be needed in the early years of the transition, but the number of vehicles per year would be low (Box 9.1) (NRC, 2008). Subsidies per vehicle would decline with fuel-cell costs, which are expected to drop rapidly with improved technology and economies of scale. By about 2025, an HFCV would cost only slightly more than an equivalent gasoline vehicle. Annual expenditures to support the commercial introduction of HFCVs would increase from about $3 billion in 2015 to $8 billion in 2023, at which point more than 1 million HFCVs could be joining the U.S. fleet annually. The cost of hydrogen also would drop rapidly, and, because the HFCV would be more efficient, it would cost less per mile to drive it than to drive the gasoline vehicle in about 2020. Combining vehicle and driving costs suggests that the HFCV would have lower life-cycle costs starting in about 2023. After that, there would be a net payoff to the country, which cumulatively would balance the prior subsidies by about 2028.

Substantial and sustained research and development (R&D) programs will be required to reduce the costs and improve the durability of fuel cells, develop new onboard hydrogen-storage technologies, and reduce hydrogen production costs. R&D investments are shown in Box 9.1. These programs should continue after

---

BOX 9.1 *Costs of Implementing Hydrogen Fuel-Cell Vehicles According to NRC (2008) Scenarios*

By 2023 (breakeven year), the government will have spent about $55 billion:
  $40 billion for the incremental cost of HFCVs.
  $8 billion for the initial deployment of hydrogen-supply infrastructure.
  $5 billion for research and development.
There would be 5.6 million HFCVs operating.

By 2050,
  There would be more than 200 million HFCVs operating.
  There would be 180,000 hydrogen stations.
  There would be 210 central hydrogen-production plants.
  There would be 80,000 miles of pipeline.
Industry would have profitably spent about $400 billion on hydrogen infrastructure.

2023 to reduce costs and improve performance further, but the committee did not estimate that funding.

The 2008 National Research Council study determined the consequent reductions in U.S. oil consumption and greenhouse gas emissions that could be expected in this scenario. HFCVs can yield large and sustained reductions in U.S. oil consumption and greenhouse gas emissions, but several decades will be needed to realize those potential long-term benefits. Figure 9.1 compares the oil consumption that would be required in this scenario with a reference case based on Energy Information Administration high oil-price projections, which include the recent increases in corporate average fuel economy standards. By 2050, HFCVs could reduce oil consumption by two-thirds. Greenhouse gas emissions would follow a similar trajectory if hydrogen produced from coal in large central stations was accompanied by carbon separation and sequestration.

The study then compared those reductions with the potential impact of alternative vehicle technologies (including conventional hybrid-electric vehicles) and

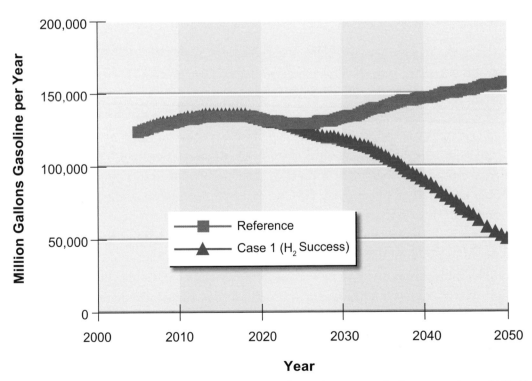

FIGURE 9.1 *Oil consumption with maximum practical penetration of HFCVs compared with reference case.*
Source: NRC, 2008.

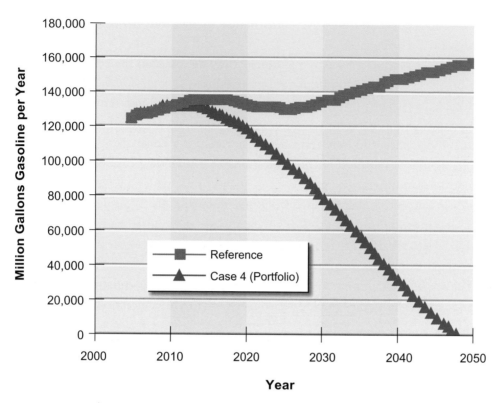

FIGURE 9.2 *Oil consumption for combined HFCVs, high-efficiency conventional vehicles, and biofuels compared with reference case.*
Source: NRC, 2008.

biofuels oil consumption and greenhouse as emissions. Over the next 2 decades, those approaches could deliver much greater reductions in U.S. oil use and greenhouse gas emissions than could HFCVs, but hydrogen offers greater longer-term potential. Thus, the greatest benefits will come from a portfolio of research and development in technologies that would allow the United States to nearly eliminate oil use in light-duty vehicles by 2050 (Figure 9.2). Achieving that goal would require substantial new energy-security and environmental-policy actions in addition to technological developments. Broad policies aimed at reducing oil use and greenhouse gas emissions will be useful, but they are unlikely to be adequate to facilitate the rapid introduction of HFCVs.

REFERENCES

Arcoumanis, C., C. Bae, R. Crookes, and E. Kinoshita. 2008. The potential of di-methyl ether (DME) as an alternative fuel for compression-ignition engines: A review. Fuel 87:1014-1030.

Economist, The. 2006. Arabian alchemy. Vol. 379, Issue 8480, 00130613, June 3.

EIA (Energy Information Administration). 2008. Natural gas consumption by end use. Available at http://tonto.eia.doe.gov/dnav/ng/ng_cons_sum_dcu_nus_m.htm. Accessed December 4, 2008.

Fackler, M. 2008. Latest Honda runs on hydrogen, not petroleum. New York Times, June 17. Available at http://www.nytimes.com/2008/06/17/business/worldbusiness/17fuelcell.html?_r=1&oref=slogin. Accessed August 28, 2008.

Fisher Scientific. 2008. Material Safety Data Sheet: Methanol. Available at http://fscimage.fishersci.com/msds/14280.htm. Accessed February 3, 2009.

Johnson-Matthey. 2006. Reducing emissions through gas to liquids technology. Available at http://ect.jmcatalysts.com/pdfs/Reducingemissionartp2-3.pdf. Accessed October 20, 2008.

Kim, M.Y., S.H. Yoon, B.W. Ryu, and C.S. Lee. 2008. Combustion and emission characteristics of DME as an alternative fuel for compression ignition engines with a high pressure injection system. Fuel 87:2779-2786.

NAS-NAE-NRC (National Academy of Sciences-National Academy of Engineering-National Research Council). 2009. America's Energy Future: Technology and Transformation. Washington, D.C.: The National Academies Press.

NRC (National Research Council). 2004. The Hydrogen Economy: Opportunities, Costs, Barriers, and R&D Needs. Washington, D.C.: The National Academies Press.

NRC. 2008. Transitions to Alternative Transportation Technologies—A Focus on Hydrogen. Washington, D.C.: The National Academies Press.

Rock, B. 2008. An overview of 2007 American 2007 natural gas vehicles. Helium, Inc. Available at http://www.helium.com/items/451632-an-overviewof-2007-american-2007-natural-gas-vehicles. Accessed September 2, 2008.

Yao, M., Z. Chen, Z. Zheng, B. Zhang, and Y. Xing. 2006. Study on the controlling strategies of homogeneous charge compression ignition combustion with fuel of dimethyl ether and methanol. Fuel 85:2046-2056.

Yborra, S. 2006. Taking a second look at the natural gas vehicle. American Gas (August/September):32-36.

Yeh, S. 2007. An empirical analysis on the adoption of alternative fuel vehicles: The case of natural gas vehicles. Energy Policy 35:5865-5875.

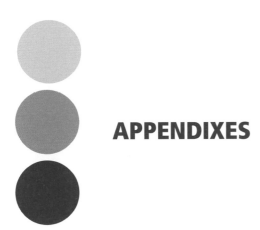

APPENDIXES

America's Energy Future Project

In 2007, the National Academies initiated the America's Energy Future (AEF) project (Figure A.1) to facilitate a productive national policy debate about the nation's energy future. The Phase I study, headed by the Committee on America's Energy Future and supported by the three separately constituted panels whose members are listed in this appendix, will serve as the foundation for a Phase II portfolio of subsequent studies at the Academies and elsewhere, to be focused on strategic, tactical, and policy issues, such as energy research and development priorities, strategic energy technology development, policy analysis, and many related subjects.

A key objective of the AEF project is to facilitate a productive national policy debate about the nation's energy future.

COMMITTEE ON AMERICA'S ENERGY FUTURE

HAROLD T. SHAPIRO, Princeton University, *Chair*
MARK S. WRIGHTON, Washington University in St. Louis, *Vice Chair*
JOHN F. AHEARNE, Sigma Xi and Duke University
ALLEN J. BARD, University of Texas at Austin
JAN BEYEA, Consulting in the Public Interest
WILLIAM F. BRINKMAN, Princeton University
DOUGLAS M. CHAPIN, MPR Associates
STEVEN CHU,[1] Lawrence Berkeley National Laboratory

[1]Resigned from the committee on January 21, 2009.

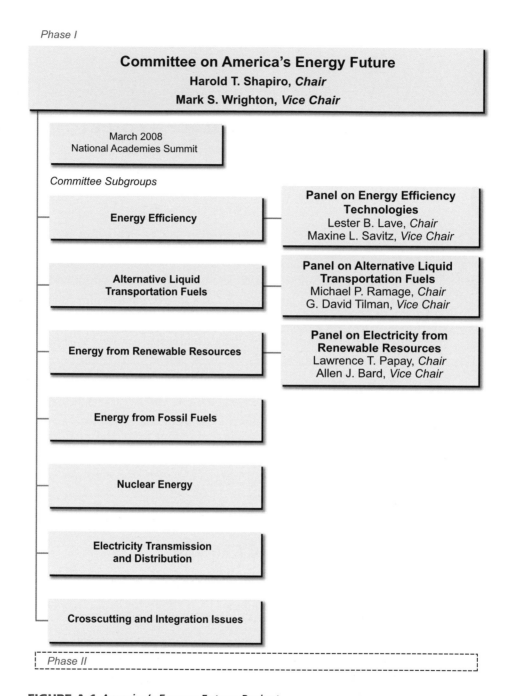

FIGURE A.1 *America's Energy Future Project.*

CHRISTINE A. EHLIG-ECONOMIDES, Texas A&M University
ROBERT W. FRI, Resources for the Future
CHARLES H. GOODMAN, Southern Company (retired)
JOHN B. HEYWOOD, Massachusetts Institute of Technology
LESTER B. LAVE, Carnegie Mellon University
JAMES J. MARKOWSKY, American Electric Power Service Corp. (retired)
RICHARD A. MESERVE, Carnegie Institution for Science
WARREN F. MILLER, JR., Texas A&M University
FRANKLIN M. ("Lynn") ORR, JR., Stanford University
LAWRENCE T. PAPAY, PQR LLC
ARISTIDES A.N. PATRINOS, Synthetic Genomics, Inc.
MICHAEL P. RAMAGE, ExxonMobil (retired)
MAXINE L. SAVITZ, Honeywell, Inc. (retired)
ROBERT H. SOCOLOW, Princeton University
JAMES L. SWEENEY, Stanford University
G. DAVID TILMAN, University of Minnesota, St. Paul
C. MICHAEL WALTON, University of Texas at Austin

PANEL ON ENERGY EFFICIENCY

LESTER B. LAVE, Carnegie Mellon University, *Chair*
MAXINE L. SAVITZ, Honeywell, Inc. (retired), *Vice Chair*
R. STEPHEN BERRY, University of Chicago
MARILYN A. BROWN, Georgia Institute of Technology
LINDA R. COHEN, University of California, Irvine
MAGNUS G. CRAFORD, LumiLeds Lighting
PAUL A. DeCOTIS, Long Island Power Authority
JAMES DeGRAFFENREIDT, JR., WGL Holdings, Inc.
HOWARD GELLER, Southwest Energy Efficiency Project
DAVID B. GOLDSTEIN, Natural Resources Defense Council
ALEXANDER MacLACHLAN, E.I. du Pont de Nemours and Company (retired)
WILLIAM F. POWERS, Ford Motor Company (retired)
ARTHUR H. ROSENFELD, E.O. Lawrence Berkeley National Laboratory
DANIEL SPERLING, University of California, Davis

PANEL ON ALTERNATIVE LIQUID TRANSPORTATION FUELS

MICHAEL P. RAMAGE, ExxonMobil Research and Engineering Company (retired), *Chair*

G. DAVID TILMAN, University of Minnesota, St. Paul, *Vice Chair*

DAVID GRAY, Noblis, Inc.

ROBERT D. HALL, Amoco Corporation (retired)

EDWARD A. HILER, Texas A&M University (retired)

W.S. WINSTON HO, Ohio State University

DOUGLAS R. KARLEN, U.S. Department of Agriculture, Agricultural Research Service

JAMES R. KATZER, ExxonMobil Research and Engineering Company (retired)

MICHAEL R. LADISCH, Purdue University and Mascoma Corporation

JOHN A. MIRANOWSKI, Iowa State University

MICHAEL OPPENHEIMER, Princeton University

RONALD F. PROBSTEIN, Massachusetts Institute of Technology

HAROLD H. SCHOBERT, Pennsylvania State University

CHRISTOPHER R. SOMERVILLE, Energy Biosciences Institute

GREGORY STEPHANOPOULOS, Massachusetts Institute of Technology

JAMES L. SWEENEY, Stanford University

PANEL ON ELECTRICITY FROM RENEWABLE RESOURCES

LAWRENCE T. PAPAY, Science Applications International Corporation (retired), *Chair*

ALLEN J. BARD, University of Texas, Austin, *Vice Chair*

RAKESH AGRAWAL, Purdue University

WILLIAM L. CHAMEIDES, Duke University

JANE H. DAVIDSON, University of Minnesota, Minneapolis

J. MICHAEL DAVIS, Pacific Northwest National Laboratory

KELLY R. FLETCHER, General Electric

CHARLES F. GAY, Applied Materials, Inc.

CHARLES H. GOODMAN, Southern Company (retired)

SOSSINA M. HAILE, California Institute of Technology

NATHAN S. LEWIS, California Institute of Technology

KAREN L. PALMER, Resources for the Future

JEFFREY M. PETERSON, New York State Energy Research and Development
    Authority
KARL R. RABAGO, Austin Energy
CARL J. WEINBERG, Pacific Gas and Electric Company (retired)
KURT E. YEAGER, Galvin Electricity Initiative

AMERICA'S ENERGY FUTURE PROJECT DIRECTOR

PETER D. BLAIR, Executive Director, Division on Engineering and Physical
    Sciences

AMERICA'S ENERGY FUTURE PROJECT MANAGER

JAMES ZUCCHETTO, Director, Board on Energy and Environmental Systems
    (BEES)

AMERICA'S ENERGY FUTURE PROJECT STAFF

KEVIN D. CROWLEY, Director, Nuclear and Radiation Studies Board (NRSB),
    Study Director
DANA G. CAINES, Financial Manager, BEES
SARAH C. CASE, Program Officer, NRSB
ALAN T. CRANE, Senior Program Officer, BEES
GREG EYRING, Senior Program Officer, Air Force Studies Board
K. JOHN HOLMES, Senior Program Officer, BEES
LaNITA JONES, Administrative Coordinator, BEES
STEVEN MARCUS, Editorial Consultant
THOMAS R. MENZIES, Senior Program Officer, Transportation Research Board
EVONNE P.Y. TANG, Senior Program Officer, Board on Agriculture and Natural
    Resources
MADELINE G. WOODRUFF, Senior Program Officer, BEES
E. JONATHAN YANGER, Senior Program Assistant, BEES

B

Statement of Task

The Panel on Alternative Liquid Transportation Fuels was tasked to examine the technical potential for reducing reliance on petroleum-based fuels for transportation, principally in automobiles and trucks but also in other vehicles and transportation modes, through the use of alternatives to traditional petroleum sources. The focus of the panel's efforts would be on liquid fuels produced from plant feedstocks and liquid fuels that can be derived from coal feedstocks. In keeping with the charge to the Committee on America's Energy Future, the panel was not to recommend policy choices but would assess the state of development of technologies.

The panel was charged to evaluate technologies on the basis of their estimated times to initial commercial deployment and to provide the following information on each:

- Initial deployment times <10 yr: costs, performance, and effects.
- 10–25 yr: barriers, implications for costs, and R&D challenges and needs.
- >25 yr: barriers and R&D challenges and needs, especially basic-research needs.

The primary focus of the study would be on the quantitative characterization of technologies whose initial deployment times would be less than 10 years. In light of existing studies and literature and the panel's own knowledge and expertise, the following should be considered for each feedstock or technology pathway chosen by the panel to the extent that existing data allow:

- For biomass-based liquid fuels, estimate the full fuel-cycle input—for example, energy, water, fertilizer, and land—needed to grow, collect and harvest, and process and convert the feedstock into a unit of fuel output. As part of its effort, the panel would also describe the implications for land use, agricultural practices, prices, externalities (such as implications for the environment), and other factors that it believes are important.
- For liquid fuels from coal, estimate the full fuel-cycle requirements—for example, mining, transport, and water—per unit of fuel produced.
- Estimate capital and operating costs per unit of output and total cost per unit of output. Costs per unit of output should be calculated on a consistent and comparable basis.
- Estimate full fuel-cycle environmental emissions per unit of fuel output—for example, carbon dioxide and other greenhouse gases, criteria pollutants, heavy metals—and land, water, and other effects identified by the panel that should be included.

It is expected that the panel would need to consider those technologies in the context of and in competition with other fuels that may enter the transportation sector during the periods examined by the panel, such as hydrogen, natural gas, electricity to power hybrid vehicles, reformulated gasoline, and petroleum-derived gasoline and diesel. The Committee on America's Energy Future, by drawing on existing National Academies and other recent comprehensive energy studies, will address the state of technology for hydrogen-fueled and hybrid electric vehicles.

Panel Members' Biographical Sketches

Michael P. Ramage *(Chair)* is retired executive vice president of ExxonMobil Research and Engineering Company. Previously, he was executive vice president, chief technology officer, and director of Mobil Oil Corporation. Dr. Ramage held a number of positions at Mobil, including research associate, manager of process research and development, general manager of exploration and producing research and technical service, vice president of engineering, and president of Mobil Technology Company. He has broad experience in many aspects of the petroleum and chemical industries. He has served on a number of university visiting committees and was a member of the Government University Industry Research Roundtable. He was a director of the American Institute of Chemical Engineers. Dr. Ramage chaired the recent National Research Council group that produced the report *The Hydrogen Economy: Opportunities, Costs, Barriers, and R&D Needs*. He is a member of the National Academy of Engineering (NAE) and has served on the NAE Council.

G. David Tilman *(Vice Chair)* is Regents' Professor and McKnight Presidential Chair in Ecology at the University of Minnesota. His research explores how to meet human needs for energy, food, and ecosystem services sustainably. He is a member of the National Academy of Sciences and the American Academy of Arts and Sciences, is a J.S. Guggenheim Fellow, and is a recipient of the Ecological Society of America's Cooper Award and its MacArthur Award, the Botanical Society of America's Centennial Award, and the Princeton Environmental Prize. He has written two books, edited three more, and published more than 200 scientific papers, including more than 30 in *Science*, *Nature*, and the *Proceedings of* *the National Academy of Sciences of the United States of America*. For the last 18 years, the Institute for Scientific Information has ranked him as the world's most-cited environmental scientist. In 2008, the emperor of Japan presented him with the International Prize for Biology.

David Gray is director of energy systems analysis at Noblis (formerly Mitretek Systems), a nonprofit consulting company. His expertise is in coal and natural-gas conversion to liquid fuels, heavy-oil and bitumen upgrading technologies, waste-to-energy conversion systems, and greenhouse gas emission and reduction analysis. Previously, he worked as a research manager at the Fuel Research Institute in South Africa on coal-to-liquid transportation-fuels production processes.

Robert D. Hall is retired general manager of Amoco Corporation. He has extensive experience in alternative-fuels R&D, in strategic planning, in R&D management, and in technology innovation. Mr. Hall held a number of positions at Amoco Corporation, including general manager of alternative-fuels development, manager of management systems and planning, director of the Amoco Oil Company R&D Department, director of the Amoco Oil Company Design and Economics Division, and supervisor of the Amoco Chemical Company Process Design and Economic Division. He has served on several National Research Council committees, including the Committee on Production Technologies for Liquid Transportation Fuels, the Committee on Strategic Assessment of the Department of Energy's Coal Program, the Committee to Review the R&D Strategy for Biomass-Derived Ethanol and Biodiesel Transportation Fuels, and the Committee on Benefits of DOE R&D on Efficiency and Fossil Energy. Mr. Hall is a past chairman of the International Council on Alternate Fuels.

Edward A. Hiler retired as the holder of the Ellison Chair in International Floriculture of Texas A&M University. He headed the Texas A&M University System Agriculture Program, which encompasses the Texas Agricultural Experiment Station, the Texas Cooperative Extension, the Texas Forest Service, the Texas Veterinary Medical Diagnostic Laboratory, and agricultural colleges at five system universities. He also served as dean of agriculture and life sciences at Texas A&M University, was head of the Department of Agricultural Engineering, and was deputy chancellor for academic and research programs and interim chancellor for the Texas A&M University System. His primary technical interests are in soil and water conservation engineering, small-watershed hydrology, irrigation and drainage engineering, and soil-plant-water-atmosphere relations in connection with irrigation management. He has been especially interested in plant response to water, nutrient, and oxygen deficits, in particular as they differ at various stages of plant growth and as they are related to irrigation and drainage management systems for minimizing these deficits. Other interests have included alternative energy sources with emphasis on biomass energy and the associated biochemical and microbiological energy-conversion processes. His career accomplishments have earned numerous honors and awards, including membership in the National Academy of Engineering, designation as a Distinguished Alumnus of Ohio State University, and presidency of the American Society of Agricultural Engineers (ASAE) in 1991–1992 and the Southern Association of Agricultural Scientists in 1999. He received the Texas A&M Faculty Distinguished Achievement Award in 1973, the ASAE Young Researcher Award in 1977, and the John Deere Gold Medal in 1991. He has served as a consultant to the U.S. Congress Office of Technology Assessment and the U.S. Department of the Interior Office of Water Research and Technology. He serves on the board of CNH Global, the world's largest manufacturer of agricultural equipment.

W.S. Winston Ho is a university scholar professor in the Department of Chemical and Biomolecular Engineering at Ohio State University. His research interests include molecular-based membrane separations, fuel-cell fuel processing and membranes, transport phenomena in membranes, and separations based on chemical reactions. In 2006, he was the recipient of the Institute Award for Excellence in Industrial Gases Technology from the American Institute of Chemical Engineers. Dr. Ho is a member of the National Academy of Engineering.

Douglas L. Karlen is a supervisory research soil scientist with the U.S. Department of Agriculture's Agricultural Research Service (USDA-ARS) and research leader in the Soil and Water Quality Research Unit of the USDA-ARS National Soil Tilth Laboratory. He is also professor in the Department of Agronomy at Iowa State University (ISU), mentor for the Graduate Program on Sustainability at ISU, and associate professor in the Department of Entomology, Soils, and Plant Sciences at Clemson University. Dr. Karlen is leading a project on sustainable agriculture and resource management and conservation and on the effects of growing crops for biofuels and bioenergy. His soil and crop management research program uses a systems approach involving producers, action agencies, nongovernment organizations, agribusiness, and other state and federal research partners to quantify the physical, chemical, and biological effects of conventional and organic farming practices. Effects of tillage, crop rotation, nutrient management, and other decision-based factors are evaluated by determining how they affect soil quality, crop productivity, plant-nutrient availability, and nutrient or soil losses in various soil types and landscape positions. Dr. Karlen has conducted a number of studies of the effects of agricultural systems and practices on nutrient loadings, biogeochemical cycles, soil and water quality, and crop production and costs. He received an MS in soil science from Michigan State University and a PhD in agronomy from Kansas State University.

James R. Katzer is an energy consultant and an affiliate professor in the Department of Chemical and Biological Engineering of Iowa State University who recently has been a visiting scholar at the Massachusetts Institute of Technology (MIT) Laboratory for Energy and the Environment and executive director of MIT's "The Future of Coal" study. He was manager of strategic planning and program analysis for the ExxonMobil Research and Engineering Company. Before that, he was vice president of technology for the Mobil Oil Corporation with primary responsibilities of ensuring Mobil's overall technical health, developing forward-looking technology scenarios, and identifying and analyzing technology and environmental developments and trends. He joined the Central Research Laboratory of the Mobil Oil Corporation in 1981 and later became manager of process research and technical service and vice president of planning and finance for the Mobil Research and Development Corporation. Before joining Mobil, he was a professor in the chemical engineering faculty at the University of Delaware and the first director of the Center for Catalytic Science and Technology. Dr. Katzer has more than 80 publications in technical journals, holds several patents, and is a coauthor or editor of several books. Dr. Katzer is a member of the National Academy of Engineering.

Michael R. Ladisch is the director of the Laboratory of Renewable Resources Engineering and Distinguished Professor of Agricultural and Biological Engineering and Biomedical Engineering at Purdue University and the chief technology officer of Mascoma Corporation. His expertise is in bioseparations, bionanotechnology bioprocess engineering, and bioenergy. His research has resulted in systematic approaches and correlations for scaling up chromatographic purification techniques from the laboratory to process-scale manufacturing systems. His work has resulted in 150 publications, a textbook on bioseparations, 14 patents, and more than 100 papers presented at national professional society meetings. He has served as a member of U.S. delegations and advisory panels to Russia, Thailand, China, and Japan to review the status of biotechnology programs. He has also chaired several National Research Council committees concerning biotechnology. He is a member of the National Academy of Engineering. He is a cofounder of Biovitesse, a startup company in pathogen detection. He serves on the scientific board of Agrivida and is a cofounder of Celsys, Inc. Both companies address technology in cellulose ethanol.

John A. Miranowski is professor of economics and director of the Institute of Science and Society at Iowa State University (ISU). Dr. Miranowski's current research is focused on economics of renewable energy and carbon policy, and he has published broadly on the economics of natural resources and environmental issues, including producer and consumer response to higher energy prices, corn and cellulosic biofuel economics, energy efficiency in agriculture, and resource conservation policy and sustainability. He previously served as chair of the Department of Economics at ISU, director of the Resources and Technology Division of the Economic Research Service of the U.S. Department of Agriculture (USDA), executive coordinator of the USDA Policy Coordination Council, and special assistant to the deputy secretary of agriculture. Dr. Miranowski also headed the U.S delegation to the Organisation for Economic Co-ordination and Development Joint Working Party on Agriculture and Environment and served as director on the Board of the Association of Environmental and Resource Economists and on the Board of the Agricultural & Applied Economics Association. Dr. Miranowski served as a member of the National Research Council Committee on Impact of Emerging Agricultural Trends on Fish and Wildlife Habitat and a panel member of the Committee on Opportunities in Agriculture. He received the USDA Distinguished Service Honor Award for Biofuels Program Development in 1993. He earned a BS in agricultural business from ISU and an AM and a PhD in economics from Harvard University.

Michael Oppenheimer is the Albert G. Milbank Professor of Geosciences and International Affairs in the Woodrow Wilson School and the Department of Geosciences at Princeton University. He is also the director of the program in science, technology, and environmental policy at the Woodrow Wilson School and faculty associate of the Atmospheric and Ocean Sciences Program and the Center of International Studies. Dr. Oppenheimer's interests include science and policy related to the atmosphere, particularly climate change and its effects. His research explores the potential effects of global warming, including the effects of warming on atmospheric chemistry, on ecosystems and the nitrogen cycle, on ocean circulation, and on the ice sheets in the context of defining "dangerous anthropogenic interference" with the climate system. Dr. Oppenheimer joined the Princeton faculty after more than 2 decades with Environmental Defense, a nongovernment environmental organization, where he served as chief scientist and manager of the Global and Regional Atmosphere Program. Recently, Dr. Oppenheimer served as a lead author of the Third Assessment Report of the Intergovernmental Panel on Climate Change. Dr. Oppenheimer was a member the National Research Council Panel on Climate Variability and Change. He received an SB in chemistry from the Massachusetts Institute of Technology and a PhD in chemical physics from the University of Chicago.

Ronald F. Probstein is Ford Professor of Engineering emeritus at the Massachusetts Institute of Technology. His research interests are in physicochemical hydrodynamics, fluid mechanics, synthetic fuels, and environmental-control technology. He was named a Guggenheim Fellow and a fellow of the American Academy of Arts and Sciences, the American Institute of Aeronautics and Astronautics, the American Physical Society, the American Association for the Advancement of Science, and the American Society of Mechanical Engineers (ASME). He was the recipient of the Freeman Award in Fluids Engineering of ASME and holds an honorary doctorate from Brown University. In addition to his research in synthetic fuels, largely in coal conversion and associated water-use minimization, he published *Synthetic Fuels*, which was reprinted by Dover Publications in 2006, and the research monograph *Water in Synthetic Fuel Production* (MIT Press, 1978). He is a member of the National Academy of Engineering and the National Academy of Sciences.

Harold H. Schobert is a professor of fuel science in the Department of Energy and Mineral Engineering at Pennsylvania State University. He also has a visiting appointment as extraordinary professor of natural sciences at North-West University in South Africa. He has published more than 100 peer-reviewed papers in coal chemistry, carbon and graphite, novel reactions in petroleum refining, and carbon dioxide capture. He has been the leader of Pennsylvania State University's coal-to-jet fuel program, which has developed a coal-based replacement for conventional jet fuels. Dr. Schobert was a member of the Energy Engineering Board at the National Research Council from 1990 to 1996.

Christopher R. Somerville is the director of the Energy BioSciences Institute in Berkeley, California. He oversees all activities at the institute, including research, communication, education, and outreach. He also chairs the institute's Executive Committee. Dr. Somerville is a professor in the Department of Plant and Microbial Biology at the University of California, Berkeley, and a visiting scientist at the Lawrence Berkeley National Laboratory. His research focuses on the characterization of proteins implicated in plant cell-wall synthesis and modification. He has published more than 200 scientific papers in plant and microbial genetics, genomics, biochemistry, and biotechnology. Dr. Somerville has served on the scientific advisory boards of many corporations, academic institutions, and private foundations in Europe and North America. He is a member of the National Academy of Sciences, the Royal Society of London, and the Royal Society of Canada.

Gregory Stephanopoulos is Willard Dow Professor of Biotechnology and Chemical Engineering at the Massachusetts Institute of Technology. The central focus of his research is metabolic engineering, the improvement of cellular properties using modern genetic tools with attention to industrial applications, and biomedical research aimed at the elucidation of key physiological differences that characterize disease states and can guide drug and therapy development. He has received numerous awards, including the American Institute of Chemical Engineers (AIChE) Wilhelm Award in Chemical Reaction Engineering (2001), the Marvin Johnson Award of the Biotechnology Division of the American Chemical Society (2000), the AIChE Food, Pharmaceutical & Bioengineering Division Award (1997), and the Technical Achievement Award of the AIChe Southern California section (1984). Dr. Stephanopoulos is a member of the National Academy of Engineering. He received a PhD in chemical engineering from the University of Minnesota, Minneapolis.

James L. Sweeney is the director of the Precourt Institute for Energy Efficiency and former chairman of the Department of Engineering–Economic Systems and Operations Research of Stanford University. He has been a consultant, director of the Office of Energy Systems, director of the Office of Quantitative Methods, and director of the Office of Energy Systems Modeling and Forecasting of the Federal Energy Administration. At Stanford University, he has been chairman of the Institute of Energy Studies, director of the Center for Economic Policy Research, and director of the Energy Modeling Forum. He has served on several National Research Council committees, including the Committee on the National Energy Modeling System and the Committee on the Human Dimensions of Global Change. He also served on the Committee on Benefits of DOE's R&D on Energy Efficiency and Fossil Energy, helping to develop the framework and method that the committee applied to evaluating benefits. His research and writings address economic and policy issues important for natural-resource production and use; energy markets, including those in oil, natural gas, and electricity; environmental protection; and the use of mathematical models to analyze energy markets. He has a BS from the Massachusetts Institute of Technology and a PhD in engineering-economic systems from Stanford University.

Presentations to the Panel

NOVEMBER 19, 2007

Robert Perlack, Oak Ridge National Laboratory
Overview of Plant Feedstock Production for Biofuel: Current Technologies and Challenges, and Potential for Improvement

Jonathan Foley, University of Wisconsin, Madison
Regional and Global Environmental Consequences of Expanding Biofuel Production from Agricultural Feedstocks: Potential Production Issues and Environmental Impacts

NOVEMBER 20, 2007

Bruce Dale, Michigan State University
Why Cellulosic Ethanol Is Nearer Than You May Think: Creating the Biofuels Future

FEBRUARY 19, 2008

Otto Doering, Purdue University
Economics of Production of Liquid Fuels from Plant Feedstocks

Robert Williams, Princeton University
Overview of the Production of Liquid Transportation Fuels from Coal Feedstocks and from Biomass Feedstocks via Gasification and Similar Technologies

Samuel Tam, Headwaters
Direct Liquefaction: Total Production Costs, Current Status of Conversion Technologies and Potential for Future Improvement, and Environmental Impacts

Sam Tabak, ExxonMobil
ExxonMobil Methanol to Gasoline

Theodore Wegner, U.S. Department of Agriculture, Forest Service
Forest Biomass for Liquid Transportation Fuels Production

FEBRUARY 20, 2008

Rich Bain and Maggie Mann, National Renewable Energy Laboratory
Thermochemical Conversion of Biomass

Amory Lovins and James Newcomb, Rocky Mountain Institute
Importance of Scale in the Production of Biofuels

Research Supporting a Landscape Vision of Production of Biofuel Feedstock

Recent field-scale precision-management studies (Kitchen et al., 2005; Lerch et al., 2005) provide a practical foundation for the panel's landscape vision of production of biofuel feedstock. In 1991, authors of those studies implemented a corn–soybean rotation in a 14.5-acre field, using mulch tillage to maintain about 30 percent residue cover. Before initiating the study, they characterized the field on the basis of georeferencing and developed an order 1 soil survey,[1] a digital elevation model, electromagnetic induction, and soil-fertility maps. They proceeded by mapping crop yield and producing profitability maps in 1993. The studies showed that the greatest yield-limiting factors were soil texture, topsoil depth, and topography (Lerch et al., 2005). All three of those factors influence soil water-holding capacity and within-field water distribution. Diminished topsoil thickness had an adverse effect on profitability (Kitchen et al., 2005) and was directly related to soil loss via erosion. Reduced soil quality is a result of poor physical and chemical characteristics of an underlying argillic claypan horizon that is not well suited for crop-root growth (Lerch et al., 2005). In a market-driven landscape vision of lignocellulosic feedstock production proposed by this panel, the precision-agriculture system devised by Kitchen et al. (2005) serves as a model because their recommendations to improve sustainability were to add more crop types and crop rotations and to use no-tillage practices tailored to specific management areas in the field based on their long-term, georeferenced database.

In another study, Williams et al. (2008) constructed a method based on

---

[1]Order 1 soil surveys are soil inventories produced for very intensive land uses that require detailed information about soils (USDA-NRCS, 2007).

the geographic information system (GIS) to delineate agroecozones and agroeco-regions that were suitable for various crops. Their procedure relied completely on digital databases and was considered more objective than methods that relied at least in part on expert opinion. The resolution of their procedure, however, was 1 km, and that leaves a spatial-resolution gap between their procedure and the approach used by Kitchen et al. (2005) for an individual field. Yan et al. (2007) described a GIS database-driven method similar to that of Williams et al. (2008). Their study, however, was conducted on a single-field scale to delineate zones requiring different management practices for a single crop.

To implement a landscape vision of bioenergy feedstock production, informa-tion should be gathered, on scales of at least 1 mile, of current land tenure and community access, drainage patterns, soil-quality status, crop-rotation and crop-distribution patterns, economic conditions, conservation practices, wildlife and human restrictions and concerns, and other pertinent factors. A potential biofuel-production scheme that increases ecosystem services might include establishing woody species (for example, *Populus)* near streams as buffers and long-term biomass sources. Next, *Miscanthus (Miscanthus x giganteus)*, reed canarygrass (*Phalaris arundinacea*), eastern gamagrass (*Tripsacum dactyloides*), or diverse mixtures of these and similar species, could be used at slightly higher landscape positions to benefit from and reduce leaching of nitrate nitrogen and to sequester carbon as soil organic matter. Slightly higher on the landscape, diverse mixtures of warm-season grasses and cool-season legumes could produce biomass and store organic carbon in soils. In fall, the perennials would be a source of biomass and thus address at least three of the landscape problems—biomass production, car-bon sequestration, and water quality. Moving up the landscape, a diversified rota-tion of annual and perennial crops would be used to meet food, feed, and fiber needs. Erosion could be partially mitigated by using cover crops or living mulches. Intensive row-crop production areas could be established by using best manage-ment practices with the awareness that if fertilizer recovery was less than desired, there would be a substantial buffer (lignocellulosic) production area lower on the landscape to capture residual nutrients and sediment. A step-by-step outline of that process is presented below:

1. Identify landscape characteristics by using georeferenced technologies and methods.
2. Identify the landscape's most important production and conservation issues.

3. Delineate critical areas that require different crops and management practices.
4. Identify suites of suitable crops, crop rotations, and conservation practices for each management area.
5. Develop a landscape-scale precision-agriculture system.
6. Apply policies, education, and programs that address social and economic concerns related to the adoption and implementation of the landscape-scale precision-agriculture systems.
7. Monitor and document the new system's performance toward production and conservation goals.
8. Re-evaluate the system and make adaptive changes to improve its performance.

In summary, the important message from the above examples and guidelines is that the technology needed to implement a sustainable landscape vision of biofuel production exists and that the practices can already be implemented efficiently and economically.

REFERENCES

Kitchen, N.R., K.A. Sudduth, D.B. Myers, R.E. Massey, E.J. Sadler, and R.N. Lerch. 2005. Development of a conservation-oriented precision agriculture system: Crop production assessment and plan implementation. Journal of Soil and Water Conservation 60:421-430.

Lerch, R.N., N.R. Kitchen, R.J. Kremer, W.W. Donald, E.E. Alberts, E.J. Sadler, K.A. Sudduth, D.B. Myers, and F. Ghidey. 2005. Development of a conservation oriented precision agriculture system: Water and soil quality assessment. Journal of Soil and Water Conservation 60:411-421.

USDA-NRCS (U.S. Department of Agriculture, Natural Resources Conservation Service). 2007. National Soil Survey Handbook, Title 430-VI. Washington, D.C.: USDA-NRCS.

Williams, C.L., W.W. Hargrove, M. Liebman, and D.E. James. 2008. Agroecoregionalization of Iowa using multivariate geographical clustering. Agriculture, Ecosystem and Environment 123:161-174.

Yan, L., S. Zhou, L. Feng, and L. Hong-Yi. 2007. Delineation of site-specific management zones using fuzzy clustering analysis in a coastal saline soil. Computers and Electronics Agriculture 56:174-186.

F

Estimating the Amount of Corn Stover That Can Be Harvested in a Sustainable Manner

The use of national average corn grain yields to estimate available amounts of corn stover for producing biofuels provides a general guideline for decision making by both industry and landowners, but it is sometimes inappropriate because of the site-specific nature of agricultural production. For example, the amount of stover needed to minimize erosion and maintain soil organic matter for a specific land area depends on several factors, including the predominant landscape (for example, rolling or flat), soil type or series, climate, tillage and crop-management practices, and yield. As a result, regardless of the type of tillage being used, some locations in a given field cannot spare any crop residue without risking degradation of soil resources. In some other locations, crops grown in rotation with corn will actually benefit from partial removal of stover.

National Agricultural Statistics Service data (USDA-NASS, 2008) on five important corn-producing states—Illinois, Iowa, Indiana, Nebraska, and Minnesota—were used to illustrate the complex relationships among seasonal weather patterns (climate), crop yields, and the multiple uses for corn stover, including mitigation of soil erosion, sustaining of soil organic matter, and biofuel feedstock. Those five states were chosen because in 2007 they accounted for 50 of the 86.5 million acres of harvested corn. In 2003–2007, average corn yields were 143–180 bushels/acre in Illinois, 157–181 bushels/acre in Iowa, 146–168 bushels/ acre in Indiana, 146–166 bushels/acre in Nebraska, and 146–174 bushels/acre in Minnesota. Variations in yield were attributed to seasonal differences in weather.

Two scenarios were constructed to help to determine a baseline amount of stover that could be harvested in a sustainable manner. First, the 5-year average yield in the five states (161 bushels/acre) was used to project 55.9 million tons of harvestable stover. Second, the average of the highest grain yields achieved in each state from 2003 to 2007, 173.8 bushels/acre, was computed for the projection. That approach increased the estimated harvestable yield to 71.2 million tons, or 94 percent of the national projection based on the 2007 average grain yield. Recognizing that the high yields occurred in different states during a given year because of weather differences, the panel discussed reducing the estimate to 70 million tons. Ultimately, the consensus was to use 76 million tons on the basis of the national corn grain yield. The panel used the high harvestable value because it took a conservative approach to estimating the amount of stover that has to be left in the field to maintain soil. The panel also assumed that crop yields increase as a result of genetic improvement to enhance a crop's stress-tolerance.

Some may consider the panel's baseline too low because many producers in Iowa and Illinois are already achieving corn yields of 208 bushels/acre, which is 20 percent higher than the 2007 average in those two states. Furthermore, if 70 percent of the corn growers in Iowa, Illinois, Indiana, Nebraska, and Minnesota grow corn continuously at that yield level, the projected amount of feedstock that could be harvested in a sustainable manner in just those states would increase to 112 million tons. That scenario justified the use of 112 million tons as the panel's 2020 estimate.

Similarly, though fewer, some producers in the five states are already achieving average yields of 230 bushels/acre by using good management practices (Elmore and Abendroth, 2008). According to the same procedure as before, that level of production could provide 135 million tons of stover per year as a biofuel feedstock. It also demonstrates the genetic potential of corn hybrids that are already commercially available. On the basis of the research and extension service reports published up to 2008, the panel chose to use 135 million tons as its projection for 2035 because it can be achieved by simply maintaining the 30-year trend of an average increase of 1.964 bushels/acre per year in the five leading corn-producing states. Achieving that nationally would increase the amount of stover that could be harvested in a sustainable manner to 232 million tons/year.

Finally, the panel computed the corn grain yield that would be needed to produce enough harvestable stover to meet the Energy Independence and Security Act of 2007 goal of at least 16 billion gallons of cellulosic biofuel. Given the conservative estimate of 50 million acres and the 70–30 distribution between continuous and rotated corn, respectively, the average grain yield would have to increase to 293 bushels/acre to meet that goal. That is not beyond some projections, but it most likely will not be required, because as feedstock demand increases, more landowners will want to participate, tillage intensity will probably be decreased, and the area that can be harvested in a sustainable manner will probably increase. In fact, achieving 228 bushels/acre on 86 million acres would also meet that goal.

REFERENCES

Elmore, R., and L. Abendroth. 2008. Are we capable of producing 300 bu/acre corn yields? Available at http://www.agronext.iastate.edu/corn/production/management/harvest/producing.html. Accessed January 14, 2009.

USDA-NASS (U.S. Department of Agriculture, National Agricultural Statistics). 2008. National Agricultural Statistics. Available at www.nass.usda.gov. Accessed June 18, 2008.

G

Life-Cycle Inputs for Production of Biomass

Nitrogen and phosphorus are the nutrients required in the largest amounts to sustain biomass-crop growth and development, and they have the greatest environmental impact. Nitrogen is leached below the crop root zone into subsurface tile drainage lines or groundwater, and phosphorus moves in runoff. Both nutrients cause eutrophication of water bodies, which contributes to such problems as hypoxia in the Mississippi River watershed and the hypoxic region ("dead zone") of the Gulf of Mexico.

Nitrogen, phosphorus, and other crop-production inputs should be applied at economically optimal rates rather than at rates that will achieve the highest yields. Phosphorus and potassium rates are generally based on soil test levels that have been optimized for each state. For Iowa, application of phosphorus pentoxide (P_2O_5) fertilizer for corn grown in soils with optimal, low, or very low soil test ratings (26–35, 16–25, and 0–15 ppm, respectively) would be at 55, 75, or 100 lb/acre. Similarly, application of potassium oxide (K_2O) rate in soils with optimal, low, or very low soil test ratings (131–170, 91–130, or less than 90 ppm, respectively) would be at 45, 90, or 130 lb/acre. For soils testing high or very high in phosphorus or potassium, fertilizer would not be recommended (Mallarino et al., 2002), assuming that corn residues are not removed.

The specific nitrogen fertilizer rate required to achieve the maximal economic net return to nitrogen (MRTN) will vary according to seasonal weather pattern, soil type, fertilizer price, management practices, and interactions among these factors. For Iowa, the MRTN for continuous corn or a corn–soybean rotation has been 175 or 125 lb/acre, respectively, for the last 10 years (Sawyer and Randall, 2008), assuming no harvest of crop residue.

Recent field studies (Karlen, 2007) indicated that nitrogen, phosphorus, and potassium removal with the cob and upper portion of the corn plant averaged 10, 2, and 13 lb/ton of dry stover, respectively. The baseline assessment is based on the assumptions that soils from which crop residues would be removed would have optimal phosphorus and potassium soil test levels and that nitrogen was being applied at the MRTN. Therefore, for a stover harvest rate of 1.5 tons/acre, the annual fertilizer requirements of nitrogen, phosphorus, and potassium would be 190, 27, and 57 lb/acre, respectively, for continuous corn or 140, 27, and 57 lb/acre for corn rotated with soybean. For 2020 projections (2.5 tons of dry stover per acre), the panel increased fertilizer input to account for nutrient removal and assumed only a slight increase (10 percent) in MRTN because of better efficiency of nitrogen use. That resulted in estimated fertilizer requirements of nitrogen, phosphorus, and potassium of 218, 29, and 69 lb/acre for continuous corn.

Inputs would be much lower for dedicated perennial biomass crops. Woody crops are rarely fertilized. Herbaceous perennial crops would be harvested when senescent, so loss of nitrogen, phosphorus, and other nutrients would be minimized because plants retranslocate such nutrients to roots in fall. The panel envisions a process in which biofuel processing facilities capture those nutrients, which can then be periodically returned to the soil. Given the ability of perennial grasses to maintain yields for many years with little or no fertilization, it might be feasible to return removed nutrients once every 3–5 years and thus reduce energy requirements for fertilizer transport and application. Commercial nitrogen fertilizer is energetically expensive but could be replaced, if needed, by growing one or more legume species with a biomass crop. Competition between the crop and the legume could be minimized, if competition occurs, by using a legume that has a different season of maximal growth from the biomass crop.

REFERENCES

Karlen, D.L. 2007. Balancing bioenergy opportunities on your natural resources base. Paper read at Indiana Crop Advisors Conference Meeting Proceedings, December 18–19, 2007, Indianapolis.

Mallarino, A.P., D.J. Wittry, and P.A. Barbagelata. 2002. Iowa soil-test field calibration research update: Potassium and the Mehlich-3 ICP phosphorus test. North Central Extension–Industry Soil Fertility Conference 18:30-39.

Sawyer, J.E., and G.W. Randall. 2008. Final Report: Gulf Hypoxia and Local Water Quality Concerns Workshop. St. Joseph, Mich.: Upper Mississippi River Sub-basin Hypoxia Nutrient Committee, American Society of Arrgricultural and Biological Engineering.

Background Information on the Economic and Environmental Assessment of Biomass Supply

The tables in this appendix present the background information and assumptions that were used in the panel's economic assessments and greenhouse gas emission analyses. They include comparisons of published and updated costs of harvest and maintenance (Table H.1), nutrient replacement (Table H.2), transportation for delivery (Tables H.3 and H.4), storage (Table H.5), and establishment and seeding (Table H.6) for different cellulosic feedstocks. Estimates of opportunity costs for cellulosic feedstocks are presented (Table H.7). The published yield values from which current and future projections were computed (Table H.8), and carbon inputs for feedstock production (Table H.9) and biomass refining (Table H.10) are also included.

TABLE H.1 Estimated Costs of Harvest and Maintenance for Cellulosic Feedstocks

Type of Feedstock	Type of Cost	Cost per Ton (cited $)	Cost per Ton (2007$)	Reference
Corn stover	Baling, stacking, grinding	26	45	Hess et al. (2007)
Corn stover	Collection	31–36	66–77	McAloon et al. (2000)
Corn stover	Collection	35–46	64–84	McAloon et al. (2000)
Corn stover	Collection	17.70	17.70	R. Perlack, Oak Ridge National Laboratory, presentation to the committee on November 19, 2007
Corn stover	Up to storage	20–21	36–39	Sokhansanj and Turhollow (2002)
Corn stover		28	36	Suzuki (2006)
Corn stover	Baling, staging	26	47	Aden et al. (2002)
Corn stover	Harvest	14	14	Edwards (2007)
Switchgrass	Collection	12–22	16–28	Kumar and Sokhansanj (2007)
Switchgrass	Harvest	32	32	Duffy (2007)
Switchgrass	Harvest	35	58	Khanna et al. (2008)
Switchgrass	Harvest, maintenance, establishment	123.5/acre	210/acre	Khanna and Dhungana (2007)
Switchgrass	Harvest	15	26	Perrin et al. (2008)
Miscanthus	Harvest	33	54	Khanna et al. (2008)
Miscanthus	Harvest, maintenance, establishment	301/acre	512/acre	Khanna and Dhungana (2007)
Nonspecific		10–30	15–45	Mapemba et al. (2007)
Nonspecific		23	38	Mapemba et al. (2008)

Note: Harvest and maintenance costs were updated by using USDA-NASS agricultural fuel, machinery, labor prices from 1999 to 2007 (USSA-NASS, 2007a,b).

TABLE H.2 Estimated Costs of Nutrient Replacement for Cellulosic Feedstocks

Type of Feedstock	Type of Cost	Cost per Ton (cited $)	Cost per Ton (2007$)	Reference
Corn stover		10.2	14.1	Hoskinson et al. (2007)
Corn stover		4.6	8.4	Khanna and Dhungana (2007)
Corn stover		7	14.4	Aden et al. (2002)
Corn stover		4.2	4.2	Petrolia (2008)
Corn stover		10	21	Perlack and Turhollow (2003)
Corn stover	Whole-plant harvest	9.7	13.3	Karlen and Birrell (Unpublished)
Corn stover	Cob, top 50% harvest	9.5	13.1	Karlen and Birrell (Unpublished)
Corn stover	Bottom 50% harvest	10.1	13.9	Karlen and Birrell (Unpublished)
Switchgrass		6.7	12.1	Perrin et al. (2008)
Switchgrass		10.8	19.77	Khanna et al. (2008)
Miscanthus		2.5	4.6	Khanna et al. (2008)

Note: Nutrient and replacement costs were updated by using USDA-NASS agricultural-fertilizer prices from 1999 to 2007 (USDA-NASS, 2007a,b).

TABLE H.3 Estimated Distance for Delivery of Cellulosic Feedstocks

Distance (miles)	Type	Reference
46–134	Round-trip	Mapemba et al. (2007)
22–62	One-way	Perlack and Turhollow (2003)
22–61	One-way	Perlack and Turhollow (2002)
50	Round-trip	Khanna et al. (2008)
50	Max one-way	English et al. (2006)
50	One-way	Vadas et al. (2008)

TABLE H.4 Estimated Costs of Transportation for Delivery of Cellulosic Feedstocks

Type of Feedstock	Type of Cost	Cost per Ton (cited $)	Cost per Ton (2007$)	Reference
Corn stover	Per ton	8.85	12.5	English et al. (2006)
Corn stover	Per ton	10.25	27	Hess et al. (2007)
Corn stover	DVC[a]	0.15	0.35	Kaylen et al. (2000)
Corn stover	Max DVC for positive NPV	0.28	0.66	Kaylen et al. (2000)
Corn stover	Per ton	10.8	10.8	Perlack (2007)
Corn stover	Per ton	13	31	Aden et al. (2002)
Corn stover	Per ton	4.2–10.5	11–27.7	Perlack and Turhollow (2002)
Corn stover	DVC	0.08–0.29	0.17–0.63	Kumar et al. (2005)
	DFC[b]	4.5	9.8	
	DFC range	0–6	0–13.3	
Corn stover	DVC	0.18	0.32	Searcy et al. (2007)
	DFC	4	7.3	
Corn stover	DVC	0.16	0.38	Kumar et al. (2003)
	DFC	3.6	8.6	
Corn stover	DVC			Petrolia (2008)
	0-25 miles	0.13–0.23	0.13–0.23	
	25-100 miles	0.10–0.19	0.10–0.19	
	>100 miles	0.09–0.16	0.09–0.16	
	DFC square bales	1.70	1.70	
	DFC round bales	3.10	3.10	
Corn stover	Per ton	10.9	13.8	Vadas et al. (2008)
Switchgrass	Per ton	14.75	14.75	Duffy (2007)
Switchgrass	Per ton	19.2–23	27–32.4	Kumar and Sokhansanj (2007)
Switchgrass	Per ton	13	28	Perrin et al. (2008)
Switchgrass	Per ton	10.9	13.8	Vadas et al. (2008)
Switchgrass or Miscanthus	Per ton for 50 miles	7.9	17.1	Khanna et al. (2008)
Nonspecific	Per ton	7.4–19.3	13.7–35.6	Mapemba et al. (2007)
Nonspecific	Per ton	14.5	31.5	Mapemba et al. (2008)
Woody biomass	Per ton		11–22	Summit Ridge Investments (2007)

Note: Transportation costs were updated by using USDA-NASS agricultural-fuel prices from 1999 to 2007 (USDA-NASS, 2007a,b).

[a]DVC, distance variable cost, per ton per mile.
[b]DFC, distance fixed cost per ton.

TABLE H.5 Estimated Storage Costs for Cellulosic Feedstocks

Type of Feedstock	Type of Cost	Cost per Ton (cited $)	Cost per Ton (2007$)	Reference
Corn stover		4.44	5.64	Hess et al. (2007)
Corn stover	Round bales	6.82	6.82	Petrolia (2008)
	Square bales	12.93	12.93	
Switchgrass		16.67	16.67	Duffy (2007)
Switchgrass		4.14	5.18	Khanna et al. (2008)
Miscanthus		4.40	5.50	Khanna et al. (2008)
Nonspecific		2	2.18	Mapemba et al. (2008)

Note: Storage costs were updated by using USDA-NASS agricultural-building material prices from 1999 to 2007 (USDA-NASS, 2007a,b).

TABLE H.6 Estimated Costs of Establishment and Seeding for Cellulosic Feedstocks

Type of Feedstock	Type of Cost	Land Rent Included	Cost per Acre (cited $)	Cost per Acre (2007$)	Reference
Switchgrass		Yes	200	200	Duffy (2007)
Switchgrass		No	25.76	46	Perrin et al. (2008)
		Yes	85.46	153	
Switchgrass	PV[a] per ton	No	7.21/ton	12.6/ton	Khanna et al.
	10-year PV per acre		142.3	249	(2008)
	Amortized				
	4% over 10 years		17.3	30.25	
	8% over 10 years		20.7	36.25	
Switchgrass		Yes	72.5–110	88.5–134	Vadas et al. (2008)
Miscanthus	PV per ton	No	2.29/ton	4/ton	Khanna et al.
	20-year PV per acre		261	457	(2008)
	Amortized				
	4% over 20 years		19	33.2	
	8% over 20 years		26.20	45.87	
Miscanthus	Total	No	1206–2413		Lewandowski
	Amortized				(2003)
	4% over 20 years		88–175	176–350	
	8% over 20 years		121–242	242–484	

Note: Establishment and seeding costs were updated by using USDA-NASS agricultural fuel and seed prices from 1999 to 2007 (USDA-NASS, 2007a,b).

[a]PV denotes present value.

TABLE H.7 Estimated Opportunity Costs for Cellulosic Feedstocks (Net Returns Forgone by Producer from Not Using Cropland to Produce Next Best Crop or Product)

Type of Feedstock	Type of Cost	Cost per Acre (cited $)	Cost per Acre (2007$)	Reference
Corn stover	Feed value 2.4 tons/acre	59.5/ton 142.8	59.5/ton 142.8	Edwards (2007)
Corn stover	Lost profits	22–58	22–58	Khanna and Dhungana (2007)
Switchgrass	Lost profits	78–231	78–231	Khanna and Dhungana (2007)
Switchgrass or *Miscanthus*	Lost profits	78	76	Khanna et al. (2008)
Miscanthus	Lost profits	78–231	78–231	Khanna and Dhungana (2007)
Nonspecific	Lost CRP[a] payments if harvest every year	35	36	Mapemba et al. (2008)
Nonspecific	Lost CRP if harvest once every 3 years	10.1	10.4	Mapemba et al. (2008)
Nonspecific	Non-CRP land crops	10/ton	10.3/ton	Mapemba et al. (2008)
Nonspecific		78	76	Khanna et al. (2008)
Woody biomass	Alternative use		0–25	Summit Ridge Investments (2007)

Note: Opportunity costs were updated by using USDA-NASS agricultural-land rent prices from 1999 to 2007 (USDA-NASS, 2007a,b).

[a]Conservation Reserve Program.

TABLE H.8 Yield Values and Ranges for Different Bioenergy Feedstocks Reported in Literature

Biomass Type	Assumptions	Estimated Yield (tons/acre)	Reference
Corn stover	Soil tolerance	2.02	Khanna and Dhungana (2007)
Corn stover		2.4	Edwards (2007)
Corn stover	2000–2005 mean yields for Wisconsin	2.31–3	Vadas et al. (2008)
Switchgrass	Iowa, Illinois field trials	2.58	Khanna and Dhungana (2007)
Switchgrass		4	Duffy (2007)
Switchgrass	Farm-scale (northern South Dakota to southern Nebraska)	2.23 (5-year average) (Range, 1.7–2.7) 3.12 (10-year average) (Range, 2.6–3.5)	Perrin et al. (2008)
Switchgrass		3.8 19.74 (10-year PV)	Khanna et al. (2008)
Switchgrass	Nitrogen level	4–5.8	Vadas et al. (2008)
Switchgrass	Research blocks	7.14 (average) 9.8 (best)	Lewandowski et al. (2003)
Switchgrass	Plot trials	3.6–8.9 (previous) 2.3–4 (own)	Shinners et al. (2006)
Switchgrass	Plot trials	6.33 4.64–8.5	Fike et al. (2006)
Switchgrass	Field trials Mean Strains: Dacotah ND3743 Summer Sunburst Trailblazer Shawnee OK NU-2 Cave-in-Rock	1.12–4.1 1.11–4.22 0.91–3.92 1.18–4.38 1.43–5.57 1.15–4.88 1.06–4.5 0.89–4.18 0.97–4.27	Berdahl et al. (2005)
Switchgrass	Plot trials Iowa Nebraska	5.2–5.6 4.7–5	Vogel et al. (2002)
Switchgrass	Peer-reviewed articles	4.46	Heaton et al. (2004a)

continues

TABLE H.8 Continued

Biomass Type	Assumptions	Estimated Yield (tons/acre)	Reference

Switchgrass	Farm trials
	Strains:
	  Alamo (1 cut)
	  Kanlow (1 cut)
	  Cave-in-rock
	  (2 cut)
		
		5.4–8.5
		5.2–6.9
		6–8.3	

Switchgrass	U.S. average	4.2	McLaughlin et al. (2002)
Grasses	County-scale in Pacific Northwest	3.4–4.1 (perennial ryegrass) 4.13–6.2 (tall fescue) 2.2–3.36 (creeping red fescue)	Banowetz et al. (2008)
Miscanthus	Simulated	8.9	Khanna and Dhungana (2007)
Miscanthus		14.5 average 12–17 range 114.58 (20-year PV)	Khanna et al. (2008)
Miscanthus	Field experiment	5.71 (14-year) 3.43–11.73 (3-year)	Christian et al. (2008)
Miscanthus		4.5–13.4	Lewandowski et al. (2003)
Miscanthus	Projection	13.36 (mean) 10.93–17.81	Heaton et al. (2004b)
Miscanthus	Peer-reviewed articles	9.8	Heaton et al. (2004a)

TABLE H.9 Carbon Inputs to Biomass Agricultural Production

Source of Input (kg CO_2 eq/ha)[a]	Corn Ethanol	Cellulosic (Switchgrass)
Nitrogen-fertilizer emissions	1638	547
Phosphorus	102	3.4
Potassium	70	2.4
Lime	228	—
Herbicide	69	10.4
Insecticide	5.4	—
Seed	—	—
Transport emissions	39	3
Gasoline	114	—
Diesel	248	341
Natural gas	46	—
Liquefied petroleum gas	61	—
Electricity	56	42
Energy used in irrigation	4	—
Labor transportation	—	—
Farm machinery	21	21
CO_2 from land-use change (kg/ha)	—	—
Total from agricultural production	2703	971
Conversion per acre (0.405 ha/acre, 2.24 lb/kg)	2452 lb CO_2 eq/ac	881 lb CO_2 eq/ac
Conversion per ton (assume 4 tons/acre)	—	220 lb CO_2 eq/ton

[a]Unless noted otherwise.
Source: Farrell et al., 2006.

TABLE H.10 Carbon Inputs to Biomass Refining, Including Transportation of Biomass

Source of Input (g CO_2 eq/L)[a]	Corn Ethanol	Cellulosic
Transport of feedstock to biorefinery	49	51
Primary energy	—	—
Diesel	—	5
Coal	885	—
Natural gas	365	—
Electricity	—	—
Biomass	—	—
Capital (plant, equipment)	8.8	29
Process water	25	19
Effluent restoration (BOD at PWTPs[b])	20	20
Transportation of chemicals to plant	—	—
Total biorefinery phase	1,353	124
Coproduct credits	525	106
Total biorefinery phase accounting for coproduct	828 g CO_2 eq/L	18 g CO_2 eq/L
Conversions:		
Initial value	828 g CO_2 eq/L	18 g CO_2 eq/L
[0.4/0.4]/0.38 L/kg	~331.2 g CO_2 eq/kg	~6.84 g CO_2 eq/kg
[8,746/8,389]/13,450 kg/ha	~2,896,675 g CO_2 eq/ha	~91,998 g CO_2 eq/ha
0.405 ha/acre	~1,173,153 g CO_2 eq/ac	~37,259 g CO_2 eq/ac
0.001 kg/g	~1,173 kg CO_2e/ac	~37.3 kg CO_2 eq/ac
TOTAL (agriculture phase + biorefinery)	4,307 lb CO_2 eq/ac	964 lb CO_2 eq/ac
Conversion per ton: (assume 4 tons/acre)	—	241 lb CO_2 eq/ton

[a]Unless noted otherwise.
[b]Biochemical oxygen demand of effluent at wastewater treatment plants.
Source: Farrell et al., 2006.

REFERENCES

Aden, A., M. Ruth, K. Ibsen, J. Jechura, K. Neeves, J. Sheehan, B. Wallace, L. Montague, A. Slayton, and J. Lukas. 2002. Lignocellulosic Biomass to Ethanol Process Design and Economics Utilizing Co-Current Dilute Acid Prehydrolysis and Enzymatic Hydrolysis for Corn Stover. Golden, Colo.: National Renewable Energy Laboratory.

Banowetz, G.M., A. Boatang, J.J. Steiner, S.M. Griffith, V. Sethi, and H. El-Nashaar. 2008. Assessment of straw biomass feedstock resources in the Pacific Northwest. Biomass and Bioenergy 32:629-634.

Berdahl, J., A. Frank, J. Krupinsky, P. Carr, J. Hanson, and H. Johnson. 2005. Biomass yield, phenology, and survival of diverse switchgrass cultivars and experimental strains in western North Dakota. Agronomy Journal 97:549-555.

Christian, D., A. Riche, and N. Yates. 2008. Growth, yield and mineral content of *Miscanthus x Giganteus* grown as a biofuel for 14 successful harvests. Industrial Crops and Products 28:320-327.

Duffy, M. 2007. Estimated Costs for Production, Storage, and Transportation of Switchgrass. Iowa State University. Available at http://www.extension.iastate.edu/agdm/crops/pdf/a1-22.pdf. Accessed April 22, 2008.

Edwards, W. 2007. Estimating a Value for Corn Stover. Iowa State University. Available at http://www.extension.iastate.edu/agdm/crops/pdf/a1-70.pdf. Accessed May 22, 2008.

English, B.C., D.G. de la Torre Ugarte, K. Jensen, C. Hellwinckel, J. Menard, B. Wilson, R. Roberts, and M. Walsh. 2006. 25% Renewable Energy for the United States by 2025: Agricultural and Economic Impacts. Knoxville: University of Tennessee.

Farrell, A., R. Plevin, B. Turner, A. Jones, M. O'Hare, and D. Kammen. 2006. Ethanol can contribute to energy and environmental goals. Science 311:506-509.

Fike, J., D. Parrish, D. Wolf, J. Balasko, J. Green, Jr., M. Rasnake, and J. Reynolds. 2006. Long-term yield potential of switchgrass-for-biofuel systems. Biomass and Bioenergy 30:198-206.

Heaton, E., T. Voight, and S.P. Long. 2004a. A quantitative review comparing the yields of two candidate C_4 perennial biomass crops in relation to nitrogen, temperature and water. Biomass and Bioenergy 27:21-30.

Heaton, E.A., J. Clifton-Brown, T.B. Voight, M.B. Jones, and S.P. Long. 2004b. Miscanthus for renewable energy generation: European Union experience and projections for Illinois. Mitigation and Adaptation Strategies for Global Change 9:433-451.

Hess, J.R., C.T. Wright, and K.L. Kenney. 2007. Cellulosic biomass feedstocks and logistics for ethanol production. Biomass, Bioproduction and Biorefining 1:181-190.

Hoskinson, R.L., D.L. Karlen, S.J. Birrell, C.W. Radtke, and W.W. Wilhelm. 2007. Engineering, nutrient removal, and feedstock conversion evaluations of four corn stover harvest scenarios. Biomass and Bioenergy 31:126-136.

Karlen, D.L., and S.J. Birrell. Unpublished. Crop Residue—What's It Worth? U.S. Department of Agriculture and Iowa State University. Available at http://www1.eere. energy.gov/biomass/pdfs/Biomass_2009_Sustainabiliy_III_Karlen.pdf. Accessed April 25, 2009.

Kaylen, M., D.L. Van Dyne, Y.S. Choi, and M. Blase. 2000. Economic feasibility of producing ethanol from lignocellulosic feedstocks. Bioresource Technology 72:19-32.

Khanna, M., and B. Dhungana. 2007. Economics of Alternative Feedstocks in Corn-Based Ethanol in Illinois and the US: A Report from Department of Agricultural and Consumer Economics. Urbana-Champaign: University of Illinois. Available at http://www.farmdoc.uiuc.edu/policy/research_reports/ethanol_report/Ethanol%20Report.pdf. Accessed April 25, 2009.

Khanna, M., B. Dhungana, and J. Clifton-Brown. 2008. Costs of producing *Miscanthus* and switchgrass for bioenergy in Illinois. Biomass and Bioenergy 32(6):482-493.

Kumar, A., and S. Sokhansanj. 2007. Switchgrass (*Panicum vigratum,* L.) delivery to a biorefinery using Integrated Biomass Supply Analysis and Logistics (IBSAL) Model. Bioresource Technology 98:1033-1044.

Kumar, A., J. Cameron, and P. Flynn. 2003. Biomass power cost and optimum plant size in western Canada. Biomass and Bioenergy 24(6):445-464.

Lewandowski, I., J. Scurlock, E. Lindvall, and M. Christou. 2003. The development and current status of perennial rhizomatous grasses as energy crops in the US and Europe. Biomass and Bioenergy 25(4):335-361.

Mapemba, L.D., F.M. Epplin, R.L. Huhnke, and C.M. Taliaferro. 2008. Herbaceous plant biomass harvest and delivery cost with harvest segmented by month and number of harvest machines endogenously determined. Biomass and Bioenergy 32:1016-1027.

Mapemba, L., F. Epplin, C. Taliaferro, and R. Huhnke. 2007. Biorefinery feedstock production on Conservation Reserve Program Land. Review of Agricultural Economics 29(2):227-246.

McAloon, A., F. Taylor, W. Yee, K. Ibsen, and R. Wooley. 2000. Determining the Cost of Producing Ethanol from Corn Starch and Lignocellulosic Feedstocks. Golden, Colo.: National Renewable Energy Laboratory.

McLaughlin, S.B., D.G. de la Torre Ugarte, C.T. Garten, Jr., L.R. Lynd, M.A. Sanderson, V.R. Tolbert, and D.D. Wolf. 2002. High-value renewable energy from prairie grasses. Environmental Science and Technology 36:2122-2129.

Perlack, R., and A. Turhollow. 2002. Assessment of Options for the Collection, Handling, and Transport of Corn Stover. Oak Ridge, Tenn.: Oak Ridge National Laboratory.

Perlack, R., and A. Turhollow. 2003. Feedstock cost analysis of corn stover residues for further processing. Energy 28:1395-1403.

Perlack, R. 2007. Overview of Plant Feedstock Production for Biofuels: Current Technologies, Challenges, and Potential Improvement. Presentation to the National Research Council Committee on Alternative Liquid Transportation Fuels. Washington, D.C., November 19, 2007.

Perrin, R., K. Vogel, M. Schmer, and R. Mitchell. 2008. Farm-scale production cost of switchgrass for biomass. Bioenergy Research 1:91-97.

Petrolia, D.R. 2008. The economics of harvesting and transporting corn stover to fuel ethanol: A case study for Minnesota. Biomass and Bioenergy 32(7):603-612.

Searcy, E., P. Flynn, E. Ghafoori, and A. Kumar. 2007. The relative cost of biomass energy transport. Applied Biochemistry and Biotechnology 137-140(1-12):639-652.

Shinners, K.J., G.C. Boettcher, R.E. Muck, P.J. Weimer, M.D. Casler. 2006. Drying, harvesting, and storage characteristics of perennial grasses as biomass feedstocks. ASABE Paper No. 061012. St. Joseph, Mich.: American Society of Agricultural and Biological Engineers.

Sokhansanj, S., and A. Turhollow. 2002. Baseline cost for corn stover collection. Applied Engineering and Agriculture 18:525-530.

Summit Ridge Investments, LLC. 2007. Eastern Hardwood Forest Region Woody Biomass Energy Opportunity. Granville, Vt: Summit Ridge Investments.

Suzuki, Y. 2006. Estimating the Cost of Transporting Corn Stalks in the Midwest. Ames: Iowa State University College of Business: Business and Partnership Development.

USDA-NASS (U.S. Department of Agriculture, National Agricultural Statistics Service). 2007a. Agricultural Prices 2006 Summary. Washington, D.C.: U.S. Department of Agriculture, National Agricultural Statistics Service.

USDA-NASS. 2007b. Agricultural Prices December 2007. Washington, D.C.: U.S. Department of Agriculture, National Agricultural Statistics Service.

Vadas, P.A., K.H. Barnett, and D.J. Undersander. 2008. Economics and energy of ethanol production from alfalfa, corn, and switchgrass in the Upper Midwest, USA. Bioenergy Research 1:44-55.

Vogel, K., J. Brejda, D. Walters, and D. Buxton. 2002. Switchgrass biomass production in the Midwest USA: Harvest and nitrogen management. Agronomy Journal 94:413-420.

Modeling of Capital and Operating Costs and Carbon Emissions of Ethanol Plants with SuperPro Designer®

The Panel on Alternative Liquid Transportation Fuels developed a model to simulate the capital and operating costs and the carbon emissions of ethanol plants. The model simulations were used to compare process economics and environmental effects in different scenarios of technological developments and improved efficiencies, with different feedstocks, and in ethanol plants of various sizes. SuperPro Designer®, chemical-process simulation software, was used by the panel to run the model simulations because it contains a set of unit procedures that can be customized to the specific modeling needs of the corn grain-to-ethanol and cellulosic biomass-to-ethanol processes. It was also used in another study (Kwiatkowski et al., 2005). The software includes a well-developed economic-evaluation package with such parameters as financing, depreciation, running royalty expenses, inflation rate, and taxes. This appendix will first discuss the composition of different biomass feedstocks, then the ethanol-plant simulation models that the panel used, and finally an example of an economic analysis generated by SuperPro Designer.

BIOMASS COMPOSITION

Feedstock Description: Poplar and High-Sugar/Glucan Biomass

Poplar woodchips were used as biomass feedstock for all initial analyses. Composition was obtained from M. Ladisch and colleagues (Purdue University) and is summarized in Table I.1. "Wet" woodchips, which are unprocessed as provided by the forestry-products industry as by-products, were used in the analyses. They

TABLE I.1 Composition of Poplar Woodchips and High-Sugar/Glucan Biomass (percentage)

	Poplar Woodchips	HGBM
Acetic acid	1.95	1.08
Ash	0.60	0.33
Cellullose	23.70	25.00
Extractives	1.95	1.08
Lignin	15.75	10.00
Water	18.00	50.00
Xylan	8.06	12.50

contain about 48 percent water, and the concentrations of sugars and lignin are 61 and 30 percent wt/wt, respectively, on a dry-weight basis. Because a high lignin content is not typical of all cellulosic biomass, the panel generated a "high-sugar/ glucan biomass" (HGBM) feedstock to analyze the effects of a different biomass composition. HGBM has sugar and lignin concentrations of 75 and 20 percent, respectively. All other components were kept at the same relative percentages as in the poplar woodchips; water content was set at 50 percent (instead of the 48 percent in poplar) for simplicity.

Cellulosic-Biomass Feedstock Alternatives

The composition of the feedstock used in the analyses could affect capital and operating costs. For example, a biorefinery that uses poplar woodchips as a feedstock has to include a burner and a steam electrical generator to burn the lignin residue for electricity generation; in contrast, wheat straw does not have enough lignin to provide any energy for the biorefinery. Therefore, the panel also assessed the process economics and environmental effects of biorefineries using different feedstocks. The different biomass compositions are shown in Table I.2. All compositions, apart from the case of poplar, are on a dry-weight basis. References obtained for these biomass compositions were inconsistent and had large ranges. The ranges of values overlap for some individual components, such as glucan or lignin. The most consistent and credible values were selected for the analysis, and they were mostly averages of the maximum and minimum for the spreads. The problem of mass closure was resolved by including a trace amount of water to reach 100 percent, and the price basis for the biomass was adjusted to reflect that. For example, if the initial price was $70/ton—(2)($35/ton of poplar woodchips on

TABLE I.2 Composition of Different Feedstocks (percentage)

	Poplar Woodchips	Wheat Straw Dry	Dry Switchgrass	Dry Corn Stover	*Miscanthus*
Acetic acid	1.9	0.0	0.3	0.0	0.0
Ash	0.6	6.4	6.0	7.0	2.0
Cellulose	23.7	39.3	32.2	35.0	38.2
Extractives	1.9	4.2	13.6	5.0	6.9
Lignin	15.7	14.5	17.3	18.5	25.0
Water	48.0	5.0	0.0	2.5	3.6
Xylan	8.1	30.6	27.9	28.0	24.3
Glucose	—	0.0%	2.7%	4.0%	

a wet-weight basis)—a 2 percent water content would reduce the price to ($70/ton)(0.98) = $68.60/ton. Another alternative would have been to augment every percentage composition proportionally so that the sum reached 100 percent. The two approaches have the same effect, but the former is more efficient.

ETHANOL-BIOREFINERY SIMULATION MODELS

Model for Corn-Grain-to-Ethanol Plants

The corn-grain-to-ethanol process is well developed and understood and is used by 130–150 ethanol plants in the United States alone; hence, it is a good starting point to evaluate the modeling method with SuperPro Designer. Because a previous study analyzed the corn-to-ethanol process with SuperPro Designer (Kwiatkowski et al., 2005), it was thought best to remodel the process with the panel's simplification constraints (discussed in Chapter 3) and any price changes in costing and to compare the results with those of the prior study. The panel's initial model would not only validate the approach but also verify that it calculated all the mass balances correctly and performed consistent energy-balance calculations for the process.

Figure I.1 shows a simple schematic of the corn-grain-to-ethanol manufacturing process, and Figure I.2 shows the corresponding schematic in SuperPro Designer. For adequate separation of concerns, the process was divided into three sections: preprocessing, production (fermentation), and recovery, including

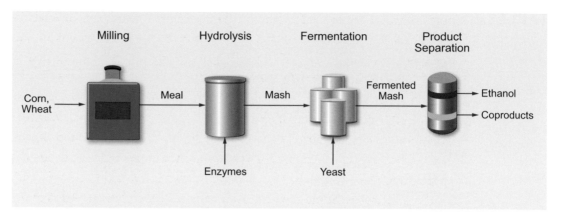

FIGURE I.1 *Schematic of processing steps for converting corn to ethanol. Source: Schwietzke et al., 2008. Reprinted with permission from IEA Bioenergy.*

recovery of ethanol as the major product, distillers dry grain solids (DGGS) as a by-product, and water (hot condensate and backset[1]). The economic evaluation report generated by SuperPro Designer is shown at the bottom of Figure I.2.

Cellulosic-Plant Model

Process Overview

There are a lot of similarities between the cellulosic-ethanol and dry-grind corn-ethanol manufacturing processes, and they share at least five main basic unit operations (Figure I.3): size reduction, saccharification, fermentation, distillation, and solids separation (centrifugation). In some plant configurations, saccharification is attempted simultaneously with fermentation, but such a design is independent of whether corn or cellulosic biomass is used as feedstock, so it is not treated as a difference between the two processes. Both systems also need some form of solids feedstock handling and storage. A variety of alternatives can be used for feedstock handling; the simplest, and the one modeled in this analysis, is a single storage bin.

One important difference between the two ethanol-manufacturing alternatives lies in the initial pretreatment of the feedstock after size reduction (grinding or chopping) for the mash to be saccharified and fermented in the later steps. Different pretreatments are used because of the difference in resilience with respect

---

[1]Backset is a portion of thin stillage.

to "liquefaction" or "softening" of the feedstock. In the case of the cellulosic-bio-mass process, there are a options to pretreat the lignocellulosic material to make the glucan and xylan or arabinan fibers available for enzyme degradation into monomers in the saccharification step.

A second difference has to do with the by-products of the process. More work and energy are used in the dry-grind corn-based case to produce DGGS of adequate quality, requiring both an evaporator and a drying step. In the case of the cellulosic alternative, only a dryer is needed to retrieve the residual solids—rich in lignin—that are then burned in a boiler to meet the energy requirements of the biorefinery. Some designs, such as that in the study of Aden et al. (2002), avoid this drying step and therefore reduce the capital cost further. However, that type of plant would depend more heavily on external sources of energy for the plant's heating and electricity needs.

A third difference is the inclusion of a lignin-based burner and boiler for the generation of steam and a steam turbine for the generation of electricity in a cellulosic-ethanol biorefinery. Those are included in the design to take advantage of the relatively high content of lignin in the feedstock and the ease with which the lignin in the residual solids can be combusted by simply providing air. The energy in the lignin is about 11.5 kbtu/lb (26.7 kJ/g).

In summary, the panel's simulation model for a cellulosic biorefinery has 10 major unit operations. They are basic and well-characterized units and modeled as shown in the SuperPro Designer schematic for the plant (Figure I.3). Even if all the units vary in their complexity (for example, a typical distillation unit includes a beer column, a rectifier column, a molecular sieve, and a stripper column), for the sake of simplicity they can be treated as simple "black boxes" that are connected to each other by one or two streams. Each unit operation is treated as a single model unit with all its details encapsulated by the single-unit "box," so that, for example, in the distillation operation, instead of six units having to be resized with all the required heat exchangers and other components, only one unit is given the value of the whole set and resized. Included are also the few heat exchangers outside the units that were difficult to model otherwise and that seem to be present in most of the currently explored configurations both in the literature and in discussions with industry. SuperPro Designer probably models some units—such as the centrifuge, the fermentor and reaction bins, and the drum dryer—after the actual physical units. The biggest simplifications are the burner, the steam turbine, and the propagation system; the first two are modeled as simple generic reaction or separation boxes. The third is modeled as a single seed fermentor. In reality, it is a

344

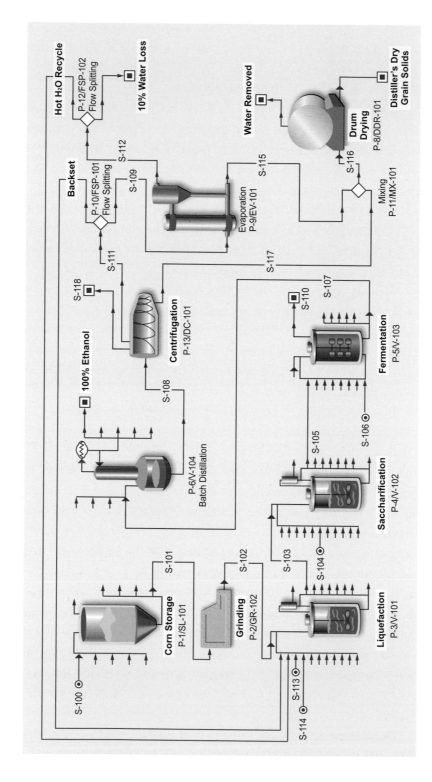

FIGURE I.2 *Schematic of processing steps for converting corn to ethanol used in SuperPro Designer®, with example (on facing page) of summary of economic evaluation report.*

Economic Evaluation Report   March 31, 2008
for

1. EXECUTIVE SUMMARY (2008 prices)

Total Capital Investment	90,886,000	$
Capital Investment Charged to This Project	90,886,000	$
Operating Cost	69,258,000	$/yr
Production Rate	119,902,339.77	kg MP/yr
Unit Production Cost	0.58	$/kg MP
Total Revenues	81,532,000	$/yr
Gross Margin	15.05	%
Return On Investment	18.28	%
Payback Time	5.47	years
IRR (After Taxes)	11.95	%
NPV (at 5.0% Interest)	62,871,000	$

MP = Flow of Component Ethyl Alcohol in Stream
100% EtOH

FIGURE I.2 *Continued*

system of many seed-fermentor units and cleaning and sterilization systems. That level of detail is encapsulated in the current unit, which, in essence, is used only for overall cost-estimation purposes.

SuperPro Designer® Network Model Description

The process is modeled in SuperPro Designer® as a batch process of 19 unit operations (including heat exchangers and flow splitters and mixers) in which some units run in continuous mode (see Figure I.4). The longest process, and therefore the one that defines the length of each batch, is fermentation, which takes 72 h. The size of the fermentors was set at 800,000 gal for all scenarios, and there can be anywhere from 5 to 23 fermentor units. The fermentors can run in a staggered fashion so that the average time to fill them and to empty the ones that have completed their fermentation is 4 h for the entire set. Thus, the whole process takes 80 h (4 h + 72 h + 4 h).

With respect to the overall batch process, it might be optimistic to consider the possibility of only 4 h in the front and back ends of the staggered fermenta-

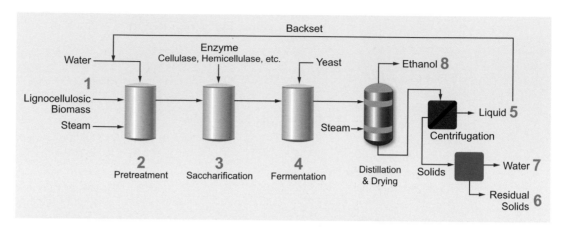

FIGURE I.3 *Simple schematic of cellulosic-ethanol manufacturing process. Courtesy of M. Ladisch and colleagues (Purdue University).*

tions in the case of 23 fermentors. However, there are only two cases in which many fermentors were used. For the sake of comparison of different batch efficiencies and sizes, the overall time was set at 80 h for all cases. Given the assumptions that the plant produces all year and that each batch takes 80 h, 109 batches could be processed per year. That number was used for all scenario analyses.

It is important to note that many unit operations in the process network have been considered to operate continuously rather than in a batched mode. The reason is that apart from the first batch, in which the completion of fermentation is the step that delays the beginning of other processes—such as distillation and drying—all the batches allow the processes to be aligned. The products of previous fermentations are fed into downstream processes in the network, so they never need to be stopped. The same is true of processes before fermentation, such as size reduction and shredding. After the first batch—for which shredding must take place before preprocessing—each batch can be shredded while the preceding batch is being processed. In this way, all units for continuous processes are sized according to their volumetric throughput per batch, assuming a run time of 80 h/batch. The processing units that have been set as continuous in the network are storage, the shredder, distillation, centrifugation, the drum dryer, the burner, the steam-turbine generator, and all mixers, splitters, and heat exchangers.

In contrast, reactor-vessel procedures, such as preprocessing and saccharification, could not be resized according to their shorter processing time with respect to fermentation. In theory, because the processing time for those steps is shorter

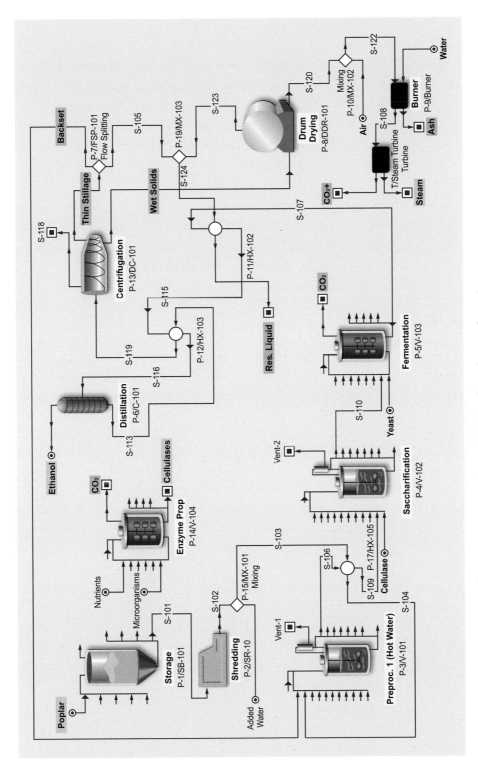

FIGURE I.4 *SuperPro Designer® network model for cellulosic-ethanol manufacturing process.*

than that for fermentation, it would be possible to have smaller vessels that are used more times for each batch while products are stored until the appropriate volume for the next fermentation is reached. The panel decided, however, to avoid the use of smaller vessels and, in essence, to use the reactor vessels themselves as the storage vessels for two reasons. First, it is usually not advisable to leave the saccharification mash idle in storage for any extended period, because the mix could spoil and create difficulties for later fermentation. Second, it is unclear whether having additional storage bins for the different pretreated or saccharified mixtures would result in substantial cost savings.

As can be seen in Figure I.3 for the case of poplar woodchips as lignocellulosic feedstock, the woodchips are deposited into a solids bin that is just big enough to hold the biomass required for each batch. The feedstock is passed through a grinder-shredder that works full-time and then is mixed with water to reach an approximate solids loading of 30 percent. After being heated by a heat-exchange element with the output stream of this step, the mixture is fed into the pretreatment vessel with the hot thin stillage backset. Of the many possible options, the pretreatment chosen for this model is the hot-water method in which the mash—now with a 21 percent solids loading after mixing with the backset—is heated with steam to 200°C for 5 min.

After pretreatment, the mash is transferred into a new reaction vessel, where it is cooled down to 65°C and mixed with cellulases at 12.6 percent wt/wt glucan. The mix is stirred for 36 h to achieve about 80–90 percent sugar yields from the total cellulose and hemicellulose in the biomass. Once that stage is complete, the mix is transferred into the fermentor, where it is cooled to 32°C and mixed with yeast at a concentration of 0.125 percent wt/wt fermentable sugars. As mentioned before, 72 h is allowed for fermentation in which 80–90 percent of the available fermentable sugar is converted into ethanol. The resulting mixture, which contains about 4–8 percent ethanol, is then passed through the distillation system. The distillation system is modeled by a single column with the assumption that 99.99 percent of the ethanol can be recovered in the distillate. In this design, the fermented mixture distilland is heated as much as possible before entering the distillation system by heat exchange with both the bottoms stream and the residual liquid stream not used as backset and by the evaporated water stream from the drying step. This approach saves as much heat energy as possible for the model.

The bottoms stream, cooled after exchanging heat with the distillation input stream, is then passed through the centrifuge. For this step, the distribution of compounds between the water stream and the solids stream is set as a percentage distribution as indicated by Ladish and colleagues, Purdue University. Fifty percent of the liquid stream, called thin stillage, is recycled as backset to be mixed with the ground or shredded feedstock for the next batch. The other 50 percent is treated as the residual liquid stream and for the model's purposes disposed of and not included in this analysis. In a real setting, that liquid stream would most likely be recovered via a water-purification system.

The residual solids coming out of the centrifuge are dried in a drum dryer to about 15 percent water content with 242°C high-pressure steam as the source of heat to allow better burning. Once dried, the solids are fed into a burner or boiler and completely burned in air to CO_2 and water. It is assumed that all compounds other than the ash already in the solids residue are hydrocarbons and are fully reacted with oxygen to CO_2 and water. Nonetheless, in the reaction enthalpy calculations, only lignin is assumed to release heat of reaction. The other compounds are not included but contribute slightly as a heat sink because they (or their products) have to be heated to the final exhaust temperature. For the sake of simplicity and because of some particularities of the program, the water to be heated to the final steam temperature used throughout the plant is mixed with the solids to be burned even though in reality these would be in separate chambers.

The generated steam is then passed through a steam turbine to generate electricity. This unit operation, however, could not be modeled adequately, because it was unclear whether and how it would be possible in the SuperPro Designer program to deduct the steam needs of the plant from the steam output of the boiler. If all generated steam were available for this unit, it would be generating at least twice the electricity that would be available to the real plant. This unit, nonetheless, was used for separating the real steam generated from the CO_2 and water stream resulting from the residual solids burn and was also used to cost the steam turbine. That was achieved after further calculations to find the real available steam for electrical-power output—and therefore the size of the unit—were carried out separately in an Excel spreadsheet.

Plant Cost Calculations

A sample detailed cost analysis for the "base-case" cellulosic plant (poplar feedstock and medium case-performance assumptions) is shown in Box I.1 at the end of this appendix.

Equipment

With respect to the major equipment specifications and freight on board (FOB) costs, there is only one unit of every unit operation (or stage)—except perhaps for one or two heat-exchanger stages that were doubled under some circumstances— in which the maximum size specifications for a unit were below the through-put for a particular case. Notable exceptions are the vessels for pretreatment, saccharificaton, and fermentation, which have constraints on how large they can be made. Therefore, although costs of all other equipment increase with size to the power of 0.6–0.7, the equipment for those three stages correlates linearly with the number of units required. The maximum and therefore the chosen working size of each vessel was selected by virtue of consolidating the decisions of industry on sizing vessels, after talks with representatives in charge of these projects, with the maximum possible size of 1 million gallons reported in the Aden et al. (2002) study. The current estimate of the base cost of these vessels was also validated in those talks.

As mentioned before, the single distillation column is a proxy for a more detailed distillation unit that to a good approximation follows single-unit comparison with Schwietzke et al.'s model (2008). The actual distillation stage would include a beer column, a rectifier column, a molecular sieve, and a stripper column, but the value and the behavior of this set of components were appropriately emulated by the single column modeled in this analysis. The overall cost of this stage was also validated by talks with industry and by the costs of such units as the centrifuge and the dryer. Scaling exponent values were also fine-tuned after discussions with industry. The scaling exponent value varies: distillation grows approximately with a scaling exponent of 0.55, and the centrifuge with 0.8. The default value was taken as 0.7.

Although the burner or boiler and steam-turbine generator help to create a more efficient biorefinery from an energy point of view, it is not obvious whether this is the most economical choice relative to the use of natural gas or coal for the energy needs of the plant. They have been included here to minimize reliance on fossil fuels. The sum of the costs of the boiler and turbine was validated independently and, on the basis of usual estimates, is 40–50 percent of total equipment costs. It varied from case to case, however, because the turbine cost for different cases was re-estimated according to the amount of available steam for electricity generation.

It should be noted that the FOB cost of equipment is about 25 percent of the total plant cost. In addition to the costs of the basic units, adding such items as piping, instrumentation, insulation, electrical facilities, buildings, and "yard improvement" (here taken as the initial landscaping needed for the construction of the facility) increases the cost. All those values are taken as percentages of the cost of the units and have been validated with industry. In addition, the percentage cost of engineering and construction and the contractor's fee and contingency have to be included. The sum is the total "direct fixed capital cost" (DFC). Finally, there are the startup costs and the initial working capital, which are expressed as percentages of the DFC (in Section 10A of the SuperPro Designer Economic Analysis Report Sample). Royalty fees, fixed at about $4 million and not based directly on the DFC, still need to be added to the DFC to provide the figure for total capital investment.

REFERENCES

Aden, A., M. Ruth, K. Ibsen, J. Jechura, K. Neeves, J. Sheehan, B. Wallace, L. Montague, A. Slayton, and J. Lukas. 2002. Lignocellulosic Biomass to Ethanol Process Design and Economics Utilizing Co-Current Dilute Acid Prehydrolysis and Enzymatic Hydrolysis for Corn Stover. Golden, Colo.: National Renewable Energy Laboratory.

Kwiatkowski, J., A.J. McAllon, F. Taylor, and D.B. Johnston. 2005. Modeling the process and costs of fuel ethanol production by the corn dry-grind process. Industrial Crops and Products 23:288-296.

Schwietzke, S., M.R. Ladisch, L. Russo, K. Kwant, T. Makinen, B. Kavalov, K. Maniatis, R. Zwart, G. Shahanan, K. Sipila, P. Grabowski, B. Telenius, M. White, and A. Brown. 2008. Gaps in the research of 2nd generation transportation biofuels. IEA Bioenergy T41:2008:2001.

BOX I.1 **Superpro Designer® Economic Analysis Report Sample**

This chart was redrawn from a formatted Microsoft Excel table generated directly by SuperPro Designer® and was augmented by further calculations related to the costing of the enzyme-propagation unit and the steam-turbine electrical generator (the augmented changes are shown in boldface).

Economic Evaluation Report
For 2008jun22 Poplar- Mid
June 24, 2008

1. EXECUTIVE SUMMARY (2008 prices)

Total Capital Investment	207923000.00 $
Capital Investment Charged to This Project	207923000.00 $
Operating Cost	81119000.00 $/yr
Production Rate	119454958.36 kg MP/yr
Unit Production Cost	0.68 $/kg MP
Total Revenues	116032000.00 $/yr
Gross Margin	30.09 %
Return on Investment	10.91 %
Payback Time	9.16 years
IRR (After Taxes)	22.27 %
NPV (at 5.0% Interest)	188837000.00 $

MP = Total Flow of Stream EtOH

2. MAJOR EQUIPMENT SPECIFICATION AND FOB COST (2008 prices)

Quantity/ Standby/ Staggered	Name	Description	Unit Cost ($)	Cost ($)
9 / 0 / 0	V-101	Stirred Reactor Vessel Volume = 783714.15 gal	669000.00	6021000.00
9 / 0 / 0	V-102	Stirred Reactor Vessel Volume = 754229.44 gal	651000.00	5859000.00
9 / 0 / 0	V-103	Fermentor Vessel Volume = 796094.41 gal	676000.00	6084000.00

1 / 0 / 0	DDR-101	Drum Dryer Drum Area = 59.88 m^2	160000.00	160000.00
1 / 0 / 0	DC-101	Decanter Centrifuge Throughput = 4400.64 L/min	2748000.00	2748000.00
1 / 0 / 0	FSP-101	Flow Splitter Size/Capacity = 207810.03 kg/h	0.00	0.00
1 / 0 / 0	SR-101	Shredder Size/Capacity = 102938.72 kg/h	182000.00	182000.00
1 / 0 / 0	SB-101	Solids Bin Vessel Volume = 12110.44 m^3	771000.00	771000.00
1 / 0 / 0	C-101	Distillation Column Column Volume = 182.20 m^3	4782000.00	4782000.00
1 / 0 / 0	V-104	Seed Fermentor Vessel Volume = 13056.19 L	7200000.00	7200000.00
1 / 0 / 0	Burner	Generic Box Size/Capacity = 1352538.12 kg/h	9290000.00	9290000.00
2 / 0 / 0	HX-102	Heat Exchanger Heat Exchange Area = 95.36 m^2	23000.00	46000.00
2 / 0 / 0	HX-103	Heat Exchanger Heat Exchange Area = 51.18 m^2	14000.00	28000.00
1 / 0 / 0	MX-101	Mixer Size/Capacity = 182244.79 kg/h	0.00	0.00
1 / 0 / 0	HX 105	Heat Exchanger Heat Exchange Area = 5488.56 m^2	90000.00	90000.00
1 / 0 / 0	MX-102	Mixer Size/Capacity = 205863.81 kg/h	0.00	0.00
1 / 0 / 0	MX-103	Mixer Size/Capacity = 127857.22 kg/h	0.00	0.00
1 / 0 / 0	Steam Turbine	Generic Box Size/Capacity = 1350432.04 kg/h	13530000.00	13530000.00
		Unlisted Equipment		0.00
			TOTAL	56791000.00

continues

3. FIXED CAPITAL ESTIMATE SUMMARY (2008 prices in $)

3A. Total Plant Direct Cost (TPDC) (physical cost)

1. Equipment Purchase Cost	56791000.00
2. Installation	8480000.00
3. Process Piping	17037000.00
4. Instrumentation	11358000.00
5. Insulation	1704000.00
6. Electrical	8519000.00
7. Buildings	11358000.00
8. Yard Improvement	1136000.00
9. Auxiliary Facilities	0.00
TPDC	116383000.00

3B. Total Plant Indirect Cost (TPIC)

10. Engineering	11638000.00
11. Construction	29096000.00
TPIC	40734000.00

3C. Total Plant Cost (TPC = TPDC+TPIC)

TPC	157117000.00

3D. Contractor's Fee & Contingency (CFC)

12. Contractor's Fee	7856000.00
13. Contingency	23552000.00
CFC = 12+13	31407000.00

3E. Direct Fixed Capital Cost (DFC = TPC+CFC)

DFC	188524000.00

4. LABOR COST - PROCESS SUMMARY

Labor Type	Unit Cost ($/h)	Annual Amount (h)	Annual Cost ($)	%
Operator	0.00	0.00	0.00	0.00
Biomass Operators	20.00	1744.00	34880.00	0.88
Shift Operator	34.00	69760.00	2371840.00	59.53
Lab Technicians	45.00	5886.00	264870.00	6.65
Maintenance	40.85	23326.00	952867.00	23.92
Plant Supervisor	60.00	5995.00	359700.00	9.03
TOTAL		106711.00	3984157.00	100.00

5. MATERIALS COST - PROCESS SUMMARY

Bulk Material	Unit Cost ($/kg)	Annual Amount (kg)	Annual Cost ($)	%
Cellulase	2.49	4026493.00	10004063.00	23.64
Poplar	0.04	897625670.00	31416898.00	74.25
Water	0.00	691548911.00	30428.00	0.07
Yeast	2.30	373412.00	858848.00	2.03
Recovered Steam	0.00	9999000000.00	0.00	0.00
Air	0.00	1552967692.00	0.00	0.00
Sugar Water	0.00	1149931.00	0.00	0.00
Trychomonas	2.30	115.00	264.00	0.00
TOTAL		13146692225.00	42310502.00	100.00

NOTE: Bulk material consumption amount includes material used as:
- Raw Material
- Cleaning Agent
- Heat Tranfer Agent (if utilities are included in the operating cost)

6. VARIOUS CONSUMABLES COST (2008 prices) - PROCESS SUMMARY

THE CONSUMABLES COST IS ZERO.

7. WASTE TREATMENT/DISPOSAL COST (2008 prices) - PROCESS SUMMARY

THE TOTAL WASTE TREATMENT/DISPOSAL COST IS ZERO.

continues

8. UTILITIES COST (2008 prices) - PROCESS SUMMARY

Utility	Annual Amount	Reference Unit	Annual Cost ($)	%	
Electricity	**56,800,000**	**kWh**	0.00	0.00	6 MW
Available Energy from Steam	**475,974,147**	**kWh**	0.00	0.00	54 MW
Usable Steam Energy (80%)	**380,779,318**	**kWh**	0.00	0.00	43 MW
Remaining Steam Energy	**323,979,318**	**kWh**	0.00	0.00	37 MW
Sales Price of Electricity	**0.05**	**$/kWh**	0.00	0.00	
Revenue from Electricity	**16,198,966**	**$**	0.00	0.00	
Steam	0.00	kg	0.00	0.00	
Steam (High P)	652362127.00	kg	0.00	0.00	
Cooling Water	63753240963.00	kg	3187662.00	99.33	
Chilled Water	0.00	kg	0.00	0.00	
TOTAL			3209010.00	100.00	

9. ANNUAL OPERATING COST (2008 prices) - PROCESS SUMMARY

Cost Item	$	%
Raw Materials	42311000.00	52.16
Labor-Dependent	3984000.00	4.91
Facility-Dependent	31616000.00	38.97
Consumables	0.00	0.00
Waste Treatment/Disposal	0.00	0.00
Utilities	3209000.00	3.96
Miscellaneous	0.00	0.00
Advertising/Selling	0.00	0.00
Running Royalties	0.00	0.00
Failed Product Disposal	0.00	0.00
TOTAL	81119000.00	100.00

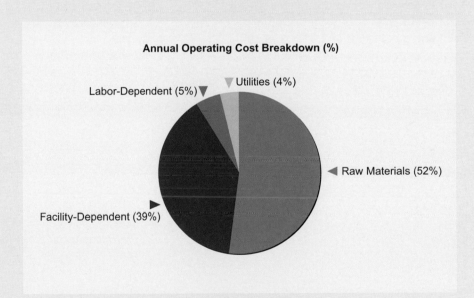

Annual Operating Cost Breakdown (%)

Utilities (4%)

Labor-Dependent (5%)

Raw Materials (52%)

Facility-Dependent (39%)

continues

10. PROFITABILITY ANALYSIS (2008 prices)

A.	Direct Fixed Capital	188524000.00 $
B.	Working Capital	4087000.00 $
C.	Startup Cost	11311000.00 $
D.	Up-Front R&D	0.00 $
E.	Up-Front Royalties	4000000.00 $
F.	Total Investment (A+B+C+D+E)	207923000.00 $
G.	Investment Charged to This Project	207923000.00 $

H.	Revenue Stream Flowrates	
	Total Flow of Stream EtOH	119454958.00 kg/yr
	Total Flow in Steam	10153761372.00 kg/yr
	Total Production of Recovered Steam	**9,999,000,000 kg/yr**
	Total Energy Content of Recovered Steam	**2,844,215,150,040 kJ/yr**
	Total Requirements of Steam (High P)	**652,362,126 kg/yr**
	Total Energy Content of Steam (High P)	**1,130,708,220,559 kJ/yr**
	Total Energy Left over for Electricity	**1,713,506,929,481 kJ/yr**

I.	Production Unit Cost	
	EtOH	0.68 $/kg

J.	Selling / Processing Price	
	Total Flow of Stream EtOH	0.84 $/kg
	Total Flow in Steam	**1.60 $/1000 kg**

K.	Revenues	
	EtOH	99833000.00 $/yr
	Steam	**16199000.00 $/yr**
	Total Revenues	116032000.00 $/yr

L.	Annual Operating Cost	81119000.00 $/yr

M.	Gross Profit (K-L)	34913000.00 $/yr
N.	Taxes (35%)	12220000.00 $/yr
O.	Net Profit (M-N)	22694000.00 $/yr

Gross Margin	30.09%
Return on Investment	10.91%
Payback Time	9.16 years

Resource Requirements for Production of Microbial Biomass

Producing biodiesel or other fuels from algae would require large-scale production of algae. This appendix discusses the resource requirements for producing algal feedstocks for production of transportation fuel.

PRODUCTION SYSTEMS

Two primary types of systems have been developed for large-scale cultivation of photosynthetic microorganisms: open systems (for example, ponds and "raceways") and closed systems (often referred to as photobioreactors).

Open Production Systems

Open production systems have been used successfully for many years for the commercial production of algae and cyanobacteria for the nutraceutical industry and have been incorporated into various fish-farming operations and wastewater-treatment facilities. The open systems use "low technology" and typically have an oval raceway configuration with a paddlewheel that mixes the culture. Data are available from numerous sources regarding the productivity of these systems, which tends to fall in the range of 25–35 g/m² per day during periods of maximum productivity (Sheehan et al., 1998; Lee, 2001; Huntley and Redalje, 2007). Simpler designs have been used for beta-carotene production; these designs feature large wind-mixed ponds and have rather low productivity.

The primary advantages of open pond systems are lower capital and operating costs. The main disadvantages of open systems are poor control of culture conditions (for example, higher susceptibility to contamination by undesired algae and predators, dilution by rain, and fouling by windborne dust and debris), high evaporative water loss, requirement for large expanses of level terrain, and high regulatory hurdles with respect to containment of recombinant strains.

Closed Production Systems

Different kinds of closed photobioreactor systems have been designed and tested. Generally, these systems are in two categories: tubular systems made of rigid or flexible plastic, and flat plate or annular reactors made of rigid materials and typically placed at upright angles to maximize use of light by the cultures. Photobioreactor design is a subject of active research in several algal-biotechnology companies. Because of high capital costs associated with rigid plastics, many of the designs being pursued are focused on tubes manufactured from flexible films. Some press releases have reported the achievement of productivity as high as 170 g/m^2 per day in novel photobioreactors (for example, GreenFuel Technologies Corporation, 2007), but it will be important to increase understanding of how the calculations were conducted to ensure valid comparisons between the various systems.

The primary advantages of closed photobioreactors are a higher degree of control over some culture conditions (for example, protection from the elements, less water evaporation and outgassing of carbon dioxide [CO_2], and delayed onset of contamination by undesired species and predators), potentially higher productivity as a result of improved use of light, and containment of recombinant strains. The overriding disadvantage of closed photobioreactors is the high capital cost associated with the construction materials, circulation pumps, and nutrient-loading systems. There are other disadvantages:

- Fouling of interior surfaces and difficulty of cleaning them.
- Accumulation of high concentrations of photosynthetically generated oxygen, which leads to photooxidative cell damage.
- Absence of evaporative cooling, which can lead to very high temperatures.

Comparison of the Two Types of Systems

Both types of systems have inherent advantages and disadvantages. It is highly unlikely that one standard system will be applicable for all strains, products, or sites, and research is being conducted on various designs. A combination of closed and open systems will probably be used in many cases—enclosed bioreactors for inoculum generation and open ponds as final production units.

To reduce the volume of water handled during cell harvesting, a flocculent (such as alum or various ionic polymers) is typically added to the cells to facilitate their concentration in a settling tank; the biomass is concentrated further with continuous centrifugation, which is an expensive process because of the capital and operating costs. For some filamentous strains, strainers or filters can be used to collect the cells. Clearly, additional research and development to improve the biomass-harvesting process will have a great effect on production costs.

Harvesting of Algae from Natural Bodies of Water

Harvesting of naturally occurring algae would eliminate the need for a photobio-reactor, but harvesting appreciable amounts of biomass would require filtering large quantities of water and extensive operating-expense outlays. In addition, environmental groups strongly resist this approach because of potential unpredictable environmental consequences of ocean and lake fertilization.

STRAINS OF MICROORGANISMS FOR BIOMASS PRODUCTION

Naturally Occurring Strains

Cultivation of the dominant strains of photosynthetic organisms in the locale of the installed system might be the easiest way to maximize the productivity of the system. However, those strains might not be optimal for biofuel production. Most fuels under development require the use of microorganisms that have high lipid content, which might not be an attribute of random local strains. Many researchers in the field therefore believe that commercial production strains will be initially selected on the basis of superior product formation and processing attributes and then developed through dedicated strain-improvement programs.

Genetically Modified Strains

Various information and tools are available for the genetic modification of photosynthetic microorganisms, including whole genome sequences, systems for gene introduction, and protocols for random and directed mutagenesis.

Genome-Sequence Information

Genome-sequence information is extremely useful for developing metabolic engineering strategies, including pathway modeling, gene-knockout strategies, and expression vector construction. Complete genome sequences are available for 11 eukaryotic microalgal species and more than 20 cyanobacterial strains. Additional genome-sequencing projects are under way.

Genetic-Engineering Tools for Cyanobacteria

The materials and methods available for genetic modification of cyanobacteria are substantially more advanced than those available for eukaryotic microalgae. DNA can be introduced into cyanobacteria via natural transformation, electroporation, or conjugation, but different strains require different methods. In some transformation systems, transgenes are included on replicating plasmids; in other cases, the foreign DNA becomes integrated at specific locations in the genome via homologous recombination. A variety of selectable marker genes have been successfully used to enable introduction of multiple foreign genes or inactivation of endogenous genes in separate steps. In addition to site-specific gene inactivation by double-crossover homologous recombination, random mutations can be generated in some species by transposon insertion or by chemical- or radiation-mediated mutagenesis.

Genetic-Engineering Tools for Eukaryotic Microalgae

Genetic engineering has been reported for a few microalgal species, including green algae, diatoms, red algae, and dinoflagellates. The limited success can be attributed in part to the small number of laboratories working in this field. In some cases, nuclear transformation was achieved, typically via random integration of the entire delivery vector into one or more chromosomes and sometimes in the form of tandem repeats. In other cases, transgenes have been successfully targeted to the chloroplast genome by the use of vectors that contain flanking regions of DNA identical with sequences found in the chloroplast. DNA introduction can be accomplished via particle bombardment, electroporation, or agitation with abrasive materials. A number of selectable markers have been used for various microalgae, including several antibiotic-resistance genes and native genes used to complement some mutations. Mutation of nuclear genes is currently limited to classical chemical- or radiation-mediated random mutation, which can be difficult with diploid organisms, such as diatoms.

Application of Genetic Tools to Production of Liquid Biofuel

There have been few efforts to use genetic-engineering tools to enhance biofuel production by photosynthetic microorganisms. In one example, the introduction of pyruvate decarboxylase and alcohol dehydrogenase genes from *Zymomonas* into the cyanobacteria *Synechococcus* and *Synechocystis* resulted in the production of small quantities of ethanol in the strains (Deng and Coleman, 1999; Fu and Dexter, 2009). Attempts have been made to enhance lipid production in the diatom *Cyclotella cryptica* by overexpressing the native acetyl-CoA carboxylase gene, but little effect was observed (Sheehan et al., 1998). A number of laboratories and companies have initiated programs to enhance biofuel production by photosynthetic microorganisms via metabolic engineering. A key goal will be to develop strains that produce large quantities of storage lipids even during periods of rapid cell division.

OTHER REQUIREMENTS FOR PRODUCTION OF MICROBIAL BIOMASS

Land

High productivity of algal or cyanobacterial cultures depends on high levels of solar radiation and an extended growing season (that is, more days with temperatures conducive to rapid culture growth). But use of land that cannot readily be used for production of food or feed crops provides cost and social advantages. Consequently, the desert regions of the southwestern United States have historically been considered the preferred site for implementation of large-scale production systems.

The culture depth of large-scale open-pond systems is typically only 20–30 cm, so precise leveling of the ground is necessary during pond construction. Because level land is needed, many regions in the United States are not suitable as production sites. Level terrain is not as important for some photobioreactor systems, however, because the growth modules tend to be less dependent on level ground and in some cases can actually benefit from the gravitational potential energy inherent in sloped land.

Water

Some strains of microalgae and cyanobacteria are able to grow in a wide variety of water types, including freshwater, saline water, brackish water, and alkaline water. Large quantities of saline groundwater that are available in the southwestern United States could be used to support the mass culture of photosynthetic microorganisms; saline water is unsuitable for crop irrigation or consumption by humans or livestock, so use of this water largely eliminates "food versus fuel" concerns that have been raised for some crop-based biofuels. It will be important to ensure that withdrawal of water from saline aquifers does not interfere with the hydrodynamics of freshwater aquifers and that the aquifers are shallow enough to avoid prohibitive pumping costs. For open-pond systems, it might be necessary to have access to freshwater for dilution of the culture medium when it becomes too saline because of evaporative water loss.

Another potential option for production facilities in coastal areas is a seawater-based culture medium. This option is probably more viable for foreign countries because much of the United States is not suitable. If cost-effective pipelines can be constructed, the number of suitable facility sites would probably increase. Recycling of nutrients and the eventual return of spent water to the ocean would probably be necessary for seawater-based production systems and would require review for regulatory compliance.

Carbon Dioxide

Large quantities of CO_2 required for biofuel production via photosynthetic processes have to be delivered to the production facility in a concentrated form. The two largest sources of CO_2 that could be tapped are coal-fired and gas-fired electric-power plants and oil wells that have been flooded with CO_2 as part of previous enhanced oil-recovery efforts (Feinberg and Karpuk, 1990). Other potential sources are fermentation facilities (such as ethanol plants), cement factories, ammonia-production plants, and oil refineries. Those sources are not all equivalent in that CO_2 is present at varied concentrations and the sources can contain different types of contaminating compounds. Purification and pressurization of the CO_2 would be necessary to reduce transportation costs and reduce contaminants (such as heavy metals, sulfur oxides, and nitrogen oxides) that can have an adverse effect on cell growth.

The CO_2 sources listed above are all point sources, so it would be highly advantageous to colocate biofuel-production facilities with CO_2-generating plants.

That will not always be possible because of unsuitable terrain or the lack of a sufficient water supply, so it will be necessary in some cases to transport concentrated CO_2 by pipeline or rail car to the production facility. Allowable transport distances will be dictated by process economics and existing market conditions for CO_2 and fuel but are not expected to exceed a few hundred miles.

Efforts to develop technology to enable cost-effective uptake and concentration of CO_2 from the atmosphere are under way. Success could have a substantial effect on the economics of biofuel production by photosynthetic microorganisms because colocation of CO_2 and biofuel-production facilities would not be necessary.

REFERENCES

Deng, M.D., and J.R. Coleman. 1999. Ethanol synthesis by genetic engineering in cyanobacteria. Applied and Environmental Microbiology 65:523-528.

Feinberg, D., and M. Karpuk. 1990. CO_2 Sources for Microalgae-based Liquid Fuel Production. Golden, Colo.: Solar Energy Research Institute and National Renewable Energy Laboratory.

Fu, P., and J. Dexter, inventors. 2009. Methods and Compositions for Ethanol Producing Cyanobacteria. USPTO Application #: 2009015587.

GreenFuel Technologies Corporation. 2007. Growth Rates of Emission-Fed Algae Show Viability of New Biomass Crop. GreenFuel Technologies Corporation Press Release. Available at http://www.greenfuelonline.com/gf_files/GreenFuel%20Growth%20Rates.pdf. Accessed April 21, 2008.

Huntley, M.E., and D.G. Redalje. 2007. Global-scale CO_2 mitigation and renewable energy from photosynthetic microbes: A new appraisal. Mitigation and Adaptation Strategies for Global Change 12:573-608.

Lee, Y.K. 2001. Microalgal mass culture systems and methods: Their limitation and potential. Journal of Applied Phycology 13:307-315.

Sheehan, J., T. Dunahay, J. Benemann, and P. Roessler. 1998. A Look Back at the U.S. Department of Energy's Aquatic Species Program: Biodiesel from Algae. Golden, Colo.: National Renewable Energy Laboratory.

Nonquantified Uncertainties That Could Influence the Costs of Carbon Storage

The estimates of potential costs of geological storage of carbon dioxide (CO_2) presented in Chapter 4 of this report are "bottom-up" and based largely on engineering estimates of expense for transport, land purchase, drilling and sequestering, and capping wells. However, ample experience suggests that the full cost of storage cannot be captured by such an approach because of various barriers to implementation that increase cost.

Historical experience with nuclear power-plant construction provides useful insights. The 2003 Massachusetts Institute of Technology study *The Future of Nuclear Power* articulated the problem: "Our 'merchant' cost model uses assumptions that commercial investors would be expected to use today, with parameters based on actual experience rather than engineering estimates of what might be achieved under ideal conditions" (MIT, 2003, Chapter 1) and "construction costs of nuclear plants completed during the 1980s and early 1990s in the United States and in most of Europe were very high. . . . The reasons for the poor historical construction cost experience are not well understood and have not been studied carefully. The realized historical construction costs reflected a combination of regulatory delays, redesign requirements, construction management and quality control problems" (MIT, 2003, Chapter 5). The study noted that the high costs were not predicted and that the experience was not being reflected in current estimates of future construction by the industry.

The issues facing storage are distinct from the problems encountered in nuclear power, but they share an uncertainty in the regulatory environment that arises from attitudes on the part of the general public and policy makers that are obscure, are not fully formed, and are likely to evolve under the influence of future events (Palmgren et al., 2004). A reliable quantitative assessment of future costs of storage would emphasize, at least qualitatively, the uncertainty arising from such attitudes, so quantitative estimates based on engineering analysis may represent a lower bound on future costs.

Storage entails a health risk associated with acute leaks and exposure of workers or populations to hazardous concentrations of CO_2 near facilities, an ecological risk to soils and groundwater due to chronic leakage, and a warming risk associated with sudden or chronic leaks that may partially or entirely vitiate the climatic value of a storage site (Anderson and Newell, 2004; Socolow, 2005). The likelihood of such acute or chronic leaks is discussed elsewhere in this report. The public and policy makers are likely to anticipate those risks and require that they be taken into account in the design, monitoring, and carbon-accounting procedures and in associated regulatory frameworks that would be part and parcel of storage (Wilson et al., 2007). Cost estimates therefore need to anticipate delay in initiating demonstration projects due to time lags in conception and development of the overall regulatory regimen for storage, as well as regulatory delay in licensing of each specific project, both in the demonstration phase and beyond. Some issues, such as liability insurance for near-term operation and for long-term site maintenance, require political resolution that may introduce additional delays (IRGC, 2008). Uncertainty in the probability of long-term leaks could translate into regulations that require the purchase of allowances equivalent to a fraction of the carbon stored by sources that are planning to sequester carbon; this requirement would increase the net cost of carbon capture and storage (CCS) compared with other alternatives.

Although there is no a priori reason for extended licensing delays to occur beyond the demonstration phase, experience with siting of a variety of industrial facilities (Reiner and Herzog, 2004) suggests that delays of a year to several years would not be unusual.

Once CCS attains full commercial-scale operation, delays could arise because of accidents that cause or threaten releases. The technologies, monitoring, and regulation of storage are likely to be closely related or even identical among sites, so interruption of operations at one site could affect operations at other sites and broadly reduce or temporarily eliminate storage; undermine credibility of the technology among investors, regulators, policy makers, and the general population; and add a substantial risk premium to investment in CCS.

Continuous storage may be subject to multiple regulatory regimens (and varied siting, licensing, and monitoring requirements) at various government levels. Moreover, storage rights to the large amount of belowground space that needs to be set aside to hold the lifetime emissions of a facility like a coal plant presumably need to be acquired at the start of a project. That involves a cost that is usually not recognized in storage-cost calculations. Depending on the details of the regulations and the degree of isolation from human settlements that is ultimately required for storage-well fields, surface-land costs may also exceed initial expectations.

One feature of CCS that improves the odds that deployment will evolve without major disruption is that many of the early CCS projects will be enhanced oil-recovery projects. These would be at sites where the general population is already familiar with and generally favorably disposed toward the oil and gas industry and where revenue streams will benefit all royalty holders, including local and state governments (Anderson and Newell, 2004; Socolow, 2005). One can expect less resistance to CCS in such instances.

Each of the aforementioned risk factors may be anticipated rationally, handled smoothly, and reflected in the cost of capital and insurance for storage operators. Or they may be ignored by all parties until experience establishes them as low risks or they cause systemic disruption of operations on a wide scale, as occurred in the United States in the case of nuclear-power-plant construction and long-term waste disposal and to a lesser extent in nuclear-power-plant operation.

There are examples of cases in which risks associated with storage were handled in the normal course of events—with smooth and reliable licensing, operation, and monitoring—and regulatory delays did not cause a serious financial burden or were appropriately recognized and incorporated in planning. CO_2 is routinely transported over long distances, injected underground, and stored without much attention being paid by the public or policy makers. Natural-gas storage and chemical storage are long-time facts of life (Reiner and Herzog, 2004), and even serious accidents and leaks do not threaten operations, at least on an industry-wide basis. But counter examples, from Bhopal to Three-Mile Island to Yucca Mountain, are also easily cited. Furthermore, the proposed scale of CO_2 storage puts it in a class by itself, and the public reaction to failure may be unique and unpredictable. Such uncertainty needs to be reflected in estimates of the cost of implementation of this technology.

REFERENCES

Anderson, S., and R. Newell. 2004. Prospects for carbon capture and storage technologies. Annual Review of Environment and Resources 29:109-142.

IRGC (International Risk Governance Council). 2008. Regulation of carbon capture and storage. Available at http://www.irgc.org/IMG/pdf/Policy_Brief_CCS.pdf. Accessed October 21, 2008.

MIT (Massachusetts Institute of Technology). 2003. The Future of Nuclear Power. An Interdisciplinary MIT Study. Cambridge: MIT.

Palmgren, C., M.G. Morgan, W. Bruine de Bruin, and D.W. Keith. 2004. Initial public perceptions of deep geological and oceanic disposal of carbon dioxide. Environmental Science and Technology 38:6441-6450.

Reiner, D.M., and H.J. Herzog. 2004. Developing a set of regulatory analogs for carbon sequestration. Energy 29:1561-1570.

Socolow, R.H. 2005. Can we bury global warming? Scientific American (July):49-55.

Wilson, E.J., S.J. Friedmann, and M.F. Pollak. 2007. Risk, regulation and liability for carbon capture and sequestration. Environmental Science and Technology 41:5945-5952.